Differentiation and Growth of Cells in Vertebrate Tissues

Differentiation and Growth of Cells in Vertebrate Tissues

Edited by
G. GOLDSPINK

Reader in Zoology,
University of Hull

CHAPMAN AND HALL
London

A HALSTED PRESS BOOK

JOHN WILEY & SONS
New York

First published 1974
by Chapman and Hall
11 New Fetter Lane, London EC4P 4EE

© *1974 Chapman and Hall Ltd*

Printed in Great Britain

Library of Congress Cataloging in Publication Data

Goldspink, G.
    Differentiation and growth of cells in vertebrate tissues.

    1. Cell differentiation. 2. Growth. I. Title.
[DLNM: 1. Cell differentiation. 2. Cells.
3. Vertebrates, QH607 G624d 1974]
QH607.G64   596'.08'761   74-4173
ISBN 0-470-31200-9

Contents

v

CONTENTS

CONTENTS

Contributors

D. BELLAMY — Department of Zoology, University College of South Wales, Cardiff, U.K.

JANE E. BROWN — Genetics Laboratory, Department of Zoology, University of Oxford, Oxford, U.K.

F. J. G. EBLING — Department of Zoology, University of Sheffield, U.K.

G. GOLDSPINK — Muscle Research Laboratory, Department of Zoology, University of Hull, U.K.

B. GONDOS — Department of Pathology, School of Medicine, University of California, U.S.A.

P. F. HARRIS — Department of Human Morphology, University of Nottingham, U.K.

J. W. HORTON — Department of Zoology, University of Durham, U.K.

M. JACOBSON — Department of Physiology and Biophysics, School of Medicine, University of Miami, U.S.A.

K. W. JONES — Department of Animal Genetics, University of Edinburgh, U.K.

MARGARET J. MANNING — Department of Zoology, University of Hull, U.K.

J. J. PRITCHARD — Anatomy Department, The Queen's University of Belfast, Belfast, N. Ireland

Preface

In recent years a new field of study has arisen called developmental biology. The term developmental biology is really a new name for embryology; it is, however, used to denote the molecular approach to the study of developing systems. In this book we have tried wherever possible to blend the older information of classical embryology and in particular organogeny with the newer concepts of developmental biology.

The original intention was to cover all the tissues of the body in this book. However, it soon became obvious that it was not possible to do this within one volume. Therefore we decided to have two general chapters, one on the basic concepts of cellular development and another on the ageing of cells (this being considered part of the normal growth process). In addition to these two general chapters we have included chapters on some of the major tissues. These were chosen not just to illustrate the points made in the general chapters but because there is enough information available on the development of these tissues for the expert in that field to present a good, readable account. It is hoped that at a later date when more information is available, we will be able to extend this work, probably as several volumes, and to include the other tissues of the body which are not dealt with in this volume.

For obvious reasons much of the recent work in developmental biology has been carried out on oocytes or cells in culture and there are many good accounts of this kind of work. Therefore it was felt that this book should concentrate on what is known about the development of the cells *in vivo* from the commencement of tissue formation and throughout the rest of life. It is hoped that the book will interest students of biological and medical sciences at both the undergraduate and postgraduate levels as well as the specialists in cellular and developmental biology.

University of Hull
August 1974   G.G.

1 Basic Concepts of Differentiation and Growth of Cells

JANE E. BROWN and
K. W. JONES

Introduction

Differentiation is the process by which the descendants of a single cell, the fertilized egg, come to differ from one another and to form tissues and organs performing specialized functions. A detailed picture of how embryos differentiate at the morphological level has been built by classical embryologists and developmental geneticists from detailed observations and intricate surgical experiments with normal and mutant organisms[1]. The central problem of developmental biology today is to understand the process in molecular terms.

In previous centuries, various 'pre-formation' theories, of which Fig. 1.1 shows an extreme example, were put forward to explain the development of a differentiated organism from a single germ cell. It was discovered in this century, however, that differentiation does not proceed inflexibly to its conclusion; that is to say, it is not a completely pre-determined process. Classical studies showed that, very early in embryogenesis, individual cells or areas of cells often retain the capacity to give rise to a complete embryo, i.e. to regulate, if separated from the dividing cell mass. Later, however they are influenced by successive inducing stimuli and thus become progressively more 'determined' to differentiate along specific pathways. These observations have led to widespread acceptance of an 'epigenetic' theory of development, in which cell interactions play a major part.

Since the discovery of the rôle of genes (DNA) in determining the synthesis of RNA and protein, it has been widely accepted that the size, shape, content and function of a cell are all ultimately the result of gene action, from which it follows that the conspicuous differences between cell types are the result of differential gene expression. Some genes are expressed only in a single cell type; for example, the genes for haemoglobin in red blood cells or for melanin synthesis in pigment cells. Other

I

genes are active in many types of cell, but may be expressed at different levels in different cell types; muscle cells have high levels of glycolytic enzymes and low levels of the enzymes of the hexose monophosphate pathway, for instance, while in adipose tissue this situation is reversed. Each differentiated cell type synthesizes a distinct pattern of enzymes, structural proteins and RNA molecules and this pattern is presumably

Fig. 1.1. A spermatozoon. *From* Hartsoeker, 1694 (Reference 2).

responsible for the specific characteristics of that cell[3]. From the totipotent zygote, these complex patterns must be built up by a combination of switching on, switching off and setting of levels of expression of many genes. Thus, to discuss differentiation is in effect to discuss the regulation of gene expression in multicellular organisms.

Modern theories of differentiation attempt to relate the observed phenomena of embryogenesis to our knowledge of gene function and regulation. The word 'programming' is often borrowed from computer jargon to describe the regulatory events which take place during differentiation. A widely supported general hypothesis runs as follows. At the outset, the fertilized egg, itself a differentiated cell, contains a set of genes programmed to direct a certain pattern of DNA, RNA and protein synthesis. As the number of cells in the embryo increases it is envisaged that polarities which were already present in the egg, together with differences in cellular environment which develop as the egg divides to form a mass of cells, exert epigenetic influences which lead to the imposition of new programmes of gene expression in different parts of the embryo. Morphogenetic movements of cells lead to the occurrence of inductive interactions between cells from different parts of the embryo and these bring more specific programmes into action and increase the commitment of the cells to specialized paths of development.

In the following pages we discuss some of the assumptions and experiments underlying this type of hypothesis and consider our knowledge of gene regulation in higher organisms. We have limited our treatment of particular examples to points which underline basic concepts of differentiation, since some of them are discussed in greater detail in other chapters.

The Control of Differentiation and Growth of Tissues

The origins of cellular diversity

If different environmental stimuli bring about gene programming in embryos we have to look for the origin of these stimuli. The cytoplasm of the zygote is the obvious place to look first. The sperm contributes little cytoplasm to the zygote and since embryos can develop without it, by parthenogenesis or by nuclear transplantation it cannot provide any elements vital for development. The egg cytoplasm, on the other hand, is voluminous in comparison with that of somatic cells and it contains many proteins, RNAs, other types of molecule and organelles of maternal origin, which might conceivably carry developmental instructions. Many eggs show a visible differentiation of their contents and it can be demonstrated that, during normal development, particular regions of the egg invariably give rise to particular structures in the adult organism. This raises the possibility that a pre-formed pattern of qualitative information is laid down in the egg and that nuclei direct differentiation according to the type of cytoplasm they come to lie in as a result of cell cleavage.

Some invertebrates have 'mosaic' eggs which develop in this way; removal of one blastomere from an early cleavage stage can result in the

3

loss of a particular tissue from the embryo and if the organization of the egg is disrupted by centrifugation before cleavage, a disorganized embryo results in which recognizable cytoplasmic regions give rise to the right tissue, but in the wrong place. Thus the organization of the egg foreshadows the organization of the adult. For the eggs of many animals, including the vertebrates, however, this is not the case. These eggs can withstand centrifugation and, for the first few cleavage divisions, separated blastomeres are capable of differentiating into complete embryos. This capacity to regulate shows that, in such embryos, the position of a cell relative to the other cells can influence its differentiation.

In regulating eggs, developmental information in the form of cytoplasmic differentiation, where it exists, must be of a relative nature. Gradients of 'animalizing' and 'vegetalizing' influences which determine the dorsoventral axis of the embryo have been demonstrated in sea urchin eggs, for instance[4]; cells differing in cytoplasmic constitution would arise from cleavage within morphogenetic gradients of this type. Diversity of cellular environment can also be imagined to arise *de novo*; after a small number of cleavage divisions, some cells begin to differ from others in their position relative to the surface of the embryo and this could affect a number of parameters, such as the rate of exchange of oxygen or waste products with the external milieu.

Experiments on the differentiation of the specialized trophoblast cells from the inner cell mass, which begins by the 8 or 16 cell stage in mouse embryos, suggest that gene expression can indeed be determined by the position of a cell relative to the rest of the developing mass[5]. Separation of the first two blastomeres in the mouse can give either two normal blastocysts or two trophoblastic vesicles lacking an inner cell mass. There is thus little evidence of pre-formed 'trophoblast-forming' or 'inner-cell-mass-forming' cytoplasm. On the contrary, the behaviour of chimaeric embryos made by the fusion of several embryos at the four or eight cell stage suggests that the cells can register whether they are on the outside or the inside of the morula and that 'outside' cells differentiate into trophoblast and 'inside' cells become inner cell mass. If a radioactively labelled embryo is completely surrounded by unlabelled ones, most or all of the label is later found in the inner cell mass of the chimaera, although two thirds of the labelled cells would have become trophoblast if they had been allowed to develop normally.

It is not known what factors signal to a cell that it is inside or outside. For mammalian embryos the influence of the maternal environment must be considered as well as factors intrinsic to the embryo. It is unlikely that

4

specific maternal factors are responsible for the trophoblast/inner cell mass differentiation, since this can take place *in vitro* in the absence of serum. Extrinsic influences may, however, be important at other stages of mammalian development; large molecules of maternal origin have been shown to enter embryos at all stages from oogenesis to implantation[6] and development of mouse embryos beyond the two cell stage *in vitro* appears to depend on their having interacted with a region of the oviduct[7]. The maternal environment is also known to exert an effect on morphological characters such as tail ring width[8] and number of lumbar vertebrae in mice[9].

No hard line can be drawn, in fact, between mosaic and regulatory eggs. Some regulation occurs in embryos from typically mosaic eggs and the amphibians, while having highly regulative early development, also provide evidence that pre-formed information can be carried by both the egg cortex and the egg cytoplasm. Part of the grey crescent which forms in the cortex at fertilization must be included in a fragment of egg or blastula for the fragment to develop differentiated structures, and grafting experiments show that this area organizes the morphogenetic movements which lead to the major events of embryonic induction. Preformation of the cytoplasm is involved in differentiation of the germ cells. In some toads and frogs, islands of yolk-free, RNA-rich 'germinal cytoplasm' appear in the vegetal (yolky) regions of the egg soon after fertilization. This cytoplasm becomes enclosed in a small number of cells in the blastula which become recognizable as primordial germ cells in the gastrula and eventually migrate from the endoderm to colonize the genital ridges. Destruction of the cells containing this cytoplasm by surgery or irradiation produces sterile adults and if these cells are replaced with primordial germ cells from another species, adults develop which produce offspring characteristic of the other species[10]. Since, however, cells histologically resembling primordial germ cells can be induced *in vitro* from gastrula ectoderm by 'vegetalizing' factor[11], it is possible that germ cell differentiation *in vivo* is not completely predetermined in the egg.

Embryonic induction

As development of a vertebrate embryo proceeds, the fates of different areas become progressively more narrowly determined. First the areas that will become ectoderm, endoderm and mesoderm are determined, then the areas which will form particular organs, then structures within the organs. At the blastula stage a complete newt can still form if half the cells are removed. A neurula embryo can compensate for the removal of an organ-forming area by recruiting surrounding cells, later embryos can only repair the loss of part of the organ and, eventually, any damage results in a defective embryo. At each stage determination can be detected by grafting experiments some time before any gross physical sign of differentiation is seen; the differentiation of distinct, histologically recognizable cell types ('cytodifferentiation') occurs at a relatively late stage in the development of an organ, although biochemical differences and differences in cell shape and surface properties can sometimes be demonstrated much earlier[12]. Studies with chimaeric mice[13] and with mammals showing mosaicism due to the inactivation of an X chromosome[14] suggest that, for some cell types at least, the number of precursor cells present at the time when the cell type is determined may be very small (10–30 cells).

From gastrulation onwards, the orderly differentiation of tissues is promoted by inductive interactions between cells, which determine which developmental paths the cells shall follow. Morphogenetic movements influence these interactions, rearranging the cells and bringing about the juxtaposition, permanently or transiently, of cells from different parts of the embryo. The basic axial organization of the embryo is established when the dorsal mesoderm comes in contact with overlying ectoderm and induces the ectoderm to form the neural tube. After this, a complex succession of secondary inductions takes place all over the embryo, some tissues being in turn induced and then themselves becoming inducers. In amphibia, for example, part of the head mesoderm induces a region of the neural tube to form forebrain and eyecups. When an eyecup grows out to meet the surface of the embryo it induces the ectoderm to form a lens, which pinches off and sinks into the head and then itself induces the overlying pigmented ectoderm to clear and become corneal epithelium.

It is not known for certain at what stage the genes of the zygote begin to be expressed; various evidence suggests that they may be largely inactive till the end of cleavage. A number of developmental mutants are known, however, which provide ample evidence of gene activity from gastrulation onwards. Some mutations, such as deficiencies in pigment formation, act autonomously, affecting only the cells in which they are expressed. Many others show how the development of one tissue can be influenced by gene expression in another. The 'creeper' (Cp) gene in chickens, a cartilage mutation which affects ossification of the long bones and also produces small, lidless eyes, acts autonomously in one tissue, but can be influenced by cell interaction in another. A Cp/Cp limb rudiment transplanted to a normal embryo develops in characteristic creeper stunted form, but a transplanted Cp/Cp eye rudiment develops normally[15]. Many mutations which disturb specific inductive interactions in embryogenesis have been studied, such as those affecting the reciprocal induction between mesoderm and ectoderm in limb formation in chicks ('wingless', 'polydactyly' and 'eudiplopodia')[15]. The various mutations in the mouse which cause abnormalities of the primitive streak and notochord and consequently disrupt primary embryonic induction and subsequent development ('Danforth's short tail', 'brachyury', 'pin tail' and 'truncate')[16] illustrate how the organization of the embryo depends on orderly and sequential expression of genes.

Very little is known about the mechanisms of embryonic induction. Induction is a localized event requiring close proximity between the interacting tissues and in some cases substances have been shown to pass from the inducer to the induced tissue. That the substances concerned are not highly specific to the inducing tissue, however, is suggested by many experiments which show that particular inductions can be effected by tissues other than the usual one, or by subjecting the target tissue to unnatural stimuli such as heat or osmotic shock or chemicals. It is possible that many different types of molecule may act as inducers. 'Neuralizing' and 'mesodermalizing' factors have been partially purified from chicken and toad tissues and can evoke many inductive steps *in vitro*[17].

It is clear that inducers have a rather generalized, activating rôle. From many grafting experiments it is known that cells must be in a state of competence before they can be induced. This state can be transitory and before it develops and after it disappears the inducer has no effect. The area of an embryo showing competence to respond to a particular inducer is often larger than the area which is actually induced. From these observations it seems likely that labile programmes of gene expression become temporarily established in response to prevailing environmental factors and that the intervention of an inducer is necessary to 'fix' them in an irreversibly determined state. The way in which regulatory powers become narrowed during development suggests that individual cells, or their progeny, may go through several 'fixing' steps, each involving a further limitation of the permitted range of gene expression.

If we can say little about the morphogenetic factors which influence, for example, the 16 cells of a mouse embryo to become either trophoblast or inner cell mass, we can say even less about the multiplicity of such factors which must presumably be involved in organogenesis. Accounting for the spatial organization of the many patterned structures which are observed, from the basic segmentation of trunk and limbs to the intricate arrangements of hairs, scales, feathers, teeth, etc., presents a particular problem. Experiments on the patterned cuticle of the bug *Rhodnius*[18] and on regeneration of hydroids[19] and of cockroach limbs[20] suggest that patterns are controlled by morphogenetic gradients, which determine the contribution of a cell to the overall pattern by inducing it to express a programme of differentiation appropriate to its position. The nature of the gradients which impart this 'positional information' to the cell is a matter for speculation. A simple model which is compatible with many observed phenomena is that of the concentration gradient formed by a diffusible substance which emanates from a source and is removed by a sink.

Hormones

While the substances regulating early differentiation are obscure, it is well known that many of the maturational steps which occur in later development, after a functional vascular system has been established, are controlled by circulating hormones. Unlike the reversible modulations of cellular activity brought about by hormones in adult cells, these changes are virtually irreversible and are initiated, and often completed, in anticipation of the environmental change or physiological need to which they adapt the cell.

Some striking examples are the many changes involved in metamorphosis in amphibia[21] and the increases in synthesis of particular enzymes which occur at birth and weaning in mammals[2]. In amphibia, the wide variety of morphological and biochemical events accompanying metamorphosis (tail resorption, skin differentiation, limb growth, etc.) are all induced by thyroxine. Environmental stimuli such as day length or temperature change probably initiate the secretion of thyrotrophic hormone which in turn evokes thyroxine. In the case of the newborn mammal, the breaking of the cord at birth abruptly deprives the fetus of its steady supply of maternal glucose and of a means of disposing of its waste products. After this has happened it must begin to synthesize its own glucose, first from intermittent supplies of milk and later from solid food. In rat liver three very eventful periods of enzyme induction occur around the time of birth and hormones connected with these inductions have been identified. The hypoglycaemia which is caused by birth

8

stimulates glucagon secretion by the newborn and this hormone can be shown to induce several enzymes of the 'neonatal' enzyme cluster prematurely if injected into fetuses *in utero*. Many enzymes of the 'late fetal' and 'late suckling' clusters can be induced prematurely by either thyroxine or hydrocortisone and the timing of secretion of these hormones *in vivo* suggests that they are probably the natural stimuli.

Hormonally-induced differentiation shares some basic features with the inductive processes which occur in earlier development and since, in the hormone controlled stages, it is often possible to identify specific gene products and to study causal relationships more closely, hormonal systems may provide valuable insight into the earlier interactions. Hormones act as triggers, setting off predisposed programmes of gene activity which may continue to unfold after the hormone has disappeared. The specificity of the reaction lies with the target cell and not with the hormone. The same hormone can elicit quite different responses from different cell types, while in a particular cell type an enzyme may rise in response to one hormone at one stage of development and to another at a later stage. As with inductions occurring in earlier development, a tissue must become competent before it can respond to hormonal inducing stimuli and in hormonally-induced systems it is possible to begin to analyse the meaning of competence in terms of biochemical events. In neonatal induction of tyrosine aminotransferase (TAT) in mammals, for example, the hypoglycaemia caused by premature delivery is only effective in bringing about premature induction of the enzyme for about one day before full term, i.e. competence to respond to this stimulus develops on the last day of gestation. Competence to respond to injected glucagon, however, appears two days before birth, and four days before birth $3':5'$-cyclic AMP (cAMP), the compound through which the action of glucagon is mediated, can evoke TAT synthesis. Thus competence to respond to the natural stimulus is built up in several stages; first the ability of liver cells to respond to cAMP, then to make cAMP in response to glucagon and finally to secrete glucagon in response to hypoglycaemia. Soon after birth these stimuli cease to be effective in causing TAT synthesis, but yet another competence then develops. Glucocorticoids, which could not induce the enzyme before birth, become the effective adult trigger for TAT synthesis.

In developing *Xenopus* larvae, spontaneous metamorphosis occurs at 80–90 days after fertilization, but competence to respond to administered triiodothyronine with typical metamorphic changes appears as early as 40–60 h after fertilization. At this time a temperature-sensitive increase in the capacity of the cells to bind triiodothyronine also occurs, suggesting that competence is conferred by the acquisition of protein receptors for the hormone. Different tissues develop different thresholds of response to thyroxine, so that as the blood level of hormone rises, changes are induced in sequence. The Mexican axolotl, *Ambystoma mexicanum*, which does not normally metamorphose in the wild, can be induced to do so in the laboratory by the administration of thyroxine; thus it shows the competence to express adult genes, but lacks the stimulus to do so. In other neotenous amphibia, the stimulus is present but the tissues fail to respond[22].

In most cases it is thought that the hormonally induced increases in enzyme activity represent increases in the amount of enzyme protein synthesized and not the stabilization or activation of enzyme already present. A number of methods have been used to show this. Assay conditions can be adjusted so that only the concentration of enzyme is limiting and the presence of naturally occurring specific inhibitors can be excluded by assaying mixtures of induced and non-induced preparations. The prevention of enzyme induction by inhibitors of protein synthesis such as puromycin and cycloheximide has also been used as evidence that induction involves *de novo* protein synthesis, although it must be borne in mind that these inhibitors may affect the synthesis, not of the induced enzyme, but of another protein which takes part in the induction process. In a few cases an increase in enzyme protein has been detected directly by immunological titration with antiserum to the purified enzyme. The breakdown of glycogen by phosphorylase, one of the neonatal activities evoked by glucagon, provides an example of an induced function which is not prevented by inhibitors of protein synthesis; in this case glucagon probably causes activation of pre-existing glycogen phosphorylase enzyme by cAMP-mediated phosphorylation, as in adult liver[2].

The way in which hormones activate cells is still only partially understood. The steroids have a different mode of action from the polypeptide hormones. It has been shown in a number of cases that when steroid hormones react with the target cell they become bound first to a cytoplasmic receptor protein and then enter the nucleus attached to a 'nuclear' receptor protein which may be a modified version of the cytoplasmic receptor[23]. For oestrogen in the uterus and progesterone in the chick oviduct, the nuclear receptor-hormone complexes have been shown to have a specific binding affinity for target cell chromatin, which suggest that steroids may act by influencing transcription. The polypeptide hormones act at the cell surface and many have been shown to stimulate the activity of adenyl cyclase in the membrane, thus causing a rise in the intracellular concentration of cAMP, which acts as the internal effector for the hormone[24,25].

Controls of growth

Many of the observed features of development in vertebrates point to the existence of controls of growth at various levels of organization. For example, the adults of most vertebrate species show characteristic size limits, even though artificial selection within a species can reveal a much wider range of possible sizes, as can be seen in the domestic dog. Size limits are genetically controlled, but they can be modified by environmental influences such as the level of nutrition or, in mammals, the maternal uterine size[26]. The genetic factors controlling whole body size are not well understood. It seems unlikely that the limits are determined by the size of the individual cells making up the organism or by a process which 'counts out' a certain number of mitoses in different parts of the animal. In salamanders, for instance, diploid, triploid, tetraploid and pentaploid larvae are all the same size, the various organs of the polyploid larvae containing fewer, but larger, cells than those of the diploid larvae[27]. In mammals, size regulation of abnormal sized embryos takes place during gestation so that, at birth, fusion chimaeras and mice derived from only one blastomere of a two cell embryo are both the same size as normal mice[7]. It is probable that the size of individual organs is genetically controlled and that the size of the adult animal is a product of these individual controls and of co-ordinate controls exerted by systemic factors, of which the most important is probably pituitary growth hormone. This hormone affects the growth of virtually every somatic tissue and extremes of overproduction or underproduction of the hormone produces the so-called pituitary 'giants' and 'dwarfs'.

During development, before cells become terminally differentiated, differential growth in the form of controlled cell proliferation contributes to the shaping of many differentiated structures[1]. 'Negative growth', or programmed cell atrophy, is also important in the shaping process, as in the regression of tadpole tails[21] or the formation of the spaces between digits at the end of a limb[1]. Controlled growth is thus an integral part of the process of differentiation and can be regarded as one of the differentiated properties of a cell type, forming part of a determined programme of gene expression. Like the stimuli which bring about the determination of cells, the stimuli which control growth in early development are difficult to identify. In later development, however, various substances are known which stimulate mitosis in specific tissues. These include many hormones which also induce the expression of specific programmes of differentiation. A particularly striking example is thyroxine, which is known to induce different cell types to undergo cell division, cell death or structural and enzymic differentiation during amphibian metamorphosis[21].

Growth controls operate at the tissue level in the adult animal as well as during development. Cycles of growth can occur in tissues such as teeth, uterine tissue and mammary glands and many organs are capable of regenerative growth. The patterns of growth shown by different adult tissues vary according to the capacity of the terminally differentiated cells of the tissue to undergo mitosis. For some differentiated cell types, particularly those in which differentiation involves the accumulation of large amounts of specialized molecules of the adoption of a highly specialized morphology, the terminally differentiated state is incompatible with cell division. This category includes nerve cells and muscle syncytia, both of which are long-lived and show no mitotic activity. Lost nerve cells are not replaced, although axon regeneration can occur. Muscles contain a small number of undifferentiated myoblasts which contribute to the regeneration of muscle tissue. A related category includes keratinizing skin cells, blood cells and various secretory cell types, in which terminal differentiation leads to cell death. These deciduous cells are short-lived and are continually replaced from a pool of precursor cells. In other cell types, differentiation and cell division are compatible, although fluctuations may occur in the synthesis of differentiated products during the cell cycle. It has been suggested that the rates of cell production and of cell loss by differentiation and death are balanced in a mitotically competent tissue by a negative feedback mechanism. Evidence has been found for tissue-specific anti-mitotic agents which have been termed chalones and it is proposed that these form part of a mass-dependent control mechanism which switches cells from proliferation to differentiation and eventual death as the tissue increases in size. Release from chalone inhibition could be the stimulus for regeneration, although activation, such as the neurotropic activation seen in amphibian limb regeneration, may also be required.

For cell types which differentiate *in vitro*, an inverse relationship between growth and differentiation has often been found. In culture medium containing high levels of serum, rapid growth with little differentiation occurs, whereas in dense cultures in which mitosis is contact-inhibited, or in medium containing low serum concentrations, the cells express their differentiated properties. Whether the change from proliferation to differentiation in these experiments is due to depletion of nutrients or of mitogenic substances from the medium, or whether chalone-like substances or specific activators of differentiation are produced by the cells in the growth-inhibiting conditions, is not known. It is possible, however, that these phenomena reflect density- or mass-dependent controls which operate *in vivo* to co-ordinate the growth and differentiation of tissues. It has been reported that the mitotic index varies inversely with cell density in the mesenchymal condensations which precede cartilage formation in chick wing morphogenesis, which would appear to be an example of such a control *in vivo*[29].

A growing number of examples show an apparent link between cyclic AMP and the balance between cell proliferation and differentiation. In cells growing rapidly *in vitro* the cAMP level is lower than in contact-inhibited cells or sparsely distributed cells grown in low concentrations of serum[30, 31]. In addition, if dibutyryl cAMP is added to the culture medium, the proliferation of various cell types is inhibited and their morphological and chemical differentiation are enhanced. This has been shown for Chinese hamster ovary cells[32], virus-transformed fibroblasts[33], neuroblastoma cells[34] and melanoblasts[35]. The significance of these findings and their relationship to the action of cAMP as a mediator of polypeptide hormone action is not yet clear. cAMP is known to be involved in various phosphorylation reactions. Phosphorylation of histones or microtubule proteins and activation of enzyme proteins by phosphorylation are among the mechanisms which have been suggested to explain some of its multifarious effects; it remains to be proved, however, whether this is its only mechanism of action. Much interest has been roused by the ubiquity of cAMP and the many effects it has on cells in both prokaryotes and eukaryotes[24, 25]. It has an important regulatory rôle in the cellular slime mould *Dictyostelium discoideum*, where it has been identified as the chemotactic and morphogenetic substance which organizes the single amoebae to form a slug[36], and in *E. coli* it is involved in the activation of certain genes under poor growth conditions (see p. 25), a situation with interesting analogies to that described above for eukaryotes.

The rôle of mitosis in cell differentiation

It has been suggested that before cells respond to a stimulus with new gene expression they must undergo a cell division in the presence of the stimulus[37]. Evidence for this has been drawn from various systems, in particular from the differentiation of the epithelial casein-producing cells of the mammary gland. These cells will differentiate *in vitro* in response to the sequential action of three hormones, insulin, hydrocortisone and prolactin[38]. Insulin stimulates a mitosis in the cells, during or after which hydrocortisone induces an increase in Golgi apparatus and rough endoplasmic reticulum. Finally, prolactin and insulin together stimulate casein production. If the insulin-stimulated cell division is inhibited, however, the milk-producing cells fail to differentiate and this has been taken to indicate that this mitosis has a special rôle in the process of differentiation[39]. A requirement for a critical or 'quantal' mitosis before terminal differentiation has been proposed for other cell types, including myoblasts, cartilage cells and erythrocyte precursors[40], and in many tissues cell division accompanies differentiation.

It might be inferred that these maturational changes in gene expression can only take place during the replicative period of the cell cycle. However, no rigorous evidence for an obligatory coupling between mitosis and differentiation has been found in any system and recent work suggests that myoblasts can differentiate *in vitro* without going through a quantal mitosis[41, 42]. It is possible, therefore, that the mitoses which accompany maturational changes are not necessarily concerned with the new gene programming. They may instead represent a growth response concerned in increasing the size of the induced cell population from a smaller pool of determined precursor cells. In frogs, for instance, the multicellular skin glands arise by cell division and cytodifferentiation, probably from a single precursor cell[43]. Both processes are stimulated by thyroxine; thyroxine also stimulates the mitotic rate in the surrounding epidermis, however, and it is likely that its effect on gland cells consists of parallel, rather than dependent, events. A connection between cell division or DNA synthesis and gene programming at some stage in the determination of cells may exist, however. Changes in the pattern of binding of histones or acidic proteins (see p. 27) might be unable to occur when newly-synthesized, 'clean' DNA is available.

The Properties of Differentiated Cells

Is differentiation reversible?

Weissman's idea, put forward in the 1890s, that organ-specifying genetic determinants contained by the egg were distributed to different parts of the embryo by cleavage, is no longer seriously entertained. It is proposed, instead, that a complete set of determinants, or genes, is received by each cell at cleavage, but that differential gene expression is brought about by environmental differences. It is still open to question, however, whether differentiation is caused by, or causes, some sort of irreversible modification of the genome.

In some exceptional examples this can be seen to be true. In the embryo of *Ascaris* and of some insects, for instance, all the cells except the potential germ cells eliminate a large number of their chromosomes[37]. Differentiation is also clearly irreversible in lens cells and mammalian erythrocytes, which lose their nuclei in the terminally differentiated state. In other cell types the genotype is unaffected by differentiation as far as can be ascertained by looking at metaphase chromosomes. The differentiated condition is very stable, however, persisting when the cells are cultured outside the body, and this stability, together with the observation of changes in nuclear morphology during the differentiation of some cells, led in the past to the suggestion that differentiation involves progressive alterations to the genetic material.

Evidence that this is not so came from nuclear transplantation experiments, which showed that a complete set of genes capable of directing the development of a whole organism is retained by differentiated cells. Nuclei from a renal tumour of adult *Rana pipiens* frogs, transplanted into enucleated frogs' eggs, can support development as far as a feeding tadpole[44], for instance, and the development of mature, fertile toads has been achieved by transplanting nuclei from *Xenopus* tadpole intestinal epithelium[45] or even cultured adult epithelial cells into *Xenopus* eggs[46, 47]. The rearing of adult toads from epithelial nuclei, albeit with a low rate of success, shows that under some circumstances nuclear differentiation can be reversed. This is not to say that nuclei of differentiated cells show no sign of having been programmed when they are transplanted. The ability of nuclei from *Rana pipiens* embryos to support advanced development falls off steadily as the embryos increase in age from blastula to neurula stage, so that few embryos derived from mid-neurula nuclei gastrulate and those that do so are severely malformed[48]. These and similar experiments show that there are limitations to the process of reversal of nuclear differentiation. These may be due to the occurrence of permanent nuclear changes in some cells, but they can also be explained in terms of failure to achieve particular conditions which favour the adjustment of the transplanted nucleus to the foreign cell. A high frequency of unspecific chromosome damage follows nuclear transplantation, and this undoubtedly contributes to the failures and abnormalities of development which have been encountered. It has been suggested that this damage is due to the superimposition of the rapid cleavage rate of the egg on a slower cell cycle in the transplanted nucleus[49]; the results obtained with *Rana* may thus reflect the progressive difficulty of bringing the two cycles into phase as the age of the donor nuclei increases rather than the occurrence of a genuinely irreversible nuclear change. They may equally well reflect the relative slowness of the changes involved in the reprogramming of differentiated nuclei. The experiments with transplanted nuclei from cultured *Xenopus* tadpole epithelial cells are most easily explained in terms of overcoming such difficulties. Only 25% of such nuclei promote development of blastulae, but nuclei from these blastulae can give rise to swimming and metamorphosing tadpoles.

Evidence for the retention of unexpressed genetic information by a differentiated cell also comes from quite a different type of experiment. The nuclei of terminally differentiated bird erythrocytes contain highly condensed chromatin and are almost completely inactive in DNA and RNA synthesis[50, 51]. If chicken erythrocyte nuclei are introduced by Sendai virus-mediated cell fusion into a mammalian cell, they recommence DNA and RNA synthesis and after a period of days synthesis of proteins coded by the chicken genome begins in the heterokaryons[50]. Synthesis of a number of functionally diverse chicken-specific proteins has been detected, (i.e. cell-surface antigens, inosinic acid pyrophosphorylase[50], interferon[52] and myosin[42]), none of which were synthesized by the erythrocytes from which the nuclei were taken. (See also p. 20.)

The stability of the differentiated state

While the experiments described above show that nuclear differentiation can be expunged and that genes can be re-programmed when a nucleus is put into a new cytoplasm, there is very little evidence that differentiation ever goes into reverse in intact cells. In fact, one of the most remarkable features of differentiation is that when a cell has been induced to move from a state of competence to a state of determination, this determined state is 'remembered' by the cell, regardless of outside influences. This inheritance of the determined state does not depend on the continued presence of the original inducing stimulus; as little as 5 min exposure to the inducing tissue is sufficient to determine the differentiation *in vitro* of brain and eye structures[53], and *in vivo* the final differentiation of a cell frequently occurs long after the contacts which induced it have been disrupted. For instance, migratory melanoblasts induced in the neural crest complete their differentiation in the hair follicles, skin and choroid layer of the eye. Furthermore, many cells continue to synthesize their characteristic differentiated products when they are cultured outside the body in standard media (e.g. kidney and liver cells, neuroblasts, chondrocytes, melanoblasts and myoblasts).

Differentiated cells sometimes fail to show their characteristic features after prolonged culturing, possibly due to their having become aneuploid. Such cells may sometimes be induced to re-express their differentiated state by manipulating the conditions of growth. An example of this has been described in the case of an amelanotic sub-line of a differentiating melanoma cell line[54]. The terminally differentiated state in some cells is incompatible with proliferation of the cells, so that, in tissue culture, differentiated products are not always formed under conditions favouring growth. Other cells in culture only show full differentiation when certain supplements are added to the medium, possibly because the cells are leaky and are unable to maintain a high level of some necessary metabolite[55]. Nevertheless, the state of determination of these cells is inherited and is re-expressed under favourable conditions.

In vivo certain types of regenerative process, such as amphibian limb regeneration, follow a course which suggests that cellular dedifferentiation and re-differentiation might be involved[56]. Among the vertebrates the amphibia show the greatest regulatory capacities, both in the embryo and in the adult, and they provide the single well authenticated example, in vertebrates, of true metaplasia, i.e. the re-differentiation of a cell as a new cell type. In many species of salamander, removal of the lens from the eye leads to the formation of a new lens from part of the iris[57]. Under the influence of inducing stimuli from the optic cup, iris cells, which are normally non-dividing, first lose their pigment and then enter a phase of cell division, during which they synthesize new RNA and develop a nucleolus. Finally, they differentiate into lens cells, becoming elongated, synthesizing lens crystallins and losing their nuclei. Recently, iris/lens metaplasia has also been demonstrated in cloned chick iris cells[58]. In other cases, such as limb regeneration, however, metaplasia has not been conclusively shown and plausible alternative interpretations are possible; for instance, that the remaining cells merely revert to a proliferative state and then re-differentiate as their original cell types, or that regenerative tissues contain multipotent precursor cells which rebuild the tissue by proliferating and differentiating *de novo*. A certain amount of developmental plasticity has been demonstrated in explants of differentiated epidermis of chicken embryos[59] and adult guinea pigs[60]. The *stratum germinativum* of the epidermis gives rise to squamous, keratinizing epithelium if the skin is cultured in vitamin A-deficient medium but to ciliated, mucous-secreting epithelium if vitamin A is added to the medium. It is unclear, however, whether both types of epithelium derive from a single precursor cell type which can adopt one of two pathways of terminal differentiation or whether two separate populations or precursors exist.

The way in which the determined state of a cell is inherited is still a mystery. From studies of cultured cells it can be shown that it is not maintained simply by the distribution among the daughter cells of molecules accumulated or synthesized at the time of induction. In some cases, differentiated cells in culture have given rise to apparently immortal cell lines which continue to exhibit differentiated properties. Unlike the cells originally taken from the animal, these lines multiply indefinitely and can be cloned and it is thus possible to show that the progeny of a single cell continue to differentiate according to the same programme even when they have become so numerous that on average each cell contains less than one hydrogen atom derived from the original parent cell. It has been suggested that determination may come about by a sequence of individually reversible steps which are irreversible in sum total because of the unlikelihood of their taking place in reverse order. Whatever the mechanism, the stability of the final state seems to be a property of the whole cell, since isolated nuclei can be completely reprogrammed (see p. 15).

Interactions between nucleus and cytoplasm

The dependence of the cytoplasm on the nucleus needs little emphasis; nuclei contain the genes which specify the structures of the cytoplasmic macromolecules and which interact to determine the shape, size and function of the whole organism. However, the proposition that clonally related nuclei in differentiated cells contain equivalent genetic information, and that differentiation is brought about through cellular interactions, bears with it the assumption that nuclear behaviour must be influenced by cues coming from the cytoplasm.

Evidence for the influence of the cytoplasm on developmental processes comes from various instances of 'cytoplasmic inheritance', in which an aspect of development is controlled by the maternal genotype and not the embryonic genotype. Thus, female *Drosophila* homozygous for a particular sterility gene are themselves fertile, but produce infertile progeny because they lay eggs with a defect in the pole plasm, the cytoplasmic area which determines the germ cells of the progeny[61]. Among vertebrates, female axolotls homozygous for the gene o produce eggs which stop developing at gastrulation, but which can be made to develop into a larva by the injection of a very small amount of cytoplasm or nucleoplasm from a normal egg[62].

Experimental evidence for cytoplasmic feedback on the nucleus is provided by the nuclear transplantation experiments described on p. 15. Similar experiments also demonstrated clearly that the three main activities of the nucleus, replication, transcription and chromosome organization, are all influenced by cytoplasmic factors[37]. *Xenopus* oocytes are very active in RNA synthesis and inactive in DNA synthesis, while activated *Xenopus* eggs can support DNA synthesis, but do not make RNA. When neurula endoderm nuclei, which synthesize much RNA, were transplanted to egg cytoplasm, their RNA synthesis was inhibited and if the embryos were allowed to develop, it did not resume until the normal developmental stage. The proportion of neurula nuclei synthesizing DNA, on the other hand, rose after the transplantation, from 30% to 70%, and when adult brain nuclei were placed in egg cytoplasm, 90% synthesized DNA within 90 min, whereas less than 0·1% would have done so in the brain. If rapidly cleaving blastula nuclei were transplanted to oocyte cytoplasm the opposite situation occurred: RNA synthesis was initiated and DNA synthesis inhibited. The cytoplasms of both the egg and the oocyte, therefore, appear to be able to impose a programme of behaviour on incoming nuclei which resembles the normal behaviour of their own nuclei. This involves both stimulation and repression of the functions of the transplanted nuclei. Chromosomal behaviour is also under cytoplasmic influence. When brain nuclei, which normally divide extremely rarely, are injected into meiosing oocytes, they rapidly form condensed chromosomes. When they are injected into eggs, they form condensed chromosomes after about 2 h (i.e. at a time when the zygote would undergo mitosis).

Similar evidence for cytoplasmic control over nuclear properties comes from experiments with artificial heterokaryons formed by the fusion, in various combinations, of HeLa cells, hen erythrocytes, rabbit macrophages and rat lymphocytes[50]. The general rule has emerged that, in heterokaryons, the combination of a cell active in the synthesis of a particular nucleic acid with an inactive one leads to the initiation of that synthesis in the inactive nucleus. In heterokaryons, suppression of synthesis is not seen even when the inactive nuclei vastly outnumber the active ones. Chromosome condensation is also under cytoplasmic control in heterokaryons[63]. When one nucleus in the heterokaryon enters mitosis, premature condensation of nuclei at earlier stages of the cell cycle occurs and the greater the proportion of mitotic nuclei the faster the premature condensation of the others. Synchronized mitoses are also observed in naturally occurring syncytia, such as mammalian megakaryocytes and cells which retain cytoplasmic interconnections after mitosis, such as vertebrate spermatocytes.

The nature of the interaction between nucleus and cytoplasm in transplantation and reactivation experiments is not well understood. It seems likely, however, that the uptake of nuclear proteins from the cytoplasm is a contributory factor. The nucleus is obligatorily non-autonomous to the extent that all the proteins concerned in nuclear function (polymerases, chromosomal proteins, nucleolar proteins, etc.) must be synthesized in the cytoplasm and transported to the nucleus. In all the above cases of nuclear reactivation, a pronounced nuclear swelling, with increase in nuclear dry weight and dispersal of condensed chromatin, accompanied the reactivation. *Xenopus* brain cells swell 60-fold before synthesizing DNA in egg cytoplasm and hen erythrocytes swell 20–30-fold in heterokaryons.

The mechanism of this swelling is not known, but it has been shown to involve the entry of proteins from the host cytoplasm. Nuclei placed in *Xenopus* egg cytoplasm containing ^{3}H-labelled proteins take up ^{3}H-label under conditions inhibitory to protein synthesis[64] and ^{125}I-labelled histones injected into eggs with a transplanted nucleus are concentrated massively in the nucleus[65]. In heterokaryons between HeLa cells and chicken erythrocyte nuclei, it has been shown most elegantly that the enlargement of the erythrocyte nucleus involves the specific uptake of human nuclear proteins[66]. By the use of specific antibodies from autoimmune patients, human nucleolar and nucleoplasmic antigens were detected in nucleoli and nucleoplasm of the red cell nuclei; human cytoplasm-specific antigens were not found in the chick nuclei, which suggests that enlargement does not occur by passive uptake of proteins

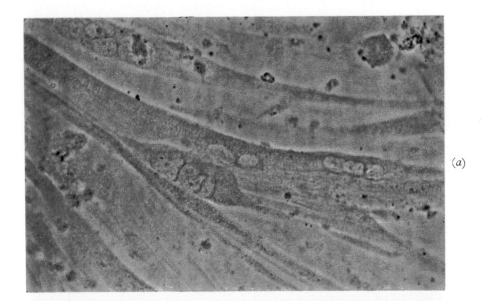

(a)

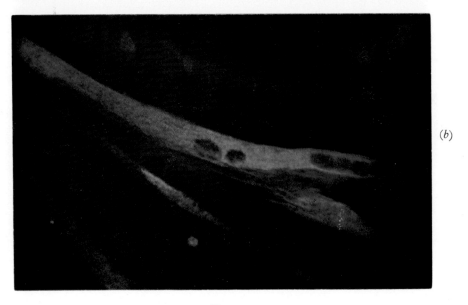

(b)

25 μm

Plate 1.1 Myotubes formed from rat myoblasts into which chicken erythrocyte nuclei had been introduced by Sendai virus-mediated fusion[42].

(a) The preparation stained by the immunofluorescence technique, using antibodies specific for chicken myosin. In the brightly fluorescing myotubes, chicken myosin synthesis, directed by the erythrocyte nuclei, has been evoked by the environment of the rat muscle cell.

(b) The same preparation seen by phase contrast microscopy.

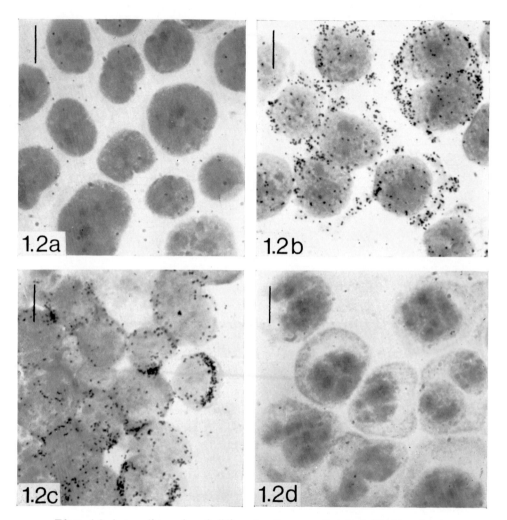

Plate 1.2 Autoradiographs of different cell types after hybridization *in situ* with cDNA prepared from mouse haemoglobin messenger RNA, using reverse transcriptase from avian myeloblastosis virus [84].

(a) Friend leukaemia cell line 707, uninduced.
(b) Friend leukaemia cell line 707, 4 days after being induced to undergo erythroid differentiation by treatment with DMSO.
(c) Cells from erythropoietic tissue (mouse fetal liver).
(d) Mouse fibroblast cell line LS.
Scale = 10 μm.

from the cytoplasm. With antibodies to chicken nucleoli it was shown that after the initial influx of human antigens, chicken antigens began to accumulate in the chicken nucleoli and were later found in the HeLa cell nuclei as well, showing that the exchange of antigens can be reciprocal. Similar experiments using rat epithelial cells have shown that the chicken nuclei compete with one another and with the host nuclei for host factors[67]. As the proportion of erythrocyte nuclei in the heterokaryons is increased, reactivation becomes slower and host RNA synthesis may be depressed. Rat cell/erythrocyte heterokaryons have also been used to show that nucleoplasmic RNA polymerase activity (see p. 27) appears in the reactivated nuclei earlier than nucleolar polymerase activity. Whether these activities are due to host enzyme or to chick enzyme is not known.

Uptake of a host enzyme by the erythrocyte nucleus is indicated in experiments on DNA repair. UV-irradiated erythrocyte nuclei do not perform DNA repair synthesis, but in heterokaryons with HeLa or mouse cells they become able to do so in the absence of erythrocyte-directed protein synthesis. That this involves the entry of host repair enzyme is borne out by the failure of the erythrocyte nuclei to repair DNA in heterokaryons with fibroblasts from *Xeroderma pigmentosum* patients, which are genetically deficient in DNA repair[68].

Transplantation and reactivation experiments show that there are considerable possibilities for interchange between the nucleus and cytoplasm and that major nuclear functions are affected by cytoplasmic factors. They also show that the expression of genes can be changed by introducing a nucleus into foreign cytoplasm. In the development of a tadpole from a differentiated cell nucleus placed in egg cytoplasm, the nucleus loses its former determination and is completely re-programmed. The range of genetic expression shown by avian red cell nuclei in heterokaryons also appears to be dictated by the host cell. When introduced into rat myotubes they begin to direct the synthesis of myosin[42] (Plate 1.1), and when introduced into mouse or HeLa cell cytoplasm they express various genes which are compatible with the state of differentiation of the host cell (see p. 16) but they fail to synthesize haemoglobin[69]. However, it is not possible in these cases to know whether the red cell nuclei have been heritably reprogrammed by the host cell, because they do not undergo cell division. The capacity to wipe out nuclear programming has thus been clearly shown only by germ cell cytoplasm.

The Primary Gene Product – Messenger RNA

Identification of mRNA

The initial step in the synthesis of region-specific or tissue-specific proteins in differentiation must be the synthesis of the corresponding messenger RNAs (mRNAs) and therefore much attention has been paid to mRNA synthesis and its distribution and utilization in embryos. In the past it has been difficult to obtain very satisfactory information on this subject because of the lack of definitive criteria for recognizing mRNA and the impossibility of identifying the mRNAs for individual proteins. On average, the base composition of mRNA was expected to be close to that of the whole DNA, as compared with the compositions of transfer RNA (tRNA) and ribosomal RNA (rRNA), and on a size basis many mRNAs were expected to fall between tRNA (4S) and the two rRNAs (18S and 28S). However, much DNA-like RNA (dRNA) was found to be degraded without leaving the nucleus and did not therefore fit the rôle of a cytoplasmic messenger[70], and preparation of mRNA on a basis of size was complicated by the presence of ribosomal RNA breakdown products.

Quantitative changes in dRNA during development have been followed using DNA–RNA hybridization and qualitative changes have been studied by the competition-hybridization method, in which the base sequence similarities between two RNA preparations are computed from the capacity of one (non-radioactive) preparation to compete with another (radioactive) one for sites on the DNA template. It has been shown that dRNA synthesis begins well before gastrulation in both sea-urchin and frog embryos[15]. It has also been shown that differentiating frog tissues contain more dRNA than 'undifferentiated' ectoderm from the same stage, that only some of the RNA present in the tail-bud stage of a frog is already present at gastrulation and that the gastrula lacks RNA which is present in neural tissue[17]. A number of other experiments show both tissue and stage specificity of dRNA synthesis[71]. These experiments can only detect gross differences, however, and are not easy to interpret. A detailed appraisal of the kinetics of DNA–RNA hybridization using conditions of RNA excess have shown that to achieve detectable levels of hybridization for RNA complementary to the non-repeated DNA sequences of the eukaryotic genome requires very high concentrations of RNA and very long hybridization times[72]. In some of the earlier experiments it is likely that the bulk of the RNA which was being studied was not in fact mRNA but RNA complementary to

22

repetitive DNA sequences and although developmental differences in this kind of RNA may prove to be of great interest, their significance is not clear at present.

Recent developments in the mRNA field promise that much more informative experiments are now possible. The specific mRNAs for a number of proteins (haemoglobin[73]; silk fibroin[74]; histones[75]; lens α-crystallin[76]; myosin[77]; immunoglobulins[78]) have been purified to a greater or lesser degree and some have been shown to direct the synthesis of their characteristic proteins in cell-free protein synthesizing systems[73, 75–78], and in some cases in living frogs' eggs[79, 80]. In addition it has been found that most mRNAs and some HnRNA have a length of polyadenylic acid (poly A) attached at the 3' end[81]. As well as having regulatory significance (see p. 30) this sequence can be exploited experimentally in several ways. First, definitive messenger RNA can be identified in, and prepared from, cytoplasm by arranging for the poly A sequences to bind to an inert substratum, either by virtue of the intrinsic 'stickiness' of the poly A in high concentrations of KCl[82] or by the binding between poly A and the synthetic polynucleotides polyuridylic acid (poly U) or polydeoxythymidylic acid (poly dT) which have previously been immobilized on the substratum[78]. Secondly, if a short length of polymerized deoxythymidylic acid (oligo dT (10)) is hybridized to the poly A sequence *in vitro* as a primer, RNA-directed DNA polymerase ('reverse transcriptase') can be used to synthesize a highly radioactive strand of DNA (cDNA) complementary to a specific purified mRNA[83]. This cDNA can then be used to detect specific mRNAs by hybridization experiments. A particularly exciting prospect is to use this copy to locate specific mRNAs *in situ* in cells, by the autoradiographic cytological hybridization method which has been used to discover the distribution of a number of DNA fractions in chromosomes. The feasibility of locating messenger RNA sequences by *in situ* hybridization has already been demonstrated by experiments in which radioactive DNA complementary to haemoglobin mRNA was hybridized to red cell precursors and Friend Leukaemia cells[84] (Plate 1.2). Radioactive poly U has also been used to detect the poly A sequences of mRNA and HnRNA *in situ* in various cell types[85].

Stability of mRNA

The use of actinomycin D (AMD), which prevents RNA synthesis by binding to the guanine bases in DNA, has shown that mRNAs may be synthesized well in advance of their utilization in protein synthesis. At low concentrations of the order of $0.05\ \mu g\ ml^{-1}$ it selectively inhibits rRNA synthesis, but at higher concentrations ($> 1\ \mu g\ ml^{-1}$) it inhibits virtually all RNA synthesis. The eggs of frogs and sea-urchins have both been shown to contain stored mRNAs which were made during oogenesis[15]. In frog oocytes, incorporation of radioactive uridine into RNA along the loops of lampbrush chromosomes can be followed by autoradiography and this incorporation ceases before fertilization. In sea-urchin embryos the addition of AMD at fertilization does not affect the rise in protein synthesis which occurs immediately after fertilization, but it inhibits a second rise which normally occurs at gastrulation[15]. This has been taken to show that both increases depend on stored RNA, the first on mRNA synthesized during oogenesis and the second on mRNA made during cleavage. Ribonucleoprotein complexes, some of which have been called 'informosomes', have been prepared from sea-urchin eggs and shown to release template RNA after trypsinization[86]. The exact nature of these particles and of their control is not understood. Recent work suggests that one of the mRNA species stored in this way codes for microtubular proteins[87].

Inhibition of transcription by AMD has been used to show that some mRNAs involved in the terminal stages of differentiation are long-lived. Haemoglobin mRNA in reticulocytes[88] and silk fibroin mRNA in silk worms[74] have lifetimes of several days and stable mRNAs, allowing differentiation in the presence of AMD, are involved in the differentiation of lens cells[89], down feather cells[90], pancreas cells[91], muscle cells[92] and others. In lens cells and reticulocytes the mRNAs continue to be translated after the cells have lost their nuclei. By measuring the decay of protein synthesis in the presence of AMD, an average lifetime of 2–4 h has been estimated for other mRNAs. However, the measurement of lifetimes by this method has been shown to be suspect, because AMD can cause the frequency of initiation of translation to be lowered and thus a low estimate of the mRNA lifetime can be obtained[93]. Recent work, using the terminal poly A sequences to quantitate mRNA directly and using no inhibitors of transcription, has shown the average lifetime of mRNA in exponentially growing HeLa cells and mouse L cells, to be 15–24 h[93, 94].

Mechanisms of Control of Gene Expression

Introduction

As yet, our knowledge of the molecular mechanisms by which differential gene action is achieved is scanty. A question which is much discussed is – to what extent is gene action controlled at source, by the direct limitation of transcription of genes into messenger RNA? Various models for transcriptional control of eukaryote gene expression, founded largely on bacterial systems, have been put forward. However, although Jacques Monod's dictum that 'what is true for *E. coli* is true for the elephant' is probably true for the molecular principles behind regulatory mechanisms, it would be rash to assume that two organisms differing so much in both structural and functional organization use identical methods for gene regulation. The bacterial control systems allow organisms which accommodate their whole repertoire of gene expression in one cell, and whose chromosomes are in direct contact with the cytoplasm, to respond rapidly and reversibly to fluctuating supplies of nutrients. The dictates of cellular economy and expediency may be very different in cells in which the chromosomes are complex structures contained within a nuclear membrane and in which regulation of gene expression occurs at several levels: first the activation and fixation of restricted gene programmes which constitute competence and determination, secondly the specific triggering of determined patterns by hormonal inducers and finally the reversible modulations of enzyme synthesis which can be induced in fully differentiated cells.

Bacterial messenger RNAs are short-lived, with an average life of the order of 1 min. Transcription, translation and degradation of the message follow one another in rapid succession[95, 96] and, as described in the next section, controls operate at the level of transcription. Eukaryotic mRNAs generally have longer lives than this (see p. 23) and it is known that genetic expression can be controlled by regulating the use of such mRNAs and their products as well as by controlling their transcription.

Bacterial models

In bacteria, highly developed genetic techniques and the development of systems for following protein synthesis *in vitro* as a DNA-dependent reaction have made it possible to characterize various regulatory mechanisms in great detail. A number of examples are described briefly here because they illustrate some general points which might be expected to apply in other systems of gene regulation; for instance, the need for protein-nucleic acid recognition, the possibility of negative or positive control and the intervention of effectors which can control a number of different genes co-ordinately.

In bacteria, specific small molecule inducers influence the rate of synthesis of enzymes by controlling transcription of the relevant enzyme structural genes into mRNA. Both positive (activating) and negative (repressing) controlling elements have been identified in E. coli. In the β-galactosidase (lac) system an 'operon', or transcriptional unit, containing three adjacent structural genes, is under negative control by a repressor protein. In the absence of inducer this protein binds to an 'operator' site on the DNA and prevents RNA polymerase (bound to a 'promoter' site) from moving along the DNA; specific inducers of the lac operon bind to the repressor protein and cause it to lose its affinity for the operator site, thereby allowing transcription to proceed[97]. The three genes of the arabinose operon appear to be regulated in a similar manner, but with an additional element of positive control. Reaction with the inducer converts the repressor protein into an activator which is required to act with another DNA site before transcription can begin[98].

The CAP locus in E. coli mediates in a general repression of many catabolic enzymes which occurs in the presence of high energy catabolites such as glucose or glucose-6-phosphate. It has been found that the CAP protein and cyclic AMP are needed for in vitro transcription of the lac operon, which is one of the loci subject to this repression. The concentration of cyclic AMP in E. coli varies inversely with that of the repressing catabolites; it seems likely, therefore, that the CAP protein exerts an integratory control over the catabolite-repressible genes, activating their transcription only when the supply of high energy substrates falls[24].

Another type of regulatory mechanism found in prokaryotes depends on the existence of multiple forms of RNA polymerase with different transcriptional specificity. This has been particularly favoured by some as a possible model for the control of expression of batteries of genes in eukaryote differentiation. In bacteria, the irreversible changes in transcription which occur during sporulation and development of lytic phage have been shown to involve various types of change in RNA polymerase[99]. These include changes in one of the subunits of the polymerase core enzyme, changes in the σ factor which binds to the core and is known to influence the specificity of initiation of transcription, and synthesis of a new polymerase altogether.

26

Transcriptional control in eukaryotes

Various attempts have been made to bring together current information about the structure and transcription of the eukaryote genome in models for transcriptional control. The main features which these models attempt to account for are the following: (1) The amount of DNA in the nucleus of a eukaryote cell may be 100–10 000 times greater than in a bacterial cell. Considered guesses as to the number of genes in eukaryotes and estimates of the mutational load which would have to be supported if a large proportion of this DNA represented structural genes have, however, led to the view that only a fraction of it, perhaps less than 5%, codes for proteins[100]. (2) Much of the DNA is composed of non-coding, repetitive sequences[101, 102]. The rôle of this DNA is not yet known; some of it is located in heterochromatic regions at the centromeres and telomeres of chromosomes and some forms 'spacer' regions separating repeated structural genes, as in the parts of the genome containing multiple copies of the genes for rRNA[103], tRNA[104], 5S RNA[105] and histone messenger RNA[106], but much of it is not accounted for. (3) There are two major classes of protein associated with the DNA. The histones are a homogeneous group of basic proteins which form a compact, supercoiled complex with DNA. The non-histone proteins are a more heterogeneous group of acidic proteins comprising about 50% of the total nuclear protein[107]. (4) Chromosomes show structural differentiation. In the giant polytene interphase chromosomes of some tissues in Diptera, dense DNA- and histone-rich bands alternate with less dense interbands, and one band and interband probably corresponds to one gene[108]. The 'lampbrush' chromosomes found in the oocytes of many organisms have a similar structure of alternating dense regions (chromomeres) and less dense regions. Coarser, species-specific banding patterns can also be seen in the metaphase chromosomes of various vertebrates using special staining methods[109]; their significance is not yet known. In metaphase chromosomes, areas of darkly staining heterochromatin can be distinguished from lighter euchromatin. Heterochromatin is inactive in RNA synthesis and tends to replicate later in the cell cycle than the active euchromatin[110]. In cases such as the inactivation of an X chromosome in female mammals or of the paternal chromosome set in the male mealy bug, chromosomes are facultatively heterochromatic, i.e. they become heterochromatic in response to their circumstances. Constitutive heterochromatin, usually found around the centromeres and sometimes in blocks elsewhere in the chromosomes, is consistently present. (5) A larger proportion of the RNA which is synthesized in the nucleus does not reach the cytoplasm, but is degraded in the nucleus[70, 111]. This included the so-called HnRNA (Heterogeneous nuclear RNA), a fraction of large RNA molecules with DNA-like base composition. Recent evidence supports the theory that HnRNA contains RNA precursors from which mRNA is released, possibly by a process similar to that which cleaves the 18S and 28S rRNAs from their common 45S precursor[103]. HnRNA from erythrocyte precursor cells has been shown to contain sequences which can direct haemoglobin synthesis[112], and in addition, poly A sequences are found attached to HnRNA molecules as well as to mRNA[80] (see p. 22).

Early models, influenced mainly by the first two points, proposed that the complexity of cells in higher organisms was controlled by hierarchies of regulatory molecules which reacted with binding sites in the DNA to regulate the transcription of single genes and of small and large batteries of genes in an integrated fashion[113, 114]. Repeated DNA sequences in these models were accounted for as loci analogous to bacterial repressor genes and promoter and operator sites, multiple copies of which were linked to many different structural genes as integrating controls.

Two more recent models, those of Crick[115] and Paul[116], acknowledge the need for hierarchical control, but lay more emphasis on the significance of the structural organization of chromatin in the control of transcription. In dipteran polytene chromosomes, differential gene action can actually be visualized[117]. Different regions of the polytene chromosomes of *Chironomus*, for instance, become puffed and active in RNA synthesis at different stages in larval development and these puffs can also be elicited by injecting moulting hormone into the larva. Both models propose separate rôles for bands and interbands, one to be the site of structural genes and the other to be a regulating site. Crick postulates that the bands are 'globular' DNA containing single-stranded (i.e. unpaired) DNA sites which are recognized by regulatory molecules which control the transcription of fibrous informational DNA in the adjacent interband. He proposes that the exposure of the recognition sites depends on the binding of chromosomal proteins to the bands. Paul's model, on the other hand, speculates that the genes are contained in the bands, but that because of the compact, probably supercoiled nature of the DNA-histone complex in the band region, RNA polymerase is prevented from initiating transcription there. He postulates that 'address' loci in the less tightly coiled interbands bind specific de-stabilizing agents, probably acidic nuclear proteins, and that this allows RNA polymerase to initiate transcription and then to continue transcription into the band region, with possibly the addition of more de-stabilizers as it proceeds. Once a portion of the genome is opened up in this way, it is suggested, that transcriptional controls of the bacterial type could operate within it. In both models large amounts of repetitive DNA are accounted for as regulatory DNA and an explanation for HnRNA precursors which are processed to give mRNA can be derived from Paul's model.

While the general application of these models might be questioned because they are founded on an atypical type of chromosome, they have the advantage of allowing for two levels of control, a coarse transcriptional control involving the masking of parts of the chromosome by chromosomal proteins and fine controls operating within specifically unmasked areas. Many of the elements of these theories are virtually impossible to test in our present state of knowledge. Evidence from various quarters supports the masking theory, however. The nuclei of different cell types show differences in the amount and distribution of condensed chromatin, which may signify the repression of different parts of the genome by the formation of supercoiled structures. The binding of histones to DNA appears to repress transcription in a non-specific manner and there is considerable support for a de-repressing rôle for non-histone proteins in regulating gene function[107]. Loci in polytene chromosomes can be shown histochemically to contain more non-histone proteins when puffed than when inactive, while the amount of histone does not change. Similarly, the transcriptionally active chromatin of sea-urchin pluteus larvae contains more non-histone proteins than the relatively inactive blastula chromatin and high levels have been found in regenerating liver cells compared with normal liver cells. There is also evidence that they can antagonize the inhibitory effect of histones on transcription *in vitro*.

So far, no factors limiting template specificity of transcription, like bacterial σ factors, have been found in eukaryotes, but in a number of animals more than one RNA polymerase has been identified[118, 119]. The enzymes are distinguishable in the basis of their α-amanitin sensitivities and optimal ionic strengths and Mn^{++}/Mg^{++} ratios. Polymerase II (or B) is nucleoplasmic and synthesizes DNA-like RNA. Polymerase III (or AIII) is also nucleoplasmic, but is α-amanitin-resistant, unlike polymerase II. Polymerase I (or AI) is restricted to the nucleolus, is α-amanitin-resistant and synthesizes rRNA. This polymerase and the mRNA which codes for it have been shown to be very labile compared with polymerase

II. It has been suggested that the effect of hormones in increasing rRNA synthesis is to stabilize polymerase I or its synthesis rather than to interact with the ribosomal genes themselves[120, 121].

A possibility for gene control at the DNA level which must also be considered is the differential amplification of genes in different cells. This occurs for ribosomal genes during the differentiation of oocytes in amphibia and possibly in other animals; the genes coding for ribosomal RNA precursors are amplified extrachromosomally more than 1000-fold and are used as templates for synthesis of rRNA for ribosomes which are stored in the egg[103]. However, there is no good evidence for gene amplification in any other systems. Recent DNA/mRNA hybridization experiments using conditions of DNA excess have shown that the number of genes coding for haemoglobin does not differ between cells committed to haemoglobin synthesis and other cell types[122] and similar results have been obtained for silk fibroin genes in the silkworm[123]. Another possibility is that genes coding for proteins required in large quantities at some stage of growth or differentiation might be present in the genome in multiple copies, as are the genes for the structural RNAs, tRNA, 5S RNA and rRNA[103-105]. So far, reiteration has been found only for the histone genes[106]. The genes coding for haemoglobin[122, 124] and silk fibroin[123] are present in very small numbers (5 or less) per haploid genome.

Post-transcriptional controls

It is a formal possibility that transcription in eukaryotes is uncontrolled and that all regulation of gene expression occurs by the interaction of controlling elements with the transcribed RNA. There are many points between the transcription of a gene and the appearance of a functional protein in the cytoplasm at which post-transcriptional controls might act. The array of mRNAs exported from the nucleus to the cytoplasm could be controlled by selective processing or transport mechanisms, while in the cytoplasm the translation of messengers and their protection from degradation could be selective. In models based entirely on post-transcriptional control it is in fact difficult to answer the question: 'how are the controlling elements themselves controlled?'. Also, there is experimental evidence for control of transcription, from dipteran polytene chromosomes and from the tissue-specificity of RNA transcribed from isolated chromatin, and it seems likely that, at least to some extent, the programming which determines which genes are to be 'on' and which 'off' in a cell operates at the chromosomal level. It is feasible, however, that a finer control of the genes that are 'on' i.e. control of the level at which they are expressed in different cells, or of the timing of their expression in development or maturation, takes place at a post-transcriptional stage.

In the nucleus, the post-transcriptional addition of poly A to the 3'-terminus of mRNA and the processing of mRNA from larger precursors are possible points for control. The significance of the poly A sequences is not known; possibly they are needed for the stable existence of mRNA either in the nucleus or in the cytoplasm. In the presence of cordycepin, an adenosine analogue thought to inhibit addition of poly A to mRNA, newly-synthesized mRNA does not appear in the cytoplasm and it has been shown that as mRNA 'ages' in the cytoplasm[125], its poly A sequences become progressively shorter[126]. Histone mRNA appears to be exceptional in not carrying a poly A sequence[80]. This fact, together with the apparent lability of histone mRNA compared with other mRNAs[127] and the finding that the histone genes are highly reiterated.[106], suggests that the expression of histone genes may be under a different type of control from other genes. Experiments on the reactivation of erythrocyte nuclei in heterokaryons have provided evidence that the export of mRNA to the cytoplasm depends on the presence of an active nucleolus[128] and recent experiments using hybridization *in situ* of radioactive poly U to detect the poly A sequences of mRNA have shown a concentration of hybridized radioactivity in the nucleolar regions of rat, mouse and human cells[85], which lends support to the idea that mRNA becomes associated with the nucleolus at some point in its history. So far, however, the lack of poly A in histone mRNA is the only evidence which has been found of selectivity in nuclear processing of mRNA.

In the cytoplasm, the number of protein molecules translated from an mRNA molecule will depend on the rate at which it is translated and on its stability. Both translation and degradation offer possibilities for the control of differential gene expression. Several models for selective translation have been provided by work on the translation of RNA phages in *E. coli*. There is evidence of differences in ribosomal binding specificity and frequency of initiation of translation between different mRNA sequences[129, 130] and also of feedback control of the translation of one sequence by the product of another[131]. Different species of the translational initiation factor f3, showing specificity for different RNAs, have been isolated from *E. coli*[132]. In eukaryotes, a requirement for tissue-specific ribosomal factors in the translation of haemoglobin and myosin mRNAs in heterologous cell-free systems has been reported[77], but a need for specific factors has not generally been found for protein synthesis *in vitro* and the ability of frog eggs and oocytes to translate purified globin, immunoglobulin and crystallin mRNAs[79, 80] also suggests that they are not of general importance. A difference in specificity between membrane-bound and free ribosomes has been inferred from the observation that cells making proteins for export are particularly rich in rough endoplasmic reticulum[133, 134]. This correlation does not hold in all cases, however. The secreted heavy and light immunoglobulin chains in mouse plasma-cell tumour cells are made on both bound and free ribosomes[135], while proliferation of rough endoplasmic reticulum, or a shift from free to bound ribosomes, is characteristic of hormonally-induced differentiation in both secretory and non-secretory cells[136]. Another possible control of translation, the regulation of the availability of different transfer RNAs in different cells, has also received some attention, but no conclusive evidence for the operation of this type of control has emerged[137]. Though there is little evidence, therefore, for translational specificity, the study of haemoglobin synthesis has shown that different mRNAs in a cell can be translated at different rates. The polysomes synthesizing haemoglobin α-chains contain fewer ribosomes than those synthesizing β-chains and it has been shown that this difference is due to slower initiation of translation of the α-chain mRNA[138].

For cells which become determined early in development and then undergo terminal differentiation in response to another stimulus at a later stage, it is possible that the synthesis of cell-type-specific mRNAs begins at determination, but that these mRNAs are not translated until after the second stimulus. Some recent experiments on erythroid differentiation have shown that no translatable haemoglobin mRNA is detectable in precursor cells before erythropoietin stimulation, although it might be present in an inactive form[139]. In Friend Leukaemia cells, which undergo characteristic erythroid differentiation when treated with dimethylsulphoxide (DMSO), no haemoglobin mRNA could be detected by *in vitro* hybridization with haemoglobin cDNA before DMSO treatment[140], although a small amount was detected by *in situ* hybridization (see Plate 1.2.)[84]. These findings suggest that haemoglobin mRNA accumulates in response to the second inducing stimulus, which may act by stimulating transcription. A post-transcriptional control of this step cannot be excluded, however. mRNA synthesized in the precursor cells may be rapidly degraded and may only rise to detectable levels when stabilized in some way. Nucleases are ubiquitous in cells and it is likely that any mRNA which is not protected in some way is attacked by them.

There is evidence that mRNA *in vivo* is always associated with protein[141], which may have a protective function, and it has also been suggested that mRNA which is not being translated is more susceptible to degradation than mRNA in polysomes[93]. If degradation automatically destroys mRNAs which are not being used, failure to detect an mRNA does not prove that it has not been synthesized.

Both translation and degradation are delayed in the case of the stored mRNAs of sea-urchin and frog eggs, whose information is not expressed until it is 'unmasked' (see p. 24). Whether the differentiation of these eggs is translationally controlled by specific unmasking of different stored mRNAs at different times or positions in the embryo, or whether the stored information is released piecemeal at fertilization, is not known. An interesting, though speculative, hypothesis implicating masked mRNAs in the activation of genes by acidic chromosomal proteins has been put forward by Paul[107]. Messenger RNAs for specific acidic proteins might be stored in the egg and selectively unmasked by specific epigenetic influences in different parts of the developing embryo. The acidic proteins translated from them would then activate specific batteries of genes, including the gene coding for their own synthesis, so that the activation would be self-perpetuating.

Experimental evidence for translational controls has come mainly from late stages of cytodifferentiation and accounts for the modulation of translation of particular mRNAs, rather than for the origin of differential gene expression. For example, translation of globin mRNA in reticulocyte cell sap can be increased by the addition of haem, although the mechanism of this stimulation is not fully understood[142]. It is thought that the reversible modulations of enzyme synthesis which can be induced in fully differentiated cells may be controlled by stimulation of mRNA translation, rather than de-repression of mRNA synthesis. A detailed study of liver tyrosine aminotransferase (TAT) induction by steroids has led to the proposal by Tomkins of a translational model for the induction[143]. The model states that, in the absence of inducer, a labile protein or RNA repressor prevents translation of TAT mRNA and promotes its degradation; the inducing steroid antagonizes the repressor in some way and TAT synthesis rises to a new level, either due to accumulation of stable message or to an increase in the initiation of translation. Removal of the inducer leads to the establishment of repression again. The critical experimental evidence for this model was the finding that the decay of TAT synthesis after removal of the inducer does not occur in the presence of AMD; this suggested that de-induction required

33

the transcription of a labile RNA. Tomkins has pointed out[144] that a number of functions other than TAT induction show a superficially similar 'paradoxical' response to AMD, which may indicate that labile controlling factors operate in a variety of systems.

A final type of post-transcriptional control mechanism which may be mentioned is the control of protein assembly. Many proteins are assembled from more than one type of monomeric subunit and some of these oligomeric proteins occur in multiple forms, known as isozymes, in the same cell[145]. The A and B subunits of lactate dehydrogenase (LDH), for example, associate randomly in binomial proportions to give five isozymes (A_4, A_3B, A_2B_2, AB_3 and B_4). The different distributions of these isozymes which can be found in different tissues in mammals can be explained by the preferential synthesis of one or the other subunit. In a number of fish, however, non-random tetramerization occurs and different non-random patterns are found in different tissues. Re-association of purified fish LDH A and B subunits *in vitro* yields the random pattern of isozymes seen in mammals. In fish, therefore, it appears that tissue-specific factors can influence the association of LDH subunits to form isozymes[146].

The genetics of regulation in eukaryotes

Whereas many details of gene regulation in bacteria (see p. 25) were initially worked out by inference from genetic experiments, it is difficult to study gene regulation in eukaryotes by a genetic approach. The many mutations affecting embryonic development are recognized primarily by morphological features and it is virtually impossible to identify the lesion at a molecular level; some may be mutations at control loci analogous to those found in bacteria, others may be biochemical lesions which alter development by interfering with important cellular inter-actions.

The diptera provide the best evidence for specific controlling genes affecting expression at the transcriptional level (see p. 27). A number of mutants exist which are unable to form puffs at certain sites and flies made heterozygous for these mutations provide evidence for both diffusible factors and chromosomally-limited factors controlling puff formation. With one type of mutation, puffs will occur on both mutant and normal homologues whether they are synapsed or not; with another, both homologues puff only if they are in synaptic contact and with a third, the failure to puff is apparently maintained autonomously by the affected homologue[147]. In the first example, the mutation is probably not

34

linked to the puff site; in the others it is closely linked. Another type of regulatory effect in diptera is the 'position effect', where translocation of a gene into, or near to, constitutive heterochromatin can repress its expression[148]. This does not necessarily provide a model for the normal regulation of the genes, but it is interesting to speculate whether investigation of the effect of heterochromatin in this case could throw any light on the models of Crick and Paul (see p. 27) which propose that gene regulation is influenced by the physical state of the chromatin. Similar position effects have been described in maize[149] and the mouse[150]. In maize some genes can be shown to be controlled by interaction between two loci, one next to the gene and possibly sited in heterochromatin and another at a distance. In the mouse, inactive X chromosome heterochromatin has a localized inactivating effect on genes at either end of a segment of autosomal euchromatin inserted into the X by translocation.

Where individual enzymes or groups of enzymes are affected by mutation it is possible to propose tentative criteria for identifying regulatory mutants in vertebrates. For example, the dominance of a mutation which prevents the induction of aryl hydrocarbon hydroxylase in mice has been taken as an indication that it probably affects a regulatory gene with a diffusible product[151]. However, it is not possible to distinguish conclusively between transcriptional and translational control from genetic experiments, without the help of cytogenetic methods such as are possible with diptera, or of the reconstruction of regulatory systems *in vitro*, as has been possible with bacteria.

A regulatory system in vertebrates which has been characterized genetically is the induction of growth and differentiation of the male sex organs from the Wolffian ducts and urogenital sinus of the mouse embryo[100]. This process requires the action of testosterone from the embryonic testes and the presence in the target tissues of the product of a single sex-linked gene. In castrated embryos or embryos possessing the testicular feminization mutation (Tfm) in the gene in question, sex organ development is prevented. So, also, are the hypertrophy and the increases in enzyme synthesis which are normally induced in kidney proximal tubules by testosterone, although uptake and metabolism of testosterone in the kidney cells can be shown to be normal. Taking into account evidence that steroids may control gene expression at a translational level (see p. 8) Ohno has suggested that the wild-type Tfm protein binds to specific mRNAs, preventing their translation. In the presence of inducing 5-α-dihydrotestosterone (DHT), an active metabolite of testosterone, it is proposed that the repressor leaves the mRNAs and moves to the nucleolus, where it initiates hypertrophy by stimulating RNA polymerase I to produce ribosomal RNA. In this scheme, the Tfm mutation would cause the translational repressor to be insensitive to DHT. Ohno has also proposed that a mutation, G^S, which produces a slightly altered β-glucuronidase which is poorly induced in kidneys, is in fact a mutation to the Tfm repressor-binding site on β-glucuronidase mRNA. The relationship of the Tfm protein to the cytoplasmic/nuclear steroid receptor has not yet been fully clarified, but the testosterone receptor does not appear to be changed by the Tfm mutation[152], which suggests that the two may not be identical.

Somatic cell genetics

In the past few years, development of techniques for performing quasi-genetic experiments on somatic cells has made it possible to look at the regulation of differentiation in quite a different way. The process of cell fusion, either spontaneously-occurring or induced by addition of inactivated Sendai virus, is used to fuse a cell expressing a particular differentiated function with one not doing so. In a proportion of the heterokaryons thus formed, the nuclei fuse at mitosis and a hybrid cell is formed. Clones of such hybrid cells can be isolated from the parental cells, by the use of culture media which specifically select against the parent cells, or by screening for clones which are different in appearance or behaviour from the parents, and the effect of fusion on the expression of the differentiated function can be analysed[153].

In a number of hybrid cells formed in this way, differentiated functions have been found to be extinguished. These include liver functions[154, 155], myogenic properties[42] and synthesis of immunoglobulin[156], melanin[157] and glycerol-3-phosphate dehydrogenase[158]. In these cases, expression of the differentiated function is suppressed by a factor or factors contributed to the hybrid by the so-called 'non-differentiated' parent cell (this cell is in fact a differentiated cell, but is not expressing the function under consideration). The suppression has been shown not to be due to physical effects of cell fusion and has also been shown to be reversible. In hybrids between mouse renal carcinoma cells and human embryo lung fibroblasts, a mouse kidney-specific esterase, ES2, ceased to by synthesized[159]. Mouse/human hybrids tend to segregate clones which have lost some of their human chromosomes, however, and some of these clones began to synthesize ES2 again. In addition, the reappearance of the enzyme could be correlated with the loss of the human C 10 chromosomes. As well as showing that the initial loss of ES2 was not due to the loss of both mouse ES2 genes, this shows that ES2 suppression in the hybrid and presumably in the human fibroblast depends on the continued presence of a particular chromosome. In some cases, the repression of a differentiated function has been partly overcome by doubling the number of chromosomes contributed to the hybrid by the differentiated parent, which suggests that the mechanism is sensitive to gene dosage[157, 158].

Some differentiated characters are not extinguished in hybrids. Of four liver-specific functions shown by rat hepatoma cells, three of them (high base-line activity of tyrosine amino transferase (TAT); inducibility of TAT by glucocorticoids[154] and liver fructose-1,6-diphosphate aldolase synthesis[155]) are repressed in hepatoma cell/mouse fibroblast hybrids, while the fourth (albumin synthesis) continues to be expressed[160]. This suggests the existence of more than one type of control mechanism for differentiated functions. In neuroblastoma cell/fibroblast hybrids, a number of neural functions (electrical responses, acetylcholine esterase synthesis and neurite formation) are also retained by the hybrids[161].

So far in cell fusion experiments the fibroblast has been used almost exclusively as the 'undifferentiated' parent cell, so that the general applicability of the findings must remain in question until other cell types have been tested. It should also be borne in mind that some of the cells used are neoplastic and aneuploid and might therefore show abnormal regulation properties. However, the fact that ES2 suppression in mouse/human hybrids depends on the presence of particular human chromosomes suggests that these experiments reveal specific control mechanisms and not merely a general disruption of cell functions caused by the mingling of two different types of cell. Two levels of control by the human contribution to these kidney/fibroblast hybrids can be distinguished, one of them diffusible and one autonomous. Synthesis of mouse ES2 is reversibly repressed by the presence of human chromosomes, but the mouse gene contribution does not become heritably re-programmed to repress ES2 production as a result of its co-existence with the human contribution. Both genomes appear to maintain their states of determination autonomously.

It is interesting to compare this autonomy with the somatic inheritance of X-chromosome inactivation in female mammals. In any one cell, either of the genetically equivalent X chromosomes can be the one which becomes inactivated, but by cloning cells it has been shown that once a particular X chromosome has been inactivated it inherits the inactive state through many cell divisions[162]. The mechanism of this inheritance is not fully understood. A single locus affecting inactivation has been described and it has been suggested that its product interacts with a number of recognition sites on one of the X chromosomes[150]. A factor which may contribute to the maintenance of the inactivation is the late replication of the inactive X^{110}. Studies have been made of auto-somal replication patterns in differentiated cells to see whether they show any differences which might indicate that differential gene activity is maintained in this way, but no differences have been found[163].

Many cell hybridization experiments have been done using parent cells from different species, in order to distinguish the products made by the hybrids on grounds of their species specificity. In two studies, the repression of a differentiated function in hybrid cells has been shown to occur in intraspecific as well as in interspecific hybrids, which confirms the validity of this approach. However, both these and the experiments with heterokaryons on pp. 18–21 raise the question of the comparability of regulatory mechanisms between species. It can be argued that the element of molecular interaction inherent in regulation would tend to mediate against evolutionary drift in such processes and support for this argument can be drawn from the evolutionary conservation shown by histones[164] and hormones[165]. There is little information on the conserva-tion of specific mechanisms, but various examples show that many basic cellular and developmental processes are probably controlled in similar ways over a wide range of animals. Viable hybrid cells can be made by fusing such bizarre combinations of cells as human and mosquito[166] (an interphylum hybridization) or hamster and Galapagos tortoise[167]. In developmental systems, chimaeric muscle syncytia will differentiate from mixtures of mouse[42], bird, rabbit or calf[168] myoblasts with rat myoblasts and many inductive interactions have been shown to be effective between tissues from different species.

Conclusion

The way in which the differentiation of cells is controlled during development still holds many mysteries. The process can, however, now be studied by new techniques which were not available to the earlier developmental biologists whose work forms the foundation for modern theories of differentiation. Improved methods for cell culture allow the study of various differentiating systems in isolation and technical advances in the field of molecular biology have made it posssible to investigate differential gene expression at the molecular level.

Of the various stages in the differentiation of a cell – the development of competence, the induction of determination, the inheritance of the determined state and the expression of the differentiated state – the last is the most clearly understood at present. A considerable amount is known about the events of terminal differentiation and the modulation of gene expression in terminally differentiated cells. Whether these processes are controlled at the level of transcription or at a post-transcriptional level may be resolved by the development of cell-free systems and the use of assays for specific messenger RNAs.

The mechanisms by which the earlier stages of differentiation are controlled are poorly understood. Further work on the acidic chromosomal proteins may show whether these proteins have a specific rôle in the induction or maintenance of determination. Experiments with somatic cell hybrids and with cells containing foreign nuclei may also provide insight into the way in which cells inherit their determined state. Our knowledge of the steps by which cells first become different from each other is still very meagre. The transitory nature of the signals and the subtlety of the cellular interactions involved mean that the processes of early embryogenesis are difficult to dissect experimentally. It is the elucidation of these processes which presents the major challenge for the future.

Summary

This chapter considers the basic concepts of differentiation and growth of cells, with particular reference to vertebrate species. At the cellular level, differentiation is considered in the wide sense of the development of differences between the cells of an organism, beginning with the zygote. At the molecular level, differentiation is equated with differential gene expression and the importance of environmental influences, cell interactions and interactions between nucleus and cytoplasm in bringing about differential gene programming in cells is stressed.

The second and third sections deal with differentiation and growth at the tissue and cell levels. The stimuli which initiate and control these processes from the fertilized egg to the neonatal animal, are described and discussed. Two sources for the initial instructions for differentiation, pre-formed information carried by the egg and environmental signals arising during cleavage, are considered. The concepts of competence and determination are defined in connection with embryonic induction and the rôle of cell interactions, the nature of inducing stimuli and the existence of gradients of information in induction are discussed. The rôle of hormones in processes of maturation is described. Controls of growth operating at the whole body level and at the tissue level are discussed. The relationship between growth and differentiation and the possibility that mitosis has a specific rôle in cell differentiation are considered.

Some basic properties of differentiated cells are outlined. Differentiation is shown to be reversible under certain circumstances, but experiments are also described which show the determined state of a cell is normally very stable and is inherited by its progeny. Examples of cell metaplasia are discussed. Evidence is given for extensive interaction between nucleus and cytoplasm in the control of nuclear functions such as DNA and RNA synthesis and mitosis.

The fourth and fifth sections are concerned with the molecular mechanisms by which differentiation is brought about. Recent advances in the study of messenger RNA, including methods for the positive identification of specific mRNA molecules and revised estimates for the stability of mRNAs in cells, are reviewed. Various mechanisms for the control of mRNA synthesis and its utilization by cells are proposed and the evidence for their occurrence is discussed. A brief description of various bacterial models for transcriptional control is given and then two models for transcriptional control in eukaryotes are described in detail and are related to the available information on the structure and function of eukaryote chromosomes. The evidence for various types of post-transcriptional control mechanisms, including the processing of mRNA, translation of mRNA and control of protein subunit assembly, is considered.

Finally, the contribution of genetic studies to the understanding of the control of differentiation is evaluated. An example of classical genetic analysis of a mammalian regulatory system is given. The effect of somatic cell fusion on the expression of differentiated functions of cells is described and the possibility of autonomous and non-autonomous controls of gene expression is discussed.

Abbreviations

DNA deoxyribonucleic acid
RNA ribonucleic acid
mRNA messenger RNA
rRNA ribosomal RNA
tRNA transfer RNA
cAMP adenosine 3' : 5'-cyclic monophosphate
AMD actinomycin D
poly A polyadenylic acid

References

1. BALINSKY, B. I. (1970), *An Introduction to Embryology*, W. B. Saunders Co.: Philadelphia and London.
2. HARTSOEKER, N. (1694), *Essai de Dioptrique*, Jean Anisson: Paris.
3. GREENGARD, O. (1971), 'Enzymic differentiation in mammalian tissues', *Essays in Biochemistry*, **7**, 159–205.
4. HÖRSTADIUS, S., JOSEFSSON, L. and RUNNSTRÖM, J. (1967), 'Morphogenetic agents from unfertilised eggs of the sea urchin *Paracentrotus lividus*', *Developmental Biology*, **16**, 189–202.
5. HILLMAN, N., SHERMAN, M. I. and GRAHAM, C. (1972), 'The effect of spatial arrangement on cell determination during mouse development', *Journal of Embryology and experimental Morphology*, **28**, 263–278.
6. GLASS, L. E. (1971), 'Transmission of maternal proteins into oocytes', *Advances in the Biosciences*, **6**, 29–61. Schering symposium on *Intrinsic and extrinsic factors in early mammalian development*. Ed. Raspé, G., Vieweg, Pergamon Press.
7. McLAREN, A. (1972), In *Reproduction in Mammals*. Eds. Austin, C. R. and Short, R. V., Part 2, Ch. 1 ,Cambridge University Press.
8. FORTUYN, A. B. D. (1939), 'A cross of gene environment as a means of studying the inheritance of some quantitative characters in *Mus musculus*', *Genetica*, **21**, 243–279.
9. McLAREN, A. and MICHIE, D. (1958), 'Factors affecting vertebral variation in mice', *Journal of Embryology and experimental Morphology*, **6**, 645–659.
10. BLACKLER, A. W. C. (1966), 'Embryonic sex cells of amphibia'. In *Advances in Reproductive Physiology*, **1**, 9–28. Ed. McLaren, A., Logos Press: London.
11. KOCHER-BECKER, U. and TIEDEMANN, H. (1971), 'Induction of mesodermal and endodermal structures and primordial germ cells in *Triturus* ectoderm by vegetalising factor from chick embryos', *Nature*, **233**, 65–66.
12. MINTZ, B. (1971), 'Clonal basis of mammalian differentiation'. In *Symposium of the Society for experimental Biology*, **25**, 345–370. Eds. Davies, D. D. and Balls, M., Cambridge University Press.

13. JONES, K. W. and ELSDALE, T. R. (1963), 'The culture of small aggregates of amphibian embryonic cells *in vitro*', *Journal of Embryology and experimental Morphology*, **11**, 135–154.

14. GARTLER, S. M. and NESBITT, M. N. (1971), 'Sex chromosome markers as indicators in embryonic development', *Advances in the Biosciences*, **6**, 225–254. Schering symposium on *Intrinsic and extrinsic factors in early mammalian development*. Ed. Raspé, G., Vieweg, Pergamon Press.

15. HAMBURGH, M. (1971), *Theories of Differentiation*, Arnold: London.

16. GRÜNEBERG, H. (1963), In *The pathology of development*, Ch. 6, Blackwell: Oxford.

17. TIEDEMANN, H. (1971), 'Induction and cell interaction in embryonic development', *Advances in the Biosciences*, **6**, 477–493, Schering symposium on *Intrinsic and extrinsic factors in early mammalian development*. Ed. Raspé, G., Vieweg, Pergamon Press.

18. LAWRENCE, P. A. (1971), 'The organisation of the insect segment'. In *Symposium of the Society for experimental Biology*, **25**, 379–390. Eds. Davies, D. D. and Balls, M., Cambridge University Press.

19. WOLPERT, L., HICKLIN, J. and HORNBRUCH, A. (1971), 'Positional information and pattern regulation in regeneration of *Hydra*'. In *Symposium of the Society for experimental Biology*, **25**, 391–415. Eds. Davies, D. D. and Balls, M., Cambridge University Press.

20. BOHN, H. (1970), 'Interkalare Regeneration und segmentale Gradienten bei den Extremitäten von *Leucophaea*-Larven (Blattaria)', *Wilhelm Roux' Archiv*, **165**, 303–341.

21. TATA, J. R. (1971), 'Hormonal regulation of metamorphosis'. In *Symposium of the Society for experimental Biology*, **25**, 163–181. Eds. Davies, D. D. and Balls, M., Cambridge University Press.

22. WEBER, R. (1967), 'Biochemistry of amphibian metamorphosis'. In *The biochemistry of animal development*. Ed. Weber, R., Vol. 2, Ch. 5, Academic Press: London.

23. In *Advances in the Biosciences*, **7** (1971). Schering symposium on *Steroid hormone receptors*. Ed. Raspé, G., Vieweg, Pergamon Press.

24. PASTAN, I. and PERLMAN, R. L. (1971), 'Cyclic AMP in metabolism', *Nature New Biology*, **229**, 5–7.

25. JOST, J-P. and RICKENBERG, H. V. (1971), 'Cyclic AMP', *Annual Review of Biochemistry*, **40**, 741–774.

26. NEEDHAM, A. E. (1964), in *The growth process in animals*, Pitman: London.

27. FANKHAUSER, G. (1945), 'The effects of changes in chromosome number on amphibian development', *Quarterly Review of Biology*, **20**, 20–78.

28. BULLOUGH, W. S. and DEOL, J. U. R. (1971). 'The pattern of tumour growth'. In *Symposium of the Society for experimental Biology*, **25**, 255–274. Eds. Davies, D. D. and Balls, M., Cambridge University Press.

29. SUMMERBELL, D. and WOLPERT, L. (1972), 'Cell density and cell division in the early morphogenesis of the chick wing', *Nature New Biology*, **239**, 24–26.

30. SEIFERT, W. and PAUL, D. (1972), 'Level of cyclic AMP in sparse and

dense cultures of growing and quiescent 3T3 cells', *Nature New Biology*, **240**, 281–283.

31. BURGER, M. M., BOMBIK, B. M., BRECKINRIDGE, B. McL. and SHEPPARD, J. R. (1972), 'Growth control and cyclic alterations of cAMP in the cell cycle', *Nature New Biology*, **239**, 161–163.

32. HSIE, A. W., JONES, C. and PUCK, T. P. (1971), 'Further changes in differentiation state accompanying the conversion of Chinese hamster cells to fibroblastic form by dibutyryl adenosine cyclic 3′,5′-monophosphate and hormones', *Proceedings of the National Academy of Sciences, U.S.A.*, **68**, 1648–1652.

33. SHEPPARD, J. R. (1971), 'Restoration of contact inhibited growth to transformed cells by dibutyryl adenosine 3′,5′-cyclic monophosphate', *Proceedings of the National Academy of Sciences, U.S.A.*, **68**, 1316–1320.

34. WAYMIRE, J. C., WEINER, N. and PRASAD, K. N. (1972), 'Regulation of tyrosine hydroxylase activity in cultured mouse neuroblastoma cells: elevation induced by analogs of adenosine-3′,5′-cyclic monophosphate', *Proceedings of the National Academy of Sciences, U.S.A.*, **69**, 2241–2245.

35. WONG, G. and PAWELEK, J. (1973), 'Control of phenotypic expression of cultured melanoma cells by melanocyte stimulating hormone', *Nature New Biology*, **241**, 213–215.

36. NEWELL, P. C. (1971), 'The development of the cellular slime mould *Dictyostelium discoideum*: a model system for the study of cellular differentiation', *Essays in Biochemistry*, **7**, 87–126.

37. GURDON, J. B. and WOODLAND, H. R. (1968), 'The cytoplasmic control of nuclear activity in animal development', *Biological Reviews*, **43**, 233–267.

38. MILLS, E. S. and TOPPER, Y. J. (1970), 'Some ultrastructural effects of insulin, hydrocortisone and prolactin on mammary gland explants', *Journal of Cell Biology*, **44**, 310–328.

39. TURKINGTON, R. W. and TOPPER, Y. J. (1967), 'Androgen inhibition of mammary gland differentiation *in vitro*', *Endocrinology*, **80**, 329–336.

40. HOLTZER, H. (1970), 'Proliferative and quantal cell cycles in the differentiation of muscle, cartilage and red blood cells'. In *Symposium of the International Society for Cell Biology*, **9**, 69–88. Ed. Padykula, H. A., Academic Press.

41. O'NEILL, M. C. and STOCKDALE, F. E. (1972), 'A kinetic analysis of myogenesis *in vitro*', *Journal of Cell Biology*, **52**, 52–63.

42. GUINNESS, F. B. (1973), 'Studies on cell fusion', Ph.D. thesis, University of Edinburgh.

43. McGARRY, M. P. and VANABLE, Jr., J. W. (1969), 'The role of cell division in *Xenopus laevis* skin gland development', *Developmental Biology*, **20**, 291–303.

44. McKINNELL, R. G., DEGGINS, B. A. and LABAT, D. D. (1969), 'Transplantation of pluripotential nuclei from triploid frog tumours', *Science*, **165**, 394–396.

45. GURDON, J. B. and UEHLINGER, V. (1966), '"Fertile" intestine nuclei', *Nature*, **210**, 1240–1241.

46. GURDON, J. B. and LASKEY, R. A. (1970), 'The transplantation of nuclei

from single cultured cells into enucleate frogs' eggs', *Journal of Embryology and experimental Morphology*, **24**, 227–248.

47. GURDON, J. B. (1972), 'Nuclear functions in embryogenesis', *Developmental Biology*, **29**, No. 2, p. f11.

48. DI BERARDINO, M. A. and KING, T. J. (1967), 'Development and cellular differentiation of neural nuclear transplants of known karyotype', *Developmental Biology*, **15**, 102–128.

49. DI BERARDINO, M. A. and HOFFNER, N. (1970), 'Origin of chromosomal abnormalities in nuclear transplants – a re-evaluation of nuclear differentiation and nuclear equivalence in amphibians', *Developmental Biology*, **23**, 185–209.

50. HARRIS, H. (1970), in *Nucleus and Cytoplasm*, Ch. 5, Clarendon Press: Oxford.

51. MADGWICK, W., MACLEAN, N. and BAYNES, Y. A. (1972), 'RNA synthesis in chicken erythrocytes', *Nature New Biology*, **238**, 137–139.

52. GUGGENHEIM, M. A., FRIEDMAN, R. M. and RABSON, A. S. (1967), 'Interferon: production by chick erythrocytes activated by cell fusion', *Science*, **159**, 542–543.

53. JOHNEN, A. G. (1956), 'Experimental studies about the relationships in the induction process', *Proceedings of the Academy of Sciences, Amsterdam, Series C*, **59**, 554–561.

54. SILAGI, S. (1969), 'Control of pigment production in mouse melanoma cells *in vitro*', *Journal of Cell Biology*, **43**, 263–274.

55. GREEN, H. and TODARO, J. G. (1967), 'The mammalian cell as differentiated microorganism', *Annual Review of Microbiology*, **21**, 573–600.

56. FLICKINGER, R. A. (1967), 'Biochemical aspects of regeneration'. In *The biochemistry of animal development*. Ed. Weber, R., Vol. 2, Ch. 6, Academic Press: London.

57. REYER, R. W. (1954), 'Regeneration of the lens in the amphibian eye', *Quarterly Review of Biology*, **29**, 1–46.

58. EGUCHI, G. and OKADA, T. S. (1973), 'Differentiation of lens from the progeny of chick retinal pigment cells cultured *in vitro*: a demonstration of switch of cell types in clonal cell culture', *Proceedings of the National Academy of Sciences, U.S.A.*, **70**, 1495–1499.

59. FELL, H. B. and MELLANBY, E. (1953), 'Metaplasia produced in cultures of chick ectoderm by high vitamin A', *Journal of Physiology, London*, **119**, 470–488.

60. BARNETT, M. L. and SZABO, G. (1973), 'Effect of vitamin A on epithelial morphogenesis *in vitro*', *Experimental Cell Research*, **76**, 118–126.

61. FIELDING, C. J. (1967), 'Developmental genetics of the mutant grandchildless of *Drosophila subobscura*', *Journal of Embryology and experimental Morphology*, **17**, 375–384.

62. BRIGGS, R. and JUSTUS, J. T. (1968), 'Partial characterisation of the component from normal eggs which corrects the maternal effect of gene o in the Mexican axolotl (*Amblystoma mexicanum*)', *Journal of experimental Zoology*, **167**, 105–115.

63. JOHNSON, R. T. and RAO, P. N. (1971), 'Nucleocytoplasmic interactions

in the achievement of nuclear synchrony in DNA synthesis and mitosis in multinucleate cells', *Biological Reviews*, **46**, 97–155.

64. MERRIAM, R. W. (1969), 'Movement of cytoplasmic proteins in nuclei induced to enlarge and initiate RNA or DNA synthesis', *Journal of Cell Science*, **5**, 333–349.

65. GURDON, J. B. (1970), 'Nuclear transplantation and control of gene activity in animal development', *Proceedings of the Royal Society, London, Series B*, **176**, 303–314.

66. RINGERTZ, N. R., CARLSSON, S-A., EGE, T. and BOLUND, L. (1971), 'Detection of human and chick nuclear antigens in nuclei of chick erythrocytes during reactivation in heterokaryons with HeLa cells', *Proceedings of the National Academy of Sciences, U.S.A.*, **68**, 3228–3232.

67. CARLSSON, S-A., MOORE, G. P. M. and RINGERTZ, N. R. (1973), 'Nucleocytoplasmic protein migration during the activation of chick erythrocyte nuclei in heterokaryons', *Experimental Cell Research*, **76**, 234–241.

68. DARŻYNKIEWICZ, Z. and CHELMICKA-SZORC, E. (1972), 'Unscheduled DNA synthesis in hen erythrocyte nuclei reactivated in heterokaryons', *Experimental Cell Research*, **74**, 131–139.

69. HARRIS, H. (1970), in *Cell fusion*, Ch. 2, Clarendon Press: Oxford.

70. HARRIS, H. (1970), In *Nucleus and Cytoplasm*, Ch. 3, Clarendon Press: Oxford.

71. CHURCH, R. B. and BROWN, I. R. (1972), 'Tissue specificity of genetic transcription'. In *Nucleic acid hybridisation in the study of cell differentiation*, **3**, 11–24. Ed. Ursprung, H., Springer-Verlag.

72. BISHOP, J. O. (1969), 'The effect of genetic complexity on the time course of ribonucleic acid – deoxyribonucleic acid hybridisation', *Biochemical Journal*, **113**, 805–811.

73. PEMBERTON, R. E., HOUSMAN, D., LODISH, H. F. and BAGLIONI, C. (1972), 'Isolation of duck haemoglobin messenger RNA and its translation by rabbit reticulocyte cell free system', *Nature New Biology*, **235**, 99–102.

74. SUZUKI, Y. and BROWN, D. D. (1972), 'Isolation and identification of the messenger RNA for silk fibroin from *Bombyx mori*', *Journal of Molecular Biology*, **63**, 409–429.

75. GROSS, K., RUDERMAN, J., JACOBS-LORENA, M., BAGLIONI, C. and GROSS, P. R. (1973), 'Cell free synthesis of histones directed by messenger RNA from sea urchin embryos', *Nature New Biology*, **241**, 272–274.

76. MATHEWS, M. B., OSBORN, M., BERNS, A. J. M. and BLOEMENDAL, H. (1972), 'Translation of two messenger RNAs from lens in a cell free system from Krebs II ascites cells', *Nature New Biology*, **236**, 5–7.

77. ROURKE, A. W. and HEYWOOD, S. M. (1972), 'Myosin synthesis and specificity of eukaryotic initiation factors', *Biochemistry*, **11**, 2061–2066.

78. MACH, B., FAUST, C. and VASSALLI, P. (1973), 'Purification of 14S messenger RNA of immunoglobulin light chain that codes for a possible light-chain precursor', *Proceedings of the National Academy of Sciences, USA*, **70**, 451–455.

79. GURDON, J. B., LANE, C. D., WOODLAND, H. R. and MARBAIX, G.

(1971), 'Use of frogs' eggs and oocytes for the study of messenger RNA and its translation in living cells', *Nature*, **233**, 177–182.

80. BERNS, A. J. M., VAN KRAAIKAMP, M., BLOEMENDAL, H. and LANE, C. D. (1972), 'Calf crystallin synthesis in frog cells: the translation of lens cell 14S RNA in oocytes', *Proceedings of the National Academy of Sciences*, **69**, 1606–1609.

81. ADESNIK, M., SALDITT, M., THOMAS, W. and DARNELL, J. E. (1972), 'Evidence that all messenger RNA molecules (except histone messenger RNA) contain poly (A) sequences and that the poly (A) has a nuclear function', *Journal of Molecular Biology*, **71**, 21–30.

82. LEE, S. Y., MENDECKI, J. and BRAWERMAN, G. (1971), 'A polynucleotide segment rich in adenylic acid in the rapidly-labelled polyribosomal RNA component of mouse sarcoma 180 ascites cells', *Proceedings of the National Academy of Sciences, U.S.A.*, **68**, 1331–1335.

83. ROSS, J., AVIV, H., SCOLNICK, E. and LEDER, P. (1972), '*In vitro* synthesis of DNA complementary to purified rabbit globin mRNA', *Proceedings of the National Academy of Sciences, U.S.A.*, **69**, 264–268.

84. HARRISON, P., CONKIE, D. and JONES, K. W. (1973), 'Timing of globin messenger RNA synthesis in foetal erythroid cell development and in Friend virus-induced cells'. In preparation.

85. JONES, K. W., BISHOP, J. O. and DA CUNHA, A. B. (1973), 'Complex formation between poly-r(U) and various chromosomal loci in *Rhyncosciara*', Chromosoma (Berl.), **43**, 375–390.

86. SPIRIN, A. S. (1966), 'On "masked" forms of messenger RNA in early embryogenesis and other differentiation systems'. In *Current Topics in Developmental Biology*, **1**, 1–36. Eds. Moscona, A. A. and Monroy, A., Academic Press.

87. RAFF, R. A., COLOT, H. V., SELVIG, S. E. and GROSS, P. R. (1972), 'Oogenetic origin of messenger RNA for embryonic synthesis of microtubule proteins', *Nature*, **235**, 211–214.

88. MARKS, P. A. and RIFKIND, R. A. (1972), 'Protein synthesis: its control in erythropoiesis', *Science*, **175**, 955–961.

89. STEWART, J. A. and PAPACONSTANTINOU, J. (1967), 'Stabilisation of mRNA templates in bovine lens epithelial cells', *Journal of Molecular Biology*, **29**, 357–370.

90. SCOTT, R. B. and BELL, E. (1964), 'Protein synthesis during development: control through messenger RNA', *Science*, **145**, 711–713.

91. WESSELLS, N. K. and WILT, F. H. (1965), 'Action of actinomycin D on exocrine pancreas cell differentiation', *Journal of Molecular Biology* **13**, 767–779.

92. SHAINBERG, A., YAGIL, G. and YAFFE, D. (1971), 'Alterations of enzymatic activities during muscle differentiation *in vitro*', *Developmental Biology*, **25**, 1–29.

93. SINGER, R. H. and PENMAN, S. (1972), 'Stability of HeLa cell mRNA in actinomycin', *Nature*, **240**, 100–102.

94. GREENBERG, J. R. (1972), 'High stability of messenger RNA in growing cultured cells', *Nature*, **240**, 102–104.

95. MORSE, D. E., MOSTELLER, R. D. and YANOFSKY, C. (1969), 'Dynamics of synthesis, translation and degradation of trp operon mRNA in *E. coli*', *Cold Spring Harbor Symposium of Quantitative Biology*, **34**, 725–740.

96. MILLER, Jnr., O. L., BEATTY, B. B., HAMKALO, B. A. and THOMAS, C. A., Jnr. (1970), 'Electron microscope visualisation of transcription', *Cold Spring Harbor Symposium of Quantitative Biology*, **35**, 505–512.

97. In *The lactose operon*. Ed. Beckwith, J. R. and Zipser, D. (1970), Cold Spring Harbor Laboratory.

98. ENGELSBERG, E., SQUIRES, C. and MERONK, F. (1969), 'The L-arabinose operon in *Escherichia coli* B/r: a genetic demonstration of two functional states of the product of a regulator gene', *Proceedings of the National Academy of Sciences*, *U.S.A.*, **62**, 1100–1107.

99. TRAVERS, A. (1971), 'Control of transcription in bacteria', *Nature New Biology*, **229**, 69–74.

100. OHNO, S. (1972), 'Gene duplication, mutation load and mammalian genetic regulatory systems', *Journal of Medical Genetics*, **9**, 254–263.

101. WALKER, P. M. B. (1971), '"Repetitive" DNA in higher organisms'. In *Progress in Biophysics and Molecular Biology*, **23**, 145–190. Eds. Butler, J. A. V. and Noble, D., Pergamon Press: Oxford.

102. FLAMM, W. G. (1972), 'Highly repetitive sequences of DNA in chromosomes'. In *International Review of Cytology*, **32**, 2–51. Eds. Bourne, G. H. and Danielli, J. F., Academic Press.

103. BIRNSTIEL, M., CHIPCHASE, M. and SPEIRS, J. (1971), 'The ribosomal RNA cistrons'. In *Progress in Nucleic Acid Research and Molecular Biology*, **11**, 351–389. Eds. Davidson, J. N. and Cohn, W. E., Academic Press.

104. CLARKSON, S., PURDOM, I. and BIRNSTIEL, M. (1973), 'Clustering of transfer RNA genes of *Xenopus laevis*', **79**, 411–429. *Journal of Molecular Biology*. In press.

105. BROWN, D. D., WENSINK, P. C. and JORDAN, E. (1971), 'Purification and some characteristics of 5S RNA from *Xenopus laevis*', *Proceedings of the National Academy of Sciences*, *U.S.A.*, **68**, 3175–3179.

106. WEINBERG, E. S., BIRNSTIEL, M. L., PURDOM, I. F. and WILLIAMSON, R. (1972), 'Genes coding for polysomal 9S RNA of sea urchins: conservation and divergence', *Nature*, **240**, 225–228.

107. PAUL, J. (1971), 'DNA/protein interactions in mammalian differentiation', *Advances in the Biosciences*, **6**, 495–509. Schering symposium on *Intrinsic and extrinsic factors in early mammalian development*. Ed. Raspé, G., Vieweg, Pergamon Press.

108. JUDD, B. H., SHEN, M. W. and KAUFMAN, T. C. (1972), 'The anatomy and function of a segment of the X chromosome of *Drosophila melanogaster*', *Genetics*, **71**, 139–156.

109. STUBBLEFIELD, E. (1973), 'The structure of mammalian chromosomes', *International Review of Cytology*. In press.

110. MITCHISON, J. M. (1971), In *The Biology of the Cell Cycle*, Ch. 4, Cambridge University Press.

111. SOEIRO, R., VAUGHAN, M., WARNER, J. and DARNELL, J. (1968), 'The

turnover of nuclear DNA-like RNA in HeLa cells', *Journal of Cell Biology*, **39**, 112–118.

112. WILLIAMSON, R., DREWIENKIEWICZ, C. E. and PAUL, J. (1973), 'Globin messenger sequences in high molecular weight RNA from embryonic mouse liver', *Nature New Biology*, **241**, 66–68.

113. BRITTEN, R. J. and DAVIDSON, E. H. (1969), 'Gene regulation for higher cells: a theory', *Science*, **165**, 349–357.

114. GEORGIEV, G. P. (1969), 'On the structural organisation of operon and the regulation of RNA synthesis in animal cells', *Journal of Theoretical Biology*, **25**, 473–490.

115. CRICK, F. (1971), 'General model for the chromosomes of higher organisms', *Nature*, **234**, 25–27.

116. PAUL, J. (1972), 'General theory of chromosome structure and gene activation in eukaryotes', *Nature*, **238**, 444–446.

117. BERENDES, H. D. (1971), 'Gene activation in dipteran polytene chromosomes'. In *Symposium of the Society for experimental Biology*, **25**, 145–161. Eds. Davies, D. D. and Balls, M., Cambridge University Press.

118. ROEDER, R. G. and RUTTER, W. J. (1970), 'Specific nucleolar and nucleoplasmic RNA polymerases', *Proceedings of the National Academy of Sciences, U.S.A.*, **65**, 675–682.

119. CHAMBON, P., GISSINGER, F., KEDINGER, C., MANDEL, J. L., MEILHAC, M. and NURET, P. (1972), 'Structural and functional properties of three mammalian nuclear DNA-dependent RNA polymerases'. In *Karolinska Symposium on Research Methods in Reproductive Endocrinology*, **5**, 222–246. Ed. Diczfalusy, E. Karolinska Institutet: Stockholm.

120. YU, F-L. and FEIGELSON, P. (1972). 'The rapid turnover of RNA polymerase of rat liver nucleolus and of its messenger RNA', *Proceedings of the National Academy of Sciences, U.S.A.*, **69**, 2833–2837.

121. SMUCKLER, E. A. and TATA, J. R. (1971), 'Changes in hepatic nuclear DNA-dependent RNA polymerase caused by growth hormone and triiodothyronine', *Nature*, **234**, 37–39.

122. BISHOP, J. O., PEMBERTON, R. and BAGLIONI, C. (1972), 'Reiteration frequency of haemoglobin genes in the duck', *Nature New Biology*, **235**, 231–234.

123. SUZUKI, Y., GAGE, L. P. and BROWN, D. D. (1972), 'The genes for silk fibroin in *Bombyx mori*', *Journal of Molecular Biology*, **70**, 637–649.

124. BISHOP, J. O. and ROSBASH, M. (1973), 'Reiteration frequency of duck haemoglobin genes', *Nature New Biology*, **241**, 204–207.

125. PENMAN, S., ROSBASH, M. and PENMAN, M. (1970), 'Messenger and heterogeneous nuclear RNA in HeLa cells: differential inhibition by cordycepin', *Proceedings of the National Academy of Sciences, USA*, **67**, 1878–1885.

126. SHEINESS, D. and DARNELL, J. E. (1973), 'Polyadenylic acid segment in mRNA becomes shorter with age', *Nature New Biology*, **241**, 265–268.

127. CRAIG, N., PERRY, R. P. and KELLEY, D. E. (1971), 'Lifetime of the messenger RNAs which code for ribosomal proteins in L cells', *Biochimica et Biophysica Acta*, **246**, 493–498.

128. DEÁK, J., SIDEBOTTOM, E. and HARRIS, H. (1972), 'Further experiments on the role of the nucleolus in the expression of structural genes', *Journal of Cell Science*, **11**, 379–391.

129. LODISH, H. F. (1970), 'Secondary structure of bacteriophage f$_2$ ribonucleic acid and the initiation of *in vitro* protein biosynthesis', *Journal of Molecular Biology*, **50**, 689–702.

130. BERISSI, H., GRONER, Y. and REVEL, M. (1971), 'Effect of purified initiation factor F3 (B) on the selection of ribosomal binding sites on phage MS2 RNA', *Nature New Biology*, **234**, 44–47.

131. EGGEN, K. and NATHANS, D. (1969), 'Regulation of protein synthesis directed by coliphage MS2 RNA', *Journal of Molecular Biology*, **39**, 293–305.

132. LEE-HUANG, S. and OCHOA, S. (1971), 'Messenger discriminating species of initiation factor F3', *Nature New Biology*, **234**, 236–239.

133. SIEKEVITZ, P. and PALADE, G. E. (1960), 'A cytochemical study on the pancreas of the guinea pig. V. *In vitro* incorporation of leucine-1-C^{14} into the chymotrypsinogen of various cell fractions', *Journal of Biophysical and Biochemical Cytology*, **7**, 619–630.

134. REDMAN, C. M. (1969), 'Biosynthesis of serum proteins and ferritin by free and attached ribosomes of rat liver', *Journal of Biological Chemistry*, **244**, 4308–4315.

135. LISOWSKA-BERNSTEIN, B., LAMM, M. E. and VASSALLI, P. (1970), 'Synthesis of immunoglobulin heavy and light chains by the free ribosomes of a mouse plasma cell tumour', *Proceedings of the National Academy of Sciences, U.S.A.*, **66**, 425–432.

136. TATA, J. R. (1970), 'Biological action of the thyroid hormones at the cellular and molecular level'. In *Biochemical actions of hormones*. Ed. Litwack, G., 58–131, Academic Press.

137. SUEOKA, N. and KANO-SUEOKA, T. (1970), 'Transfer RNA and cell differentiation'. In *Progress in Nucleic Acid Research and Molecular Biology*, **10**, 23–55. Ed. Davidson, J. N. and Cohn, W. E., Academic Press.

138. LODISH, H. F. (1971), 'Alpha and beta globin messenger ribonucleic acid: different amounts and rates of initiation of translation', *Journal of Biological Chemistry*, **246**, 7131–7138.

139. TERADA, M., CANTOR, L., METAFORA, S., RIFKIND, R. A., BANK, A. and MARKS, P. (1972), 'Globin messenger RNA activity in erythroid precursor cells and the effect of erythropoietin', *Proceedings of the National Academy of Sciences, U.S.A.*, **69**, 3575–3579.

140. ROSS, J., IKAWA, Y. and LEDER, P. (1972), 'Globin messenger RNA-induction during erythroid differentiation of cultured leukaemia cells', *Proceedings of the National Academy of Sciences, U.S.A.*, **69**, 3620–3623.

141. PERRY, R. P. and KELLEY, D. E. (1968), 'Messenger RNA-protein complexes and newly synthesised ribosomal subunits: analysis of free particles and components of polyribosomes', *Journal of Molecular Biology*, **35**, 37–59.

142. LEGON, S., JACKSON, R. J. and HUNT, T. (1973), 'Control of protein synthesis in reticulocyte lysates by haemin', *Nature New Biology*, **241**, 150–152.

143. TOMKINS, G. M., GELEHRTER, T. D., GRANNER, D., MARTIN, D., Jnr, SAMUELS, H. H. and THOMPSON, E. B. (1969), 'Control of specific gene expression in higher organisms', *Science*, **166**, 1474–1480.

144. TOMKINS G. M., LEVINSON, B. B., BAXTER, J. D. and DETHLEFSEN, L. (1972), 'Further evidence for posttranscriptional control of inducible tyrosine aminotransferase synthesis in cultured hepatoma cells', *Nature New Biology*, **239**, 9–13.

145. MASTERS, C. J. and HOLMES, R. S. (1972), 'Isozymes and ontogeny', *Biological Reviews*, **47**, 309–361.

146. ROSENBERG, M. (1971), 'Epigenetic control of lactate dehydrogenase subunit assembly', *Nature New Biology*, **230**, 12–14.

147. ASHBURNER, M. (1970), 'A prodromus to the genetic analysis of puffing in *Drosophila*', *Cold Spring Harbor Symposium of Quantitative Biology*, **35**, 533–538.

148. BAKER, W. K. (1968), 'Position effect variegation'. In *Advances in Genetics*, **14**, 133–169. Ed. Caspari, E. W., Academic Press.

149. MCCLINTOCK, B. (1965), 'The control of gene action in maize', *Brookhaven Symposium of Biology*, **18**, 162–184.

150. CATTANACH, B. M., POLLARD, C. E. and PEREZ, J. N. (1969) 'Controlling elements in the mouse X chromosome', *Genetical Research, Cambridge*, **14**, 223–235.

151. NEBERT, D. W., GOUJON, F. M. and GIELEN, J. E. (1972), 'Aryl hydrocarbon hydroxylase induction by polycyclic hydrocarbons: simple autosomal dominant trait in the mouse', *Nature New Biology*, **236**, 107–110.

152. DREWS, U., ITAKURA, H., DOFUKU, R., TETTENBORN, U. and OHNO, S. (1972), 'Nuclear DHT-receptor in Tfm/Y kidney cell', *Nature*, **238**, 216–217.

153. EPHRUSSI, B. (1972), In *Hybridisation of somatic cells*, Princeton University Press.

154. WEISS, M. C. and CHAPLAIN, M. (1971), 'Expression of differentiated functions in hepatoma cell hybrids: reappearance of tyrosine aminotransferase inducibility after the loss of chromosomes', *Proceedings of the National Academy of Sciences, USA*, **68**, 3026–3030.

155. BERTOLOTTI, R. and WEISS, M. C. (1972), 'Expression of differentiated functions in hepatoma cell hybrids II. Aldolase', *Journal of Cell Physiology*, **79**, 211–224.

156. COFFINO, P., KNOWLES, B., NATHENSON, S. G. and SCHARFF, M. D. (1971), 'Suppression of immunoglobulin synthesis by cellular hybridisation', *Nature New Biology*, **231**, 87–90.

157. DAVIDSON, R. L. (1972), 'Regulation of melanin synthesis in mammalian cells: effect of gene dosage on the expression of differentiation', *Proceedings of the National Academy of Sciences, U.S.A.*, **69**, 951–955.

158. DAVIDSON, R. L. and BENDA, P. (1970), 'Regulation of specific functions of glial cells in somatic hybrids. II. Control of inducibility of glycerol-3-phosphate dehydrogenase', *Proceedings of the National Academy of Sciences, U.S.A.*, **67**, 1870–1877.

159. KLEBE, R. J., CHEN, T. R. and RUDDLE, F. R. (1970), 'Mapping of a human genetic regulator element by somatic cell genetic analysis', *Proceedings of the National Academy of Sciences, U.S.A.*, **66**, 1220–1227.

160. PETERSON, J. A. and WEISS, M. C. (1972), 'Expression of differentiated functions in hepatoma cell hybrids: induction of mouse albumin production in rat hepatoma-mouse fibroblast hybrids', *Proceedings of the National Academy of Sciences, U.S.A.*, **69**, 571–575.

161. MINNA, J., NELSON, P., PEACOCK, J., GLAZER, D. and NIRENBERG, M. (1971), 'Genes for neuronal properties expressed in neuroblastoma x L cell hybrids', *Proceedings of the National Academy of Sciences, U.S.A.*, **68**, 234–239.

162. MIGEON, B. R., DER KALOUSTIAN, V. M., NYHAN, W. L., YOUNG, W. J. and CHILDS, B. (1968), 'X-linked hypoxanthine-guanine phosphoribosyl transferase deficiency: heterozygote has two clonal populations', *Science*, **160**, 425–427.

163. MARTIN, P. G. (1966), 'The pattern of autosomal DNA replication in four tissues of the Chinese hamster', *Experimental Cell Research*, **45**, 85–95.

164. DELANGE, R. J., FAMBROUGH, M., SMITH, E. L. and BONNER, J. (1969), 'Calf and pea histone IV', *Journal of Biological Chemistry*, **244**, 5669–5679.

165. BARRINGTON, E. J. W. (1964), In *Hormones and Evolution*, English Universities Press: London.

166. ZEPP, H. D., CONOVER, J. H., HIRSCHHORN, K. and HODES, H. L. (1971), 'Human mosquito somatic cell hybrids induced by UV inactivated Sendai virus', *Nature New Biology*, **229**, 119–121.

167. GOLDSTEIN, S. and LIN, C. C. (1972), 'Somatic cell hybrids between cultured fibroblasts from the Galapagos tortoise and the golden hamster', *Experimental Cell Research*, **73**, 266–269.

168. YAFFE, D. and FELDMAN, M. (1965), 'The formation of hybrid multinucleated muscle fibres from myoblasts of different genetic origin', *Developmental Biology*, **11**, 300–317.

2 Differentiation and Growth of Nerve Cells

MARCUS JACOBSON

Introduction

The diversification of the neuronal stem cells into thousands of anatomically and functionally different types of neurons is termed differentiation. Neural differentiation is initiated by the process of determination or specification which programmes the cells for their future course of development. Because the key to understanding the morphological differentiation of nerve cells is to be found in the early events occurring during neural development, the latter deserve to receive a considerable amount of our attention in this chapter. By contrast, I shall give less attention to the overt structural changes in differentiating young neurons which many authors have vested with great developmental significance. These fine structural changes in the developing nerve cells are interesting, but they are merely expressions of earlier steps in cell differentiation and cannot be understood without first considering the developmental programmes which precede visible changes in cellular phenotypes. The large number of different neurophenotypes are the terminal branches of a developmental tree which stems from the genotype and is shaped by epigenetic factors such as hormones and cellular interactions which modify the development of embryonic nerve cells. We should also try to see how these neurophenotypes represent evolutionary solutions to environmental challenges. These challenges were successfully met by molecular, developmental and behavioural modifications which have led to the evolution of nervous systems that are found in the species that have survived to this time.

There is much evidence that the developmental programme of nerve cells is under the control of genes and that gene expression is regulated by molecules whose function is the selective activation of sets of genes that control specific programmes of development. These regulator molecules may either be intrinsic to the cells, and transmitted directly from mother to daughter cells during mitosis, or the regulator molecules may be free to diffuse from cell to cell through intercellular junctions. Thus, they may exert differential actions on cells at different positions. In the

53

latter mechanism there is a field of differential distribution of signals and the cells are then labelled according to their position, whereas in the former, cell lineage mechanism, cells are labelled with reference to the temporal order of their origin.

These two modes of determination have been studied by a number of experiments in which embryonic cells have been re-located or re-combined in various patterns. If neuronal determination or specification is the result of differential cell division, the developmental programme carried out by each cell is independent of the programmes executed by neighbouring cells. This is called mosaic development. In other cases, in which intercellular communication occurs, cellular interactions play a predominant rôle in regulating the cellular developmental programmes. This is what is meant by regulative development. In such cases cells can be made to differentiate in numerous ways depending on their positions within the multicellular context. In most animal tissues, development is partly mosaic and partly regulative, and as a rule the capacity to regulate is maximal early in development and is lost at a later stage. There is no evidence of mosaic development in the nervous system of vertebrate embryos, in which regulative mechanisms predominate during the early stages when the developmental programmes are initiated. Thus no abnormalities result from loss of cells or experimental translocation of the neuroepithelial stem cells when these operations are performed early in embryonic development. However, after the proliferative activity of the stem cells has been completed, usually during larval or fetal stages, nerve cells cannot be replaced, and damage to the central nervous system results in irreparable structural damage and a corresponding loss of function.

Programming of Embryonic Nerve Cells

From these introductory remarks it will be gathered that the nerve cell passes through several intermediate states before arriving at the final neurophenotype. Development of nerve cells can be regarded as a three stage process, namely neural induction, neural determination or specification and neural differentiation and growth. These stages of nerve cell development are carried out according to a developmental programme initiated at the time of neural induction in the gastrula. We may summarize the various states through which the developing nerve cell passes as follows[1]. First, *neural induction* transforms ectodermal cells to neurectodermal cells as a result of the interactions between the archenteron roof and the overlying ectoderm[2]. Second, the cells in the neurectoderm acquire position-dependent properties. These appear initially in the anteroposterior (craniocaudal) axis of the neurectodermal cell population during the early phases of gastrulation and arise a few hours later in the mediolateral axis of the neurectodermal cell population[1, 3]. The neurectodermal cell population comprises the neural plate covering the dorsal surface of the embryo. These cells acquire developmental potentials which are related to their relative positions within the boundaries of the cell population[4]. The position-dependent properties constitute cellular 'addresses' which might at first be properties of the cell-population as a whole, and might later become regionalized to discrete sub-populations before finally becoming uniquely related to individual cells within the population. It is likely that this progression from tissue-specific to cell-specific properties may not go to completion in all cases. That is, developmental programmes become restricted initially to groups of cells and finally to some individual cells that ultimately will form discrete subdivisions of the nervous system. The Mauthner neuron provides a striking example of precocious restriction of a unique developmental programme to a single nerve cell[5, 6]. One Mauthner neuron develops on each side of the medulla in fish and larval amphibians, but it fails to develop if a specific small region of neural plate is excised. Specification of this cell occurs some days before the Mauthner cell can be recognized histologically.

Transplantation or isolation of small fragments of the embryo, or dissociation of embryonic tissues into small groups of cells or single cells, are means of testing whether individual nerve cells or small neuronal aggregates can differentiate autonomously and these techniques thus show whether the parts have been determined or whether their differentiation can be altered by interactions with the surrounding tissues[7, 8]. Such experiments have shown that a piece of the neural plate always becomes nervous tissue after being transplanted outside the nervous system or explanted to tissue culture. For example, if the region of the neural plate that is destined to give rise to the eye is excised and transplanted to the lateral ectoderm of another embryo of the same age, the transplant differentiates into a fully developed eye which contains all the neuronal types normally found in the retina. Even a small part of the eye rudiment excised from the early neurula may give rise to an entire small eye. Moreover, if a second eye rudiment is implanted in contact with a normal eye-forming region in the neural plate, the two fuse completely to produce a single eye in which part is derived from the implant and part from the original eye-forming cells. Whether the retina of such an eye is equivalent to a normal eye (that is whether it contains a full, reduced or augmented set of cellular specificities found in the normal retina) is not known.

One can also try to discover when the earliest stable programme for neural differentiation has been initiated in the neurectoderm by removing pieces of presumptive neural tissue from a series of amphibian embryos at progressively earlier stages of development and culturing them in balanced solution for up to fifteen days. The results differ according to the developmental age of the explanted ectoderm; presumptive neural ectoderm from the early or mid-gastrula does not differentiate as neural tissue but in some cases neurectoderm derived from the late gastrula gives rise to small aggregates of nerve cells. The complexity of neural differentiation in such isolated fragments of neurectoderm increases as the age of the embryo increases. After the rise of the neural folds at the sides of the neural plate, isolated pieces of the neural plate always give rise to histotypic aggregates of nerve cells which can be identified as particular parts of the brain.

These experiments show that in the very earliest stages of formation of the neurectoderm a programme for neural differentiation is initiated which may even be expressed in tissue culture, resulting in the differentiation of various classes of neurons. However, the programme is modifiable to some extent: if the cells are translocated within the neural ectoderm they interact with neighbouring cells to regulate their developmental programmes in accord with their new positions and their new relationships within the context of the population of interacting cells. These cellular interactions are possible only as long as the appropriate communication channels are open between the developing cells[9]. After the neural cells withdraw from the mitotic cycle their capacity for cellular interactions diminishes greatly, probably because the junctions between the cells diminish or disappear[10].

The cellular addresses endow some individual neurons with a *locus specificity* which pre-disposes each outgrowing axon to project to a specific position in the brain and to form synaptic connections there with selected neurons[11, 12]. Outgrowth of dendrites and axons and formation of synaptic connections are the hallmarks of neuronal differentiation. It is only after synapses have developed and nerve circuits have formed that reflex action is initiated. Then the animal becomes capable of responding to external stimulation. With further development of interneuronal circuits, central nervous integration and learning become possible. Adaptive modification of neurons or 'fine tuning' of neuronal circuits may then occur as the result of experience and learning, and these changes are the final stage of nerve cell differentiation.

The Developmental Programme of Retinal Ganglion Cells

Delineation of these states of development has so far been attempted only for a single neurophenotype: the ganglion cells of the vertebrate retina. The retinal ganglion cells have long axons, which grow out of the eye into the optic stalk and then into the brain, and ultimately form synaptic connections very selectively with specific groups of neurons, at various places, collectively termed the visual centres. These retinal ganglion cells exhibit all the main developmental changes that are known to occur in other nerve cells and thus they will be used as the prime example of such determinative and differentiative changes[12]. For no other type of neuron can we give such a detailed history, following it from its stem cell, through a succession of cell divisions, its withdrawal from the mitotic cycle, its overt cytodifferentiation, growth of its axons and dendrites and formation of specific synaptic connections with nerve cells within the retina and with those in the visual centres. At this point one should note that there are five different kinds of nerve cells in the vertebrate retina, and hundreds of other kinds in the brain of any species of vertebrate. The ways in which the developmental programmes of these different neuronal phenotypes differ from each other are not known, nor have many of the steps in the expression of the developmental programme resulting in the differentiation and growth of any type of nerve cell been studied.

In the neural plate we can demarcate an eye field which is a small group of neurectodermal cells that will ultimately, if left undisturbed, form the neural retina. These cells have already, at the time of neurulation, embarked on the initial steps in the developmental programme and they already show positional differences. Thus the entire cell population of the neural plate is axially polarized first in the anteroposterior axis and later in both the anteroposterior and the mediolateral axis[3]. This polarization can be operationally demonstrated by excising the neural plate, and replacing it back-to-front (180° rotated), whereupon the whole nervous system develops in correspondingly inverted orientation. This result implies that the neural plate, from which the central nervous system originates, already has axial cues that are aligned with the major axes of the embryo. These axes are the property of the entire neural plate for if a small fragment of the neural plate is rotated *in situ* it does not show axial independence but takes on the axes of the surrounding parts of the neural plate. However, such a small fragment may already have embarked on a programme that results in the development of a distinct region of the brain, such as the forebrain or eye, so that when it is transplanted to another part of the body or explanted to tissue culture it develops autonomously as forebrain or as an eye. But the axes of such a fragment are reversible – when the fragment is inverted within the neural plate or even when it is transplanted to another site on the embryo, it interacts with the surrounding tissues to adopt the axes of the dominating surroundings. Thus, the original axes of the small population of cells that was inverted are substituted by new axes that are aligned with the surrounding tissue. However, when the same number of cells is permitted to develop in isolation, the original position-dependent properties exert themselves and the programme is completed that was set up in the cells' original position. A similar result is found at later stages of development of the embryonic eye rudiment. If the early eye rudiment is excised and rotated 180°, a normal set of position-dependent properties is substituted for the original set and the ganglion cells that develop send their nerve fibres into the brain to form completely normal connections with the visual centres[13]. However, if the early eye rudiment is excised, placed in tissue culture and allowed to develop to an advanced stage before it is re-implanted in an inverted position in the orbit, the eye forms an inverted pattern of nerve connections with the brain[12, 14]. This shows that a programme for the differentiation of position-dependent properties in the retinal ganglion cells is initiated at an early stage of eye development, but is reversible or replaceable when the eye is inverted *in situ*.

The reversal or replacement of the position-dependent properties must have resulted from the interaction between the eye and the surrounding tissues because it did not occur when the eye was removed to tissue culture. It seems as if the alignment of the axes of the eye field with those of the embryo depends upon a cellular interaction between the retinal cell population and the surrounding cell population. This cellular interaction appears to occur only when the retinal cells are in a process of proliferation, that is, when they are neurectodermal stem cells. As soon as the first retinal cells withdraw from the mitotic cycle they immediately lose the ability to re-negotiate their position-dependent properties after their positions are changed. This can be shown by

58

inverting the eye after the first retinal ganglion cells have withdrawn from the mitotic cycle. It is then found that the connections formed between the eye and the brain are inverted, indicating that they have been programmed on the basis of the position-dependent properties of the ganglion cells that were obtained before inversion of the eye. The change from a reversible to an irreversible state proceeds step-wise: if the eye is inverted during the period of withdrawal of ganglion cells from the mitotic cycle only the anteroposterior components of the positional-dependent properties is irreversible and fixed, but the dorsoventral component is replaced with another dorsoventral axis aligned with the body axis. If the eye is inverted a few hours later both the dorsoventral and the anteroposterior axis are permanently inverted. The neuron is said to be completely specified when it has attained a stable irreversible state. The change from a reversible to an irreversible state has also been shown to correlate with the disappearance of intercellular junctions called gap junctions which are presumed to mediate the intercellular communications that is necessary for the re-alignment of the axes of the eye after it has been inverted[12, 15].

The position-dependent properties which develop within the ganglion cell population of the retina are essential pre-requisites for the subsequent development of the synaptic connections at the proper locations in the visual centres. Thus the trajectory of growth of the nerve fibre, the selection of a locus at which it is to terminate and the operation of forming a synapse, are all probably predicated upon the acquisition of a specific cellular property dependent upon each cell's position at a particular time in its early development. As all these developmental events unfold gradually during the life span of the nerve cell, which may be many years, it seems necessary for the positional-dependent properties to have a remarkable stability.

Stability of the Differentiated State of Nerve Cells

In general we wish to know the stability of a given state of differentiation. This is of particular importance in the case of nerve cells, in which diversification of cell types have gone far beyond that found in all other systems combined, and in which the long-term memory functions of the system are apparently dependent on the structural and functional integrity and stability of the nerve cells. The most fine-grained properties of the nervous system are the locus specificities of the individual neurons.

It is, thus, of great interest to assay the stability of these locus specificities. This has been done in the case of the retinal ganglion cells by two methods. First, the eye, just after specification, may be excised from a donor embryo, isolated *in vitro* for six to ten days, or grafted in an ectopic position on the side of the body for 30 days, and then returned to the orbit of a carrier embryo[14]. The eye forms synaptic connections with the brain of the carrier on the basis of the set of specificities gained in the original donor orbit. This shows that the locus specificities programmed in the original donor were stable for at least ten days *in vitro* or 30 days in an ectopic position on the embryo. A second method of assaying the stability of the specified state in the retinal ganglion cell population, is to graft an already specified eye in an inverted position into a carrier embryo at an earlier stage of development. This experiment shows that the original set of locus specificities in the retinal cell population is not lost or modified.

These experiments on the eye rudiment are the only ones in which stability of the developmental programme has been assayed in a population of embryonic nerve cells. Therefore, the result should not be generalized too freely, and while it seems probable that the large neurons (Class I) have great phenotypic stability, it is also probable that other types of neurons, especially Class II nerve cells with short axons, are more labile, or retain the capacity to regulate their developmental programmes in response to changes in extrinsic conditions. One indication, however, of the stability of the differentiated state in all neurons is the great rarity of primary neoplasms of nerve cells, whereas neoplasms of glial cells are quite common. Another indication of the stability of the specified state of the neuron is that differentiated nerve cells, once they have withdrawn from the mitotic cycle, cannot be induced to divide. Finally the invariance of neuronal cytoarchitectonics – the pattern of nerve cell locations and interconnections is apparently almost identical in all individuals of the same species – is another indication of the stability of the neurophenotype.

In making the point about invariance of neuronal cytoarchitectonics we should add the qualification that certain types of neuron show considerable variability, and it is important to make a distinction between the invariant and the variable classes of neurons[11, 16, 17]. The highly invariant neurons, tend to be large, with long axons, and comprise the primary projection pathways of the nervous system (Class I neurons). These should be contrasted with Class II neurons, which are small nerve cells with short axons and variable morphologies. The Class I neurons become post-mitotic early in embryonic development and their developmental programme is fixed at about that time. As a result, their growth and maturation, including the formation of synaptic connections, is under the aegis of a developmental programme that is initiated in the early embryo and is usually completed before the organism is exposed to the vagaries of a variable environment. Class I neurons include spinal motor neurons, the sensory neurons of the spinal ganglia, retinal ganglian cells, pyramidal cells of the cerebral cortex, cerebellar Purkinje cells, and all the other large, principal neurons of the major nuclei of the central nervous system whose axons constitute the peripheral nerves and main long tracts of the central nervous system. These Class I neurons have great structural stability and invariance. By contrast, the Class II neurons are small neurons with short axons which synapse with other neurons in their own immediate vicinity. They originate at a much later time than the Class I neurons with which they are associated in any part of the brain. Class II neurons are very variable in their morphology and are probably less stable and therefore more modifiable than Class I neurons. More experimental data are needed before these hypotheses concerning the stability of the differentiated state of the two classes of neurons can be elevated to a greater degree of certainty. Experiments such as those that have shown the stability of the specified state in retinal ganglion cells, need to be carried out to examine the stability of the specified state in other types of neurons, particularly in Class II neurons. For example, the effect of heterotopic grafting of Class II nerve cells or of their isolation in tissue culture needs to be given more attention. Classical transplantation experiments need to be done to show whether the isolated neurons have stable, invariant phenotypes or whether they exhibit modifications of their phenotypes. Other experiments need to be designed to assay the plasticity or modifiability of identified Class I or Class II neurons, for example, by studying the permanent changes produced during learning.

Another aspect of the stability of the neurophenotype is demonstrated by studying the capacity of the developing system to recover from injury. Excision of parts of the developing nervous system allows one to determine whether there is any functional compensation or recovery or whether a permanent deficit results. Recovery may occur as a result of regeneration from the residual parts or by regulation (in which the neighbouring parts take over the developmental programme of the missing piece. As a rule, the later the stage of development the less the restitution of the excised neural rudiments, and the larger the excised fragment the less complete its restitution. Another general rule is that the structures and functions that develop earliest are the most vulnerable to injury whereas the structures and functions that develop late, for example language in man, are able to recover from injury inflicted at relatively late stages of development.

Structural Changes Associated with Nerve Cell Differentiation

The mature nerve cell is polarized structurally and functionally. At one pole are cytoplasmic processes, the dendrites, that assume various branching patterns, as characteristic of each type of nerve cell as branching patterns are typical of each species of tree. At the opposite pole is a single cytoplasmic process the axon or axis cylinder, ranging in diameter from 0·1 to 15 μm in various types of neurons, and also arborizing at its terminals. The cell body, containing the nucleus and the greater part of the synthetic mechanisms is interposed between dendrites and axon. The morphologies of dendrites, axons, and cell bodies are so varied that they cannot be described briefly, but a glance at the figures illustrating a textbook of neurohistology will convince the reader that the diversification of neuronal phenotypes is a major unsolved problem in developmental biology.

The functional polarization of the neuron is not readily seen but is a basic aspect of nerve cell development. Electrical activity flows from the dendrites to the cell body, and then cellulifugally to the axonal terminals. However, products synthesized in the nerve cell body flow proximodistally at rates ranging from 1 mm per day to several hundred millimeters per day into the axon as well as into the dendrites. Dendrites and axonal processes cannot synthesize enough material to sustain their own growth and materials must be produced in the cell body and be transported actively into the outgrowing cytoplasmic extensions. This transport is maximal during nerve cell growth but continues in the adult neuron in order to provide materials required for normal turnover in the dendritic and axonal branches[18, 19]. Some materials can enter the growing nerve fibre by pinocytosis from the extracellular fluid[20, 21]. This 'retrograde' flow[22] of materials may provide signals that regulate the differentiation and growth of the nerve cell.

At the time of its origin the young nerve cell has no clearly distinguishable axon and dendrites, but these usually appear after a latent period of hours to days, the axonal outgrowth usually preceding that of the dendrites. The cytoplasmic outgrowths of the nerve cell are amoeboid. The tip of the outgrowing axon or dendrite is termed the *growth cone* whose mobility is due to contractile cytoplasmic microfilaments[23]. These are fibrous macromolecules with a diameter of 90 to 100 Å and a length of several microns. Mechanical stability to the elongating nerve processes is provided by intracellular microtubules, which have a diameter of about 240 Å and which may run the entire length of the axon or dendrite. Neurotubules probably also form part of the mechanism of transport of materials in the nerve cell[24].

Colchicine is a drug which binds specifically with microtubules. When colchicine is added to a culture of outgrowing axons, in a concentration sufficient to disrupt the neurotubules ($2 \cdot 5 \times 10^{-6}$M), axons elongation stop within 30 min. The axons then shorten and may retract completely into the cell body. By contrast, the growth cone remains active for a longer time. Growth cone mobility may be inhibited within a few minutes by addition of cytochalasin, a drug that affects cytoplasmic microfilaments but not microtubules[23, 24].

One of the most striking features of nerve cell differentiation is the *outgrowth of the axon*. This cytoplasmic elongation occurs at the rate of about 20 to 40 μm per hour at 20 °C and at about 100 to 170 μm per hour at 34 °C. The vagaries of axonal growth – production of branches, their elongation, retraction and degeneration – makes the process appear to be a matter of trial and error. It is not known why many axonal branches fail to survive, and the evidence is inadequate to either confirm or refute any of the hypotheses of chemical or mechanical guidance of axonal growth[1].

The axon constitutes the presynaptic element of the neuronal nexus. Very little is known about the way in which the axonal terminal selects an appropriate nerve cell with which to form a synaptic connection. Probably some type of chemoaffinity between the pre- and post-synaptic elements is necessary[25], involving molecules at the external surface of the cell membrane. Indeed, so many activities of the developing nerve cell involve cellular interactions between cells in contact, that the cell membrane obviously plays a cardinal rôle in nerve cell development. *Deposition of myelin sheaths* around the axon is another membrane activity. Although the morphology of myelin and the process of its development have been described and studied for more than thirty years[28], the mechanism of the interactions and cellular affinities between Schwann cells and peripheral nerve fibres or between oligodendroglial cells and central nerve fibres are not understood[29]. The developmental mechanics of myelination, which include cell affinities, interactions, mobility, synthesis and transport are so protean as to constitute a separate essay in developmental biology.

Studies of *growth of dendrites* has not gone beyond the descriptive stage[26, 27]. That each type of nerve cell has a distinctive pattern of dendritic branching indicates a large measure of genetic control, but dendritic morphology may be grossly distorted (although the pattern rarely becomes unrecognizable) by a variety of extrinsic conditions acting on the developing dendrites. Their growth is stunted and they become deformed if deprived of the presy-naptic axon terminals that normally connect with them[30, 31]. A similar, but less severe dendritic pathology results from severe undernutrition during the neonatal period when the outgrowth of dendrites occurs in most regions of the brain[1, 32]. The dendrites provide more than 90% of the post-synaptic membrane area of the neuron. Thus, any reduction in dendritic development inevitably interferes with the formation of synaptic connections.

Studies of *development of synapses* have also been greatly assisted by the electron microscope. These descriptive anatomical studies have shown that new synapses continue to develop in the mammalian cerebral cortex during the post-natal period. The number of synapses in the cerebral cortex of the rat increases approximately tenfold between the twelfth and thirtieth day after birth[33, 34]. Some observations indicate that synaptogenesis may be influenced by sensory stimulation[35, 36, 37]. For example, the excitatory synapses on pyramidal cells of the cerebral cortex are largely restricted to specialized post-synaptic structures, the dendritic spines. The number of dendritic spines increases from birth to adulthood in the mammalian cerebral cortex, with the maximal increase occurring in the neonatal period. In the mouse, the increase in dendritic spines in the visual cortex can be greatly reduced by raising the animals in the dark[39]. Less drastic visual disturbances, such as monocularly sewing the eyelids together, or a monocular squint, in cats during the fourth to sixth weeks after birth, resulted in severe alterations in the functional properties of nerve cells in the visual cortex[39, 40]. These changes are largely irreversible. They show that adequate stimulation during a short, critical period after birth is a necessary requirement for the postnatal development and growth of some types of nerve cells.

Summary

The perplexities of nerve cell differentiation and growth are the same as those that we face in considering the mechanisms of development of other types of cells but the complexities are considerably greater in the nervous system. Genetic control is necessary for differentiation and growth of nerve cells, but it is usually insufficient, and other factors are also required for the normal maturation of neurons. Intercellular communication and cellular interactions are most important at all stages of neural ontogeny, and the development of the nerve cell can only be studied in the multicellular context. During the stages of development in the egg or in the uterus the nervous system is shielded from extrinsic influences, and factors controlling nerve cell development are almost entirely confined within the developing organism. However, after birth the interaction between the organism and the external environment plays a very important rôle in regulating the final stages in nerve cell differentiation and growth. Thus sensory deprivation has catastrophic effects on the development of nerve cells in the newborn mammal. The cellular mechanisms of these effects are virtually unknown.

References

1. JACOBSON, M. (1970), *Developmental Neurobiology*, Holt, Rinehart and Winston: New York.
2. SAXÉN, L. and TOIVONEN, S. (1962), *Primary Embryonic Induction*, Logos: London.
3. ROACH, F. C. (1945), 'Differentiation of the central nervous system after axial reversals of the medullary plate of *Amblystoma*', *Journal of Experimental Zoology*, **99**, 53–77.
4. JACOBSON, C. O. (1959), 'The localization of the presumptive cerebral regions in the neural plate of the axolotl larva', *Journal of Embryology and Experimental Morphology*, **7**, 1–21.
5. STEFANELLI, A. (1951), 'The Mauthnerian apparatus in the Ichthyopsida; its nature and function and correlated problems of histogenesis', *Quarterly Review of Biology*, **26**, 17–34.
6. JACOBSON, C.-O. (1964), 'Motor nuclei, cranial nerve roots, and fibre pattern in the medulla oblongata after reversal experiments on the neural slate of Axolotl larvae. I. Bilateral operations', *Zoologisk Bidrag. Uppsala*, **36**, 73–160.
7. KALLEN, B. (1958), 'Studies on the differentiation capacity of neural epithelium cells in chick embryos', *Zeitschrift für Zellforschung und mikroskopische Anatomie. Abteilung Histochemie*, **47**, 469–480.

8. CORNER, M. A. (1964), 'Localization of capacities for functional development in the neural plate of *Xenopus laevis*', *Journal of Comparative Neurology*, **123**, 243–256.

9. LOEWENSTEIN, W. R. (1968), 'Communication through cell junctions. Implications in growth control and differentiation', *Developmental Biology, Supplement*, **2**, 151–183.

10. JACOBSON, M. (1968), 'Cessation of DNA synthesis in retinal ganglion cells correlated with the time of specification of their central connections', *Developmental Biology*, **17**, 219–232.

11. JACOBSON, M. (1969), 'Development of specific neuronal connections', *Science* **163**, 543–547.

12. JACOBSON, M. and HUNT, R. K. (1973), 'The origins of nerve cell specificity', *Scientific American*, **228**, 26–35.

13. JACOBSON, M. (1968), 'Development of neuronal specificity in retinal ganglion cells of *Xenopus*', *Developmental Biology*, **17**, 202–218.

14. HUNT, R. K. and JACOBSON, M. (1973), 'Specification of positional information in retinal ganglion cells of *Xenopus*: Assays for analysis of the unspecified state', *Proceedings of the National Academy of Sciences, U.S.A.*, **70**, 507–511.

15. DIXON, J. S. and CRONLY-DILLON, J. R. (1973), 'The fine structure of the developing retina in *Xenopus laevis*', *Journal of Embryology and Experimental Morphology*, **28**, 659–666.

16. JACOBSON, M. (1970), 'Development, specification and diversification of neuronal connections'. In *The Neurosciences: Second Study Program*. Ed. Schmitt, F. O., 116–129, Rockefeller University Press: New York.

17. JACOBSON, M. (1973), 'A plenitude of neurons'. In *Studies on the Development of Behavior and the Nervous System*. Ed. Gottlieb, G., Vol. 2, pp. 151–166, Academic Press: New York.

18. LASEK, R. J. (1970), 'Protein transport in neurons', *International Review of Neurobiology*, **13**, 289–324.

19. OCHS, S. (1972), 'Fast transport of materials in mammalian nerve fibers', *Science*, **176**, 252–260.

20. HOLTZMAN, E. and PETERSON, E. R. (1969), 'Uptake of protein by mammalian neurons', *Journal of Cell Biology*, **40**, 863–869.

21. HOLTZMAN, E. (1971), 'Cytochemical studies of protein transport in the nervous system', *Philosophical Transactions of the Royal Society*, London, *B*, **261**, 407–421.

22. KRISTENSSON, K. and OLSSON, Y. (1971), 'Retrograde axonal transport of protein', *Brain Research*, **29**, 363–365.

23. YAMADA, K. M., SPOONER, B. S. and WESSELLS, N. K. (1971), 'Ultrastructure and function of growth cones and axons of cultured nerve cells', *Journal of Cell Biology*, **49**, 614–635.

24. SCHMITT, F. O. (1968), 'Fibrous proteins – neuronal organelles', *Proceedings of the National Academy of Sciences, U.S.A.*, **60**, 1092–1101.

25. SPERRY, R. W. (1963), 'Chemoaffinity in the orderly growth of nerve fiber patterns and connections', *Proceedings of the National Academy of Sciences, U.S.A.*, **50**, 703–710.

26. MOREST, D. K. (1969), 'The growth of dendrites in the mammalian brain' *Zeitschrift für Anatomie und Entwicklungsgeschichte*, **128**, 290–317.
27. HINDS, J. W. and HINDS, P. L. (1972), 'Reconstruction of dendritic growth cones in neonatal mouse olfactory bulb', *Journal of Neurocytology*, **1**, 169–187.
28. BUNGE, R. P. (1968), 'Glial cells and the central myelin sheath', *Physiological Reviews*, **48**, 197–251.
29. FRIEDE, R. L. (1972), 'Control of myelin formation by axon caliber (with a model of the control mechanism)', *Journal of Comparative Neurology*, **144**, 233–252.
30. SIDMAN, R. L. (1968), 'Development of interneuronal connections in brains of mutant mice'. In *Physiological and Biochemical Aspects of Nervous Integration*. Ed. Carlson, F. D., pp. 163–193, Prentice-Hall: New Jersey.
31. RAKIC, P. (1972), 'Extrinsic cytological determinants of basket and stellate cell dendritic pattern in the cerebellar molecular layer', *Journal of Comparative Neurology*, **146**, 335–354.
32. HALTIA, M. (1970), 'Postnatal development of spinal anterior horn neurones in normal and undernourished rats', *Acta physiologica Scandinavica, Supplement*, **352**, 1–70.
33. EAYRS, J. T. and GOODHEAD, B. (1959), 'Postnatal development of the cerebral cortex in the rat', *Journal of Anatomy*, **93**, 385–402.
34. AGHAJANIAN, G. K. and BLOOM, F. E. (1967), 'The formation of synaptic junctions in developing rat brain: a quantitative electron microscopic study', *Brain Research*, **6**, 716–727.
35. CRAGG, B. G. (1968), 'Are there structural alterations in synapses related to functioning?' *Proceedings of the Royal Society*, London, *B*, **171**, 319–323.
36. CRAGG, B. G. (1969), 'Structural changes in naive retinal synapses detectable within minutes of first exposure to daylight', *Brain Research*, **15**, 79–96.
37. MÖLLGAARD, K., DIAMOND, M. D., BENNETT, E. L., ROSENZWEIG, M. R. and LINDNER, B. (1971), 'Quantitative synaptic changes with differential experience in rat brain', *International Journal of Neuroscience*, **2**, 113–128.
38. RUIZ-MARCOS, A. and VALVERDE, F. (1969), 'The temporal evolution of the distribution of dendritic spines in the visual cortex of normal and dark raised mice', *Experimental Brain Research*, **8**, 284–294.
39. HUBEL, D. H. and WIESEL, T. N. (1970), 'The period of susceptibility to the physiological effects of unilateral eye closure in kittens', *Journal of Physiology*, **206**, 419–436.
40. HIRSCH, H. V. B. and SPINELLI, D. N. (1971), 'Modification of the distribution of receptive field orientation in cats by selective visual exposure during development', *Experimental Brain Research*, **13**, 509, 527.

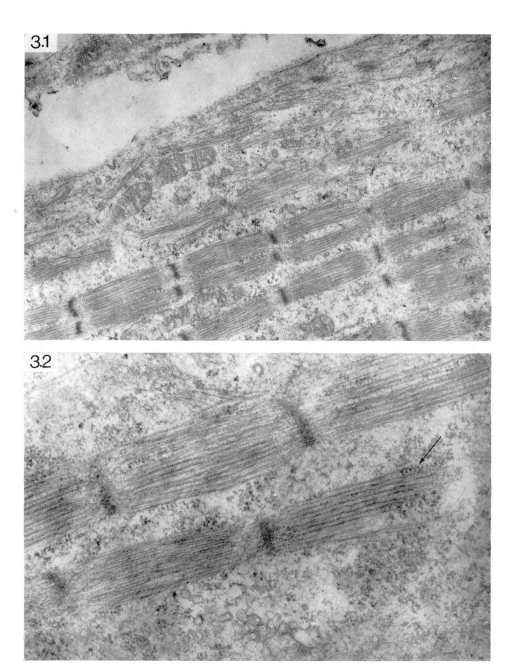

Plates 3.1 and 3.2 Electron micrographs of longitudinal sections of myotubes showing myofibrillogenesis. In Plate 3.1 numerous free filaments can be seen especially at the periphery of the myotube. Some of the filaments are apparently in the process of being assembled into myofibrils. Plate 3.2 is an electron micrograph at higher magnification of a myotube and shows polyribosomes associated with the myosin filaments. The ribosomes are in a spiral configuration. (*From Williams and Goldspink*[25].) Plate 3.1 × 16 000. Plate 3.2 × 32 000.

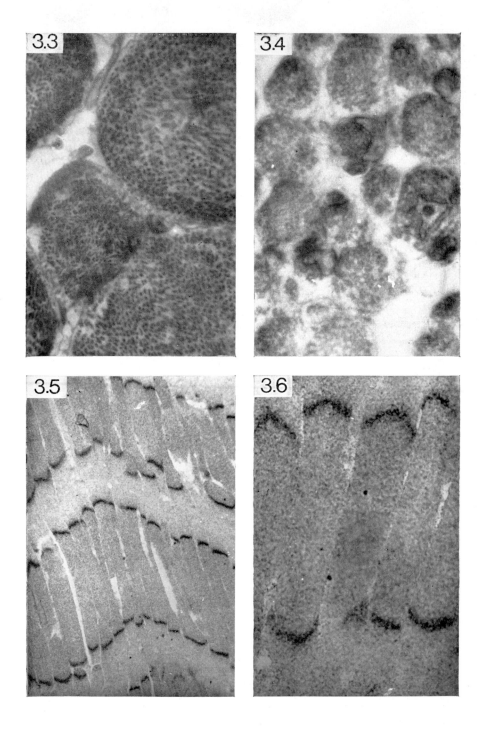

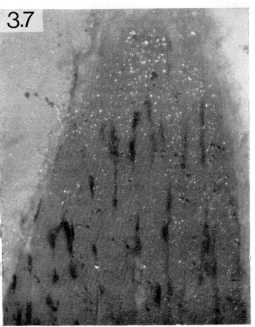
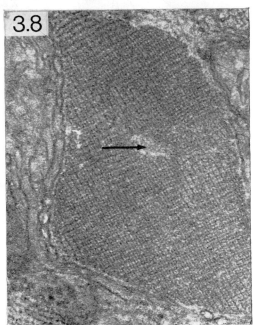

Plates 3.3 and 3.4 Light microscope photographs of transverse sections of mature muscle (3.3) and developing muscle (3.4). Note the centrally placed nuclei in the myotube and the relatively few myofibrils in the developing muscle. The section of the mature muscle shows fibres in the two phases of development. Note that in the large fibres the myofibrils are longer and more abundant than in the small fibre. × 1080.

Plates 3.5 and 3.6 Autoradiographs of amphibian muscle (tail muscle of the African clawed toad, *Xenopus Laevis*) that has been reared in a medium containing ³H-leucine. Note the bands of heavy labelling near the ends of each fibre. This is the region of longitudinal growth of the myofibrils. Plate 3.6 is a higher magnification of part of Plate 3.5. Plate 3.5 × 90. Plate 3.6 × 225.

Plate 3.7 An autoradiograph prepared from a longitudinal section of a biceps brachii muscle taken from a young mouse which had been injected with 75 μCi of ³H-adenosine over a period of 3 days. The photograph was taken using Leitz Ultrapak optics therefore the photographic grains showing the localization of the isotope appear as white spots. The localization of the isotope at the end of the muscle indicates that this is the growth region of the muscle where new sarcomeres are being produced during longitudinal growth. × 500 (*From* Griffin, Williams and Goldspink [42].)

Plate 3.8 Electron micrograph of a transverse section of a Z disc which has just commenced to rip. The rips are seen to begin as small holes in the centre and they then presumably extend across the Z disc with the direction of the weave. × 800 000. (*From* Goldspink [62].)

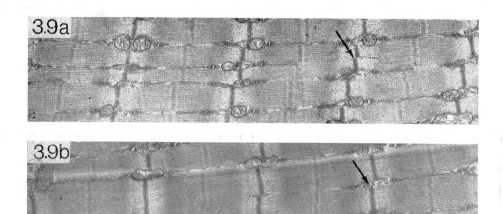

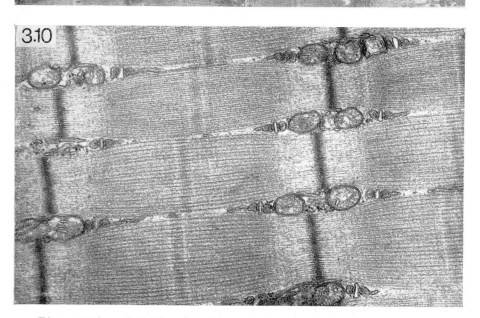

Plates 3.9 a and b. Electron micrographs of longitudinal sections showing myofibrils from the biceps brachii muscle of the mouse that are in the process of splitting. The splits are denoted by the arrow. × 13 500. (*From* Goldspink[62].)

Plate 3.10 Electron micrograph of a longitudinal section from a biceps brachii muscle fixed during contraction; sarcomere length 2·0 μm. Note that the peripheral actin filaments run obliquely from the Z disc and that the peripheral myosin filaments are bowed. This is due to a mismatch in the actin and myosin filament lattices. × 36 000. (*From* Goldspink[62].)

3 Development of Muscle

G. GOLDSPINK

Introduction

In recent years interest in muscle development has increased considerably. One reason for this is that muscle affords a good example of tissue differentiation and growth, in that it is a tissue that is apparently programmed to produce considerable quantities of rather specialized proteins in order to carry out its contractile function. Muscle therefore provides the cellular and molecular biologist, who is interested in cellular differentiation and growth, with an excellent system for study.

The study of muscle development is also of considerable importance in medicine and in agriculture. From the medical point of view it is important that we gain knowledge of muscle development in order to understand the derangement of the normal processes which occur with muscular dystrophy and other related diseases. From the agricultural standpoint it is important that we understand more about muscle, because striated muscle, as meat, provides a substantial proportion of the protein in our diet. One of the most serious problems regarding the food supply for an increasing world population is the relative scarcity of protein, particularly protein such as meat, with its high content of essential amino acids. It is therefore very desirable that we study the way in which muscle grows and the factors that influence its growth. When this is fully understood it may be possible to increase the efficiency and perhaps even the method by which meat is produced. Also, as muscle is the most abundant tissue in the body it is important that we understand the way in which it develops in order to appreciate the growth of the body as a whole. In this respect muscle development should not be studied in isolation but in relation to the development of other tissue particularly nerve and bone.

69

Differentiation of Muscle Cells

Origin of muscle tissue

With one or two exceptions, skeletal muscle is derived from lateral plate and paraxial somitic mesoderm. The main mass of muscle arises from the inner part of each myotome. Several other tissues, such as cartilage bone and dermis, are also derived from the inner part of the myotome. One of the unsolved problems is how presumptive cells of muscle, cartilage, bone and dermis, which are so close together in the myotome, can give rise to these different tissues. In other words, how is it that one cell may give rise to muscle whilst an adjacent cell may give rise to cartilage. Some experiments were carried out quite some time ago in which prospective muscle-forming areas were transplanted into the mid-ventral belly region or grown in epidermal jackets, and it was found that the cells did not form muscle but developed into pronephric-like tubules[1]. These experiments indicate that there are some environmental factors that induce the cells to develop into muscle rather than cartilage, dermis or kidney. The search for these factors has implicated the noto-chord[2] and the spinal cord[3]. However, as Muchmore[4] has pointed out, this cannot be the whole answer to the induction of muscle differentia-tion, because somitic mesoderm can, in certain circumstances, develop into muscle tissue in the absence of these tissues. The fact that muscle can differentiate in the absence of the nervous system is amply borne out by tissue-culture studies, for example, Yaffe[5] has kept several lines of isolated myoblasts propagating for over two years, and these have retained their capacity to differentiate into muscle. Although the early development of muscle is independent of neurogenic control, the further development of muscle is very much influenced by its nerve supply.

Proliferation of myoblasts

In the chick the wing bud mesenchyme appears as a separate cell con-densation in the embryo at stage 14–16 (24 somite embryo) and the development of the leg bud occurs in the 16–17 stage embryo (26 somite embryo). The enzymes required to synthesize chondroitin sulphate can be detected at stage 19[6], whilst myosin or other contractile proteins cannot be detected until stage 22[7]. However, if the chondrogenic tissue is taken at stages as late as 23 and 24 and transplanted into the non-chondrogenic soft tissue of the limb, 84% of the transplants develop into muscle or connective tissue and not into cartilage. If the reverse experiment is carried out in which the soft tissue is transplanted into the chondrogenic region, then 95% of the transplants develop into cartilage. This indicates that the determination of the tissue is not complete at this stage in development.

The first discernible step in myogenesis is the accumulation of mitotically active cells at sites which are destined to form muscle. The presumptive myoblasts become spindle shaped, and can therefore be distinguished from the polygonally shaped mesenchymal cells. A few of these cells may become very long and produce myofibrils; however, most of the myoblasts divide several times before fusing with other myoblasts or myotubes, and it is not until fusion has taken place that they produce myofibrils. These mononucleate cells with myofibrils do not seem to go on to develop into muscle fibres, as they do not incorporate ^{3}H-thymidine into their DNA, and therefore they probably remain mononucleate throughout their existence, which is probably relatively short, as the mononucleate cells with myofibrils are only very rarely seen in advanced embryos.

The time sequence of myoblast development has been well established for chick muscle. The rate of proliferation of myoblasts in the leg myogenic tissue is quite high at the end of the fourth day of leg development and reaches its maximum around the sixth or seventh day. On the seventh day of leg development about 70% of the myoblasts are in some phase of the replication cycle. This is compared with 12·5% for the surrounding mesenchymal cells[8]. The number of myogenic cells involved in replication, however, drops from 70% to 20% by the eleventh day. Holtzer[9] suggests that myoblasts undergo two types of cell division; proliferative mitoses, which just increase the number of cells and do not lead to any diversification, and quantal mitoses, which result in a slightly different kind of cell following the division. He suggests that myoblasts have to undergo a certain number of quantal mitoses before they can fuse and form myotubes. His suggestion is that each quantal mitoses would segregate a cytoplasmic factor and/or de-repress a portion of the genome.

Investigations of myoblast development using the electron microscope have not been very fruitful. In electron micrographs it is difficult to recognize myoblasts with any degree of certainty. Some cells can be assumed to be myoblasts because of their position, in other words they are found adjacent to myotubes or muscle fibre with which they are presumably going to fuse. Myoblasts can usually be distinguished from fibroblasts, as they contain many membrane bound ribosomes, whereas myoblasts have very little rough endoplasmic reticulum. Also in myoblasts there are often many thin filaments called cortical filaments. These have a diameter of approximately 5 nm and bind heavy meromyosin[10] which indicates they are actin-like in composition. However, these filaments are not unique to myogenic cells, being found in cells which undergo cytoplasmic elongation, amoeboid movement or cytokinesis. Other types of filaments have also been found in myoblasts, some of which do not bind heavy meromyosin; the function of these filaments is unknown at the present time.

Positive identification of myoblasts in culture has been achieved by transferring cells from the culture vessel to another vessel. Using this technique, Konisberg[11] showed that it is the bipolar spindle-shaped cells that give rise to muscle tissue whereas the flatter triangular or polygonal cells give rise to fibroblasts colonies. The difference in shape of the cell may reflect the difference in the cell membrane properties and in particular it may be associated with the change in the cell membranes prior to fusions. The change in the membrane properties of the myoblasts which permit fusion could be regarded as the first step that is unique to muscle differentiation.

Myotube formation

There were two theoretical possibilities for the formation of the multinucleated myotubes. One possibility is for the single nucleus of the myotube to undergo many mitoses and for a concomitant development of the cytoplasm of the cell. The second possibility was for the myotube to be formed by the fusion of several myoblasts.

All the evidence suggests that this second possibility represents the true situation. There is in fact overwhelming evidence from light and electromiscroscopy, quantitative DNA measurements by microspectrophotometry, autoradiographic studies of ^{3}H-thymidine incorporation and inhibition studies in which DNA synthesis is blocked, that the nuclei of the developing myotubes do not undergo mitosis. This evidence is also supported by the results of some very sophisticated experiments carried out by Mintz and Baker[12]. These workers aggregated blastomeres from two genetically distinct strains of mice to form 'chimeric' embryos. After uterine implantation of these embryos the mice were raised to maturity and the muscles found to contain two types of nuclei, which could be distinguished because the two strains of mice had different subunits of the enzyme isocitric dehydrogenase which could be easily identified by gel electrophoresis. In the chimeric mice a third intermediate isoenzyme band which had both types of subunits was found in muscle but not in the other tissues, therefore it must be concluded that the muscle fibres were formed by the fusion of cells which were originally from the two mice of different strains.

Therefore, it seems safe to conclude that after a certain number of mitoses the myoblasts fuse with one another to form myotubes or with myotubes to form longer myotubes. This process apparently continues throughout the growth of the animal with myoblasts fusing with muscle fibres although in this case the term satellite cell is used instead of myoblast (p. 80). These satellite cells or residual myoblasts are also believed to be very important in muscle regeneration following injury.

Development of myotubes into muscle fibres

The development of myotubes into muscle fibres does not occur synchronously. Also the parts of myotubes develop at different rates. The central region is the first to develop and the end regions the last to develop. The most noticeable change associated with the development of the myotubes is the rapid production of the myofibrillar proteins. There is also a considerable increase in the numbers of mitochondria. The myofibrils first form around the periphery of the myotube cytoplasm (Plate 3.1) whilst the nuclei tend to be centrally placed. As the myotube develops the central region becomes filled with myofibrils and most of the nuclei either migrate or are pushed by the myofibril mass to the periphery, where they occupy a position just under the sarcolemma (Plates 3.1 and 3.2).

The newly formed myofilaments and other structures exhibit an axial alignment to the long axis of the myotube. Unfortunately there is little information on how the myofilaments are aligned and assembled into myofibrils. However, at least three structures may be implicated in the determination of intracellular symmetry; the extracellular collagen, microtubules, and certain cytoplasmic filaments. The relative importance of the rôle that each may play is not known at the present time.

In skeletal muscle the bulk production of myofibrillar proteins only seems to occur when DNA synthesis has finished. In cardiac muscle, however, DNA synthesis can be demonstrated in myotubes which have well organized myofibrils[13]. On the other hand DNA synthesis is not observed in adult cardiac muscle fibre even during the hypertrophy of the tissue in response to functional overload[14].

It is quite possible that contractile proteins are synthesized at a low level in the stage before myotube formation. It is in fact likely that low levels of the contractile proteins exist in most cell types, indeed actin has been isolated from axons and myosin has been isolated from fibroblast cultures. However, as far as muscle is concerned it does appear that no bulk synthesis of contractile proteins occurs until myotube formation has commenced. Therefore the problem is not to pinpoint the first appearance of myosin and actin but rather to elucidate the factors which switch on the bulk synthesis of the contractile proteins.

As already stated the initial differentiation of myogenic cells is independent of any neuronal influence. However, it appears that innervation is necessary for the future development of the fibres, because if muscle fibres are denervated they degenerate. The innervation of myotubes has been studied[15]. It appears that the neuron usually branches several times and the individual branches make contact with the myotubes at special sites. Following this the terminal part of the nerve branch develops into a motor end plate and the myotubes begin to exhibit electrical activity and are sensitive to tubocuramine[16].

Problems of muscle differentiation

The asynchronous development of myogenic cells is one of the major problems that is encountered when studying muscle differentiation. Because of this, biochemical measurements tend to be rather meaningless as at any one time there are cells at several stages of development. Also the biochemical methods for estimating muscle proteins are not sensitive enough to determine the very small quantities of contractile and other typical muscle proteins that are initially produced in developing myogenic cells. It is therefore difficult to decide just when a presumptive muscle cell becomes a definite muscle cell. Several attempts have been made to block the various steps of muscle development in order to obtain more homogeneous cell population. Some success has been obtained with nucleic acid base analogues; FUdR (5-flurodeoxyuridine) for instance will block DNA synthesis by inhibiting the enzyme thymidylate synthetase. With this inhibitor it is, therefore, possible to selectively destroy the mononucleate cells in cultures without affecting the myotubes or muscle fibres[17]. Another base analogue BUdR (bromodeoxyuridine) if used in concentrations of 10^{-5} to 10^{-6} will prevent myotube formation without affecting the proliferation of the mononucleate cells. It seems that before muscle cell differentiation can be understood in detail it will be necessary to develop some sensitive methods for determining the cell specific proteins of muscle, e.g., micro

74

methods for myosin, actin, myoglobin creatine kinase, etc., and to find some way of synchronizing or separating out the various stages of development. Only when these methods become available we will apparently be able to elucidate the factors that control the differential gene expression which results in the bulk synthesis of muscle proteins.

Synthesis of Contractile Proteins

The control of protein synthesis during differentiation and growth is one of the most interesting problems in the field of molecular biology. At the present time we know that the genetic information is contained in the DNA of each cell. We also know that all the cells of the same animal have the same information; however, we do not know why some cells become muscle cells or other cells become, for example, liver or nerve cells. It seems almost certain that the differentiation of cells into different types is the result of muscle cells using different genes to liver cells. However, the control of differential gene expression in vertebrate cells

still very much a mystery. Certainly in muscle cells very little is known about the transcription of the DNA code into messenger RNA. However, some information is now available about the next step, namely the translation of the messenger RNA into protein, Heywood and co-workers[18] have developed a reproducible method of isolating polyribosomes from embryonic muscle (Fig. 3.1). Using radioactively labelled amino acids they have shown that the large polyribosomes which contain 50–60 ribosomes are capable of synthesizing myosin. The newly formed myosin was identified by acrylimide gell and most of the radioactive label that had been incorporated was located in this fraction. A similar study has been carried out on polyribosomes of other sizes and it has shown that some of these synthesize the other contractile proteins such as actin and myosin; the size of the polyribosome cluster corresponding to the molecular weight of the protein. These findings indicate that the different contractile proteins are synthesized independently and not produced from the one polycistronic strand of mRNA. This in turn means that the rate of synthesis of the different proteins may be independent of one another and therefore may be controlled by separate mechanisms. Heywood and Nwagwu[19] have now been able to isolate the messenger RNA of myosin from the large polyribosome clusters and have shown that the size of the mRNA corresponds to the large subunit of the myosin molecule which has a molecular weight of about 200 000.

Studies by Sarkar and Cooke[20] have shown that the low molecular weight components of myosin are made by another size class of polyribosome.

Although it is relatively easy to measure the rate of accumulation of the different contractile proteins during growth, it is much more difficult to estimate their rates of synthesis and turnover. There are several problems associated with the measurement of the rates of synthesis *in vivo*

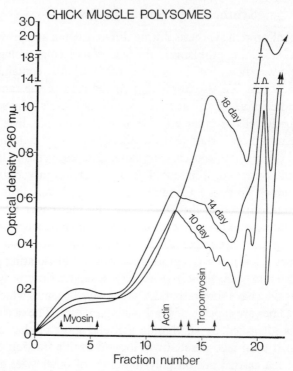

Fig. 3.1 Shows the polyribosome preparations from chick muscle of different ages. These were prepared by sucrose gradient analyses of cytoplasmic extracts from 0·7 g embryonic chick leg muscle at different ages of embryological development. The polyribosomes responsible for synthesizing myosin actin and tropomyosin occur in different size ranges and can be separated using this technique. (*From* Heywood and Rich[18].)

using labelled amino acids as precursor. These include the fact that the various proteins have different rates of turnover and the fact that the size of the amino acid pools, the permeability of the plasma membrane and the blood supply to different muscles differs at different ages.

In spite of the difficulties experienced in studying the factors which influence the rate of protein synthesis *in vivo*, some information has been obtained about protein synthesis in muscle during growth. Several workers have noted that the total RNA and the ribosomal RNA in muscle decreases during post-natal growth[21]. As well as the decrease in the total ribosome content, the percentage of the ribosomes in poly-ribosome clusters also decreases during growth[22, 23]. Polyribosome aggregates are certainly much more frequently observed with the electron microscope in embryonic or neonatal muscles than in adult tissue. In developing tissue, the polyribosomes are frequently seen in coiled configurations and they often seem to be actually coiled around individual myosin filaments[24, 25] (Plates 3.1 and 3.2). In adult muscle, the ribosomes are not so obvious, because although the fibres probably have nearly the same total number of ribosomes[22], they are distributed over a greater area, and there is also a tendency for them to be hidden by the mass of myofibrils and glycogen granules.

The use of cell-free systems to study the mechanism of protein synthesis during growth has also been tried by several other workers[22, 26, 27]. Breuer and Florini[22] found that the ability of the cell-free systems in synthesizing proteins was relatable to their polyribosome content; in other words, their messenger RNA content. The polyribosome content of the total ribosome fraction was found to be 83% in the muscle of young rats and 72% in the muscle of older rats. Similar results have also been obtained for chick muscle. Although the concentration of ribosomes decreases with age, it seems therefore that the production of mRNA becomes the limiting factor in myofibrillar production because the decrease in the percentage of ribosomes in polyribosomes that occurs during growth is probably a reflection of the reduction in mRNA synthesis. This has been verified by Srivastava[23], who demonstrated that the protein synthetic activity of muscle ribosome preparations could be stimulated by synthetic mRNA and that the extent to which they were stimulated depended on the age of the muscle. It is difficult to obtain information about mRNA in muscle or any other tissue, because it represents only a small percentage of the total RNA and it is readily subjected to hydrolysis by the ribonuclease of the tissue.

In addition to these quantitative changes in the myofibrillar proteins, there are suggestions that the structure of the proteins is altering during this early post-natal growth period. The characteristics of the myosin molecules change in several ways. There is evidence that the specific activity of the myosin ATPase increases and that the inhibitory effect of EGTA on the myofibrillar ATPase system also increases during this period[28, 29]. There are other differences, for example in the 3-methyl-histidine content of the myosin, that have been reported[30] which indicate fetal myosin is different to adult myosin. This is supported to some extent by the finding that the immunochemical response of the myosin changes during development, although its molecular weight and hydrodynamic parameters are apparently unchanged. It has been suggested by Perry[31] that myosin may exist in the form of two (or more) isoenzymes with different ATPase activity, one being the fetal type and the other the adult type. As the animal develops, the fetal type is replaced by the adult type, although the ratio of fetal to adult would vary from muscle to muscle, depending on whether the muscle was fast or slow.

Striated muscle offers many advantages for studies on protein synthesis because it produces considerable quantities of unique types of proteins. It is very likely, therefore, that this problem will attract the attention of many more biochemists and cell biologists, with the result that we shall arrive at a more complete understanding of the molecular processes involved in protein synthesis and the control of protein synthesis during growth.

Future Development of Muscle Fibres

Muscle fibre length

The limbs of most species of animals increase in length by several times during post-natal growth. The increase in limb length is accompanied by an increase in the fibres of individual muscles of the limb and it has been shown to be associated mainly with an increase in the number of sarcomeres along the fibres[25, 32]. As well as the increase in sarcomere number there may also be an increase in the length of the individual sarcomeres in both invertebrate[33–35] and vertebrate muscles[32]. Although in some invertebrates there is evidence that the filaments increase in length during growth[33], in vertebrates the increase in sarcomere length does not involve a change in filament lengths and the change in sarcomere length during differentiation and growth is presumably due to the difference in the rate of muscle growth and the rate of sarcomere production. The pulling out of the sarcomeres is apparently not uniform along the muscle fibre length as the terminal sarcomeres are invariably shorter than those in the middle of the fibres. However, the contribution made by the increase in sarcomere length to the increase in fibre length is relatively small. The main factor is the increase in the number of sarcomeres along the length of the fibres. The number of sarcomeres in series has been counted in individually teased fibres from the soleus and biceps brachii muscles of mice at different ages[25] and during post-natal growth the number of sarcomeres was found to increase from a few

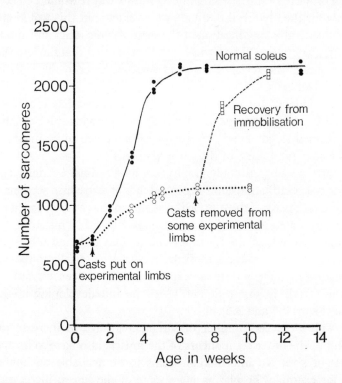

Fig. 3.2 Graph showing the rate of sarcomere production along the length of the myofibrils during growth in normal soleus muscles (●) and in soleus muscles that have been immobilized by means of plaster casts (○). Recovery of the muscles after removal of plaster casts is also shown (□) (from Williams and Goldspink[25]).

hundred to over two thousand with most of the increase occurring shortly after birth (Fig. 3.2).

The way in which new sarcomeres are added to the myofibrils is a fascinating problem. The point or points at which the sarcomeres are added to the myofibril has been a controversial matter for some time. Some authors[36, 37] have suggested that the myofibrils grow interstitially, in other words new sarcomeres are added to the myofibrils at points along their length. The evidence for this theory is based on the fact that the sarcomeres of adjacent myofibrils are often out of register because of slight differences in sarcomere length. Therefore, for a given length of muscle fibre some of the myofibrils will have an additional sarcomere and this is taken as evidence that a sarcomere has been inserted. However, in order to insert new sarcomeres in this way it would be necessary not only for the myofibril to divide transversely but this would also involve considerable modification of the sarcoplasmic reticulum and transverse tubular system. Other workers[38–41], suggested that the lengthening process entails the serial addition of sarcomeres onto the ends of the existing myofibrils.

Direct evidence for serial addition of sarcomeres has recently been obtained by injecting tritiated adenosine into growing mice to label the newly formed sarcomeres[25, 42]. Adenosine is known to be incorporated into the structural ADP of the actin filaments.

By carrying out autoradiography on longitudinal sections and by carrying out scintillation counting in batches of transverse sections taken from different parts of the muscle it was shown that most of the label had been incorporated into the end regions, see Fig. 3.7. This strongly suggests that the ends of the fibres are the regions of longitudinal growth and that new sarcomeres are most probably added serially to the ends of the existing myofibrils. Similar experiments have been carried out by the author on the incorporation of ^{3}H-leucine into developing amphibian muscle (Plates 3.5 and 3.6).

Not only does the number of sarcomeres have to increase during growth but the other components of the muscle also have to increase in number or size. No information seems to be available on how some parts, for example the plasma membrane of the fibres, increases with growth. Williams and Goldspink[25] have counted the total number of fibre nuclei in individual teased muscle fibres while Moss[43] has counted nuclei in sections in muscle of different ages. They found that the fibre nuclei continue to increase in number beyond the stage at which there was no further increase in muscle fibre length. They also found that the larger diameter fibres possessed more nuclei, therefore it must be concluded that the increase in nuclei is associated with an increase in the girth of the fibres as well as the increase in length. The increase in fibre nuclei does not result from mitosis of the existing nuclei. As mentioned on p. 73 certain cells known as satellite cells or residual myoblasts fuse with the fibres and donate nuclei to the growing fibres[44, 45]. These cells can be seen to be associated with the muscle fibres particularly in young muscles and they actually lie under the basal membrane of the fibres.

The percentage of satellite cell nuclei with respect to muscle fibre nuclei has been shown to decline in rat muscle by a factor of about 8 times between birth and maturity[46].

Certain physiological experiments have shed light onto the factors that may be involved in stimulating muscle fibres to produce more sarcomeres. Surgical modification of the distance that the muscle has to contract is known to radically affect the longitudinal growth of the fibre[47, 48]. For example, immobilization of limbs of growing mice using plaster casts results in a considerable reduction in the number of sarcomeres in series and therefore in shorter muscle fibres[25]. It has also shown that the nervous system has some influence on longitudinal growth[49]. This has been reinforced by the studies of Tabary et al.[50] on the longitudinal growth of muscle in children afflicted with cerebral palsy. In these children the defect in the central nervous system in some way prevents the muscle fibres producing enough sarcomeres for the muscle to attain its correct length. However the influence of the nervous system is apparently not a direct one, because denervated muscles are still able to produce more sarcomeres in series when immobilized in the lengthened position. Further studies will be necessary to elucidate the link between the mechanical event of contraction and the molecular processes involved in the synthesis and assembly of sarcomeres; however this is certainly an aspect of muscle which is worthy of more extensive investigation.

Muscle fibre number

In theory, muscle may increase in girth during growth by either an increase in muscle fibre number and/or an increase in muscle fibre size. It is now generally accepted that in mammalian muscles the number of fibres does not increase once the embryonic differentiation of the tissue is complete[51–55] and that the number of muscle fibres does not increase during the post-natal growth except perhaps for a short period after birth when the muscle fibres are still being formed. A very considerable post-natal increase in fibre number has, however, been reported as occurring in muscles of the marsupial Setonix brachyurus (the Australian quokka)[56]. This marsupial is born in a very immature condition and it seems that the first stage of muscle growth involves the rapid development of some of the fibres; just enough to enable the newborn animal to climb to its mother's pouch. The full complement of fibres in the muscles is then attained at about 100 days after birth. This is a special adaptation and should be regarded as a protraction of the embryonic differentiation of the tissue.

As far as the two sexes are concerned, there is no difference in the total number of fibres between the same anatomical muscles although fibre size is greater in males than in females[55]. There are, however, considerable differences in fibre number between different genetic strains and of course between different species. Small mammals tend to have small fibres and large mammals tend to have larger fibres, but the difference in muscle size between species is attributable mainly to differences in fibre number[57]. From a physiological point of view it is not feasible to have fibres developing beyond a certain size because the distance from the outside to the centre would be too great for oxygen diffusion and possibly too great for the transmission of the impulse down the T system to the centremost myofibrils.

A modification of fibre number has also apparently resulted from the artificial selection of animals. Staun[58] estimated the number of fibres in the muscles of different breeds of pig and found that fibre number was characteristic for a particular breed. This finding has also been verified for different strains of inbred mice by Luff and Goldspink[59] who measured the number of fibres in the muscles of mice that were originally from the same strains but which had subsequently been selected for largeness and smallness over several generations. The difference in muscle size between the large line and the small line could be attributed to a difference in the total number of fibres in the muscle and not to a difference in fibre size.

The fact that fibre number becomes fixed before birth or shortly after birth and that it can only be normally altered by natural or artificial selection, suggests that it is under direct or indirect genetic control. A high fibre number has also been correlated with increased meat producing potential and therefore the possibility of selecting meat-producing animals based on the fibre number of a muscle taken out by biopsy, seems to merit further investigation. It may also be possible to modify fibre complement by experimental means, particularly if the experiments are carried out on the fetus or the neonatal animal during the time that the myoblasts are still proliferating. For instance, a slight modification in the rate of mitosis would presumably result in a considerable alteration in fibre number. The cellular mechanism involved in producing a certain complement of fibres for a given muscle are unfortunately not understood at the present time.

Muscle fibre girth

It has long been recognized that muscle fibres increase in size during the post-natal growth of the animal. Several workers have measured the overall increase in fibre size during growth[51, 53, 54]. However, few workers have attempted to study the way in which the fibres increase in size or the ultrastructural and biochemical changes that accompany the post-natal increase in fibre size. Work on mouse muscles[53, 55] which are particularly suitable for histological studies because the fibres usually run from tendon to tendon and because of their reasonably small size, has indicated that the muscle fibres grow in a discontinuous way rather than in a gradual and continuous manner. Shortly after birth all the fibres are about the same size, 15–20 μm in diameter. In some muscle such as the soleus and extensor digitorum longus the fibres grow very little and more or less stay at this basic level of development throughout the life of the animal. However, in other muscles, such as the biceps brachii, some of the fibres undergo further growth or hypertrophy to a size of about 40 μm in diameter. The conversion of small fibres into larger fibres can be detected during growth by plotting fibre size distribution for muscles at different ages (Fig. 3.3). In the mouse soleus and extensor digitorum longus muscles the modes and shape of the distribution plots do not change to any great extent during growth. However, in the mouse biceps brachii and anterior tibialis muscle the appearance of the large fibres is discernible by three weeks of age when a fibre size peak appears at about 35–40 μm. As the animal grows the number of fibres in this peak increases. In the mature biceps brachii this population of large fibres is approximately equal to the number in the small phase. The proportions of fibres in the large or small phase can be altered considerably by exercise or changing the level of nutrition in the animal. Recent work has shown that there is, however, a population of small oxidative fibres in the mouse biceps brachii which do not change size in response to changes in the level of nutrition of exercise. Because of their high levels of oxidative enzymes it is suspected that they play a postural rôle, in other words maintaining the tone of the muscle.

The physiological reason for the conversion of the small phase fibres into large phase fibres becomes apparent when the ultrastructure of fibres in these two phases of development is examined[60]. The large phase fibres possess larger myofibrils and a greater number of myofibrils. Indeed the increase in girth of the fibre can be explained almost entirely by the increase in the number and size of the myofibrils. In larger muscles which grow at a slower rate the discontinuous growth of the fibres apparently is not readily detectable by constructing distribution histograms[66]. In these muscles the fibres probably undergo several increases in size over a relatively long period of time before they attain

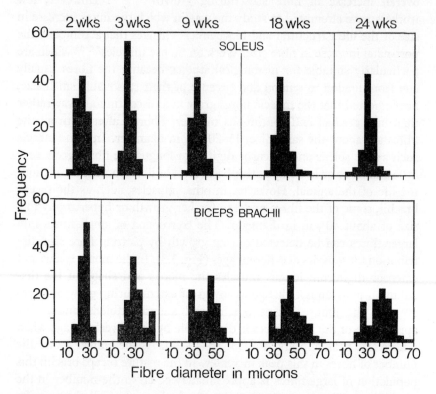

Fig. 3.3 Shows distribution histograms of fibre size for the soleus and biceps brachii muscles at different ages. Note that the soleus muscle fibres show a unimodal distribution at all ages. However the biceps brachii develops a bimodal distribution because of the rapid development of some of the fibres. In the adult biceps brachii, therefore, there are fibres in both the small and large phases of development. (*From* Rowe and Goldspink[55].)

their ultimate size. This is contrasted with the situation in the mouse muscle where the fibres undergo only one post-embryonic size increase to reach their ultimate size and this occurs in a relatively short time. A general scheme for the different stages of muscle fibre development is presented in Fig. 3.4.

The stimulus that causes a fibre to undergo further growth of hypertrophy may be the intensity of the work-load to which the fibre is subjected. As the body weight of the animal increases during growth, the work-load on most of the skeletal muscles must also be considerably increased. It seems reasonable to suggest that when the 'work-load threshold' for an individual fibre is exceeded this, in some way, triggers off the production of more myofibrils and causes the fibre to increase in size. However there are other influences such as the effects of androgens, which also influence fibre development.

The possible mechanism of myofibril production is of great interest as

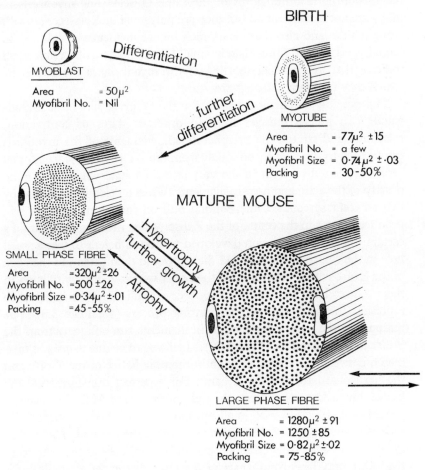

Fig. 3.4 Shows a generalized scheme of muscle development for mouse muscle. The last set of arrows indicates the further increases in size that may occur in the muscles of larger animals. Note that the myofibril production is associated with the development of the myotubes into small but proper muscle fibres and also with the hypertrophy of these fibres into large phase fibres.

it is perhaps the most obvious ultrastructural change associated with muscle fibre growth. The number of myofibrils within an individual muscle fibre of the mouse may increase during growth by more than 15 times[60]. Evidence obtained by examining muscle fibres at different stages of growth strongly suggests that this proliferation of myofibrils is the result of longitudinal splitting of the existing myofibrils once they attain a certain size (Plates 3.8–3.10). The evidence for this includes the observations that the myofibrils in some fibres are about twice the size of the myofibrils in other fibres of the same muscle. The myofibrils of fibres are not usually round but they are polygonal in cross-section with straight sides and also there is a much higher incidence of splitting in large hypertrophied fibres than in small fibres[60]. These observations all indicate that longitudinal splitting of the myofibrils is the means by which they proliferate during growth.

The possible mechanical reason for the splitting has been investigated[62]. In mouse muscles fixed in different states of contraction, relaxation and stretch the peripheral actin filaments of the myofibrils can be seen to run slightly obliquely from the Z disc. This is true for all sarcomere lengths except in the very stretched state when there is no overlap of the actin and myosin filaments. When tension is developed by two adjacent sarcomeres the oblique pull on the actin filaments will produce a stress in the centre of the Z disc. When the myofibrils attain a certain thickness the tension developed will be sufficient to tear a small hole in the centre of the disc and the rip would then extend to the edge of the Z disc with the direction of the lattice weave. Z discs with holes in their centres have frequently been observed and these are believed to represent the first stage of the splitting process (see Plate 3.8). The reason why the more peripheral actin filaments run obliquely from the Z disc to the A band (see Plate 3.10) is believed to be due to a slight mismatch between the square or slightly rhombic lattice of the Z disc and the hexagonal lattice of the A band. For a perfect transformation the lattice dimensions are critical; the ratio of the Z and M lattice spacings should be $1:1.5$. However, in all but the very stretched myofibrils, the ratio is nearer to $1:2$ (Z:M) which means that the peripheral actin filaments are always displaced more than the central ones as they run from the Z disc into the A band. It seems therefore that in vertebrate muscles there is this built-in mechanism for ensuring that the myofilamental mass as it increases, is subdivided into smaller units. If the myofilamental mass was built up as one uninterrupted mass this would present problems as far as the activation of the contractile apparatus is concerned; however, the splitting mechanism allows the sarcoplasmic reticulum and the transverse tubular system to invade the mass and to develop at about the same rate as the myofibril content of the fibres[62].

It is well known that the myofibrils of slow muscles tend to be less discrete and more irregular in cross-section than those of fast muscles. The different appearance of the myofibrils in these two kinds of muscles have been termed 'Felderstruktur' in the case of slow muscle and 'Fibrillenstruktur' in the case of fast muscle[63]. In the chicken, the Felderstruktur appearance of the slow fibres is known to develop during growth[64] and it is believed to be the result of the less frequent and incomplete splitting of the myofibrils. To explain this, it has been suggested[60, 61] that the rate at which tension developed by the myofibril is perhaps more important than the total amount of tension developed in myofibril splitting.

It would seem that the method of myofibril proliferation during growth involves an increase in the number of actin and myosin filaments in the myofibril and the splitting of the myofibril once it has attained a critical size. This method of myofibril production may, of course, be different in many respects from that by which myofibrils are first produced during early embryonic development. The way in which more actin and myosin subunits are added to the myofibrils is not known. However, recent radio-autography in conjunction with electron microscopy has indicated that the newly formed contractile proteins are added to the outside of the myofibrils[64]. This certainly does seem reasonable as it is difficult to imagine the new filaments being produced at or being attached to the centre of the myofibril. Also in adult muscle it is unlikely that whole filaments are added as the myofibrils are enclosed by the sarcoplasmic reticulum and it is difficult to see how the filaments, which are known to be semi-rigid, could get through the sarcoplasmic reticulum in order to be latched onto the myofibril. It seems more feasible that the myofibrils are built up by the addition of individual protein subunits of actin, myosin and the other myofibrillar proteins. Presumably the subunits, once they have been synthesized by the ribosomes, will attach themselves to the nearest 'receptive' myofibril. The self assembly of myosin and actin molecules has been shown in vitro by Huxley[65] and it seems probable that this process operates in vivo. Auber[35] has studied the addition of filaments in developing insect flight muscle and noted that myosin filaments on the periphery of the myofibrils are often smaller in diameter thus indicating that the construction of these peripheral filaments is incomplete. Auber's observation therefore lends support to the idea of the *in situ* assembly of the myofilaments.

As the myofibrils increase in number, concomitant changes must occur in the sarcoplasmic reticulum and transverse tubular systems. In rat muscle the sarcoplasmic reticulum undergoes some striking changes in the first two weeks after birth[66-68]. The sarcoplasmic reticulum in the new born rat consists of a few irregularly orientated tubules which do not completely surround the myofibrils. At this stage there are also very few contacts between the sarcoplasmic reticulum and the transverse tubular system. During the first two weeks after birth the sarcoplasmic reticulum develops into its mature form of a well-organized system of tubules which envelop the myofibrils. The transverse tubular system, which arises from the plasma membrane, penetrates deep into the muscle fibres and makes contact with the sarcoplasmic reticulum to form triads at the level of the AI junction. The number of triads observed in sections during this period increases very considerably during the first few weeks after birth. The transverse tubular system and the sarcoplasmic reticulum must presumably continue to extend as long as the fibre is producing myofibrils.

Factors Which Influence Muscle Fibre Development

Activity

Hypertrophy of striated muscle fibres as a result of exercise has been an accepted fact for many years[69, 70, 71]. More recent biochemical studies[72] and cytological studies on striated muscle[73, 74, 75] and on cardiac muscle[76] clearly indicate that hypertrophy is usually associated with a large increase in the myofibrillar material of the fibre. However, this does not preclude the possibility that under certain conditions of exercise some hypertrophy of the fibres may be due partly or wholly to an increase in the mitochondrial and sarcoplasmic proteins[77].

It is well known from observations on athletics that exercise of endurance leaves the size of the musculature relatively unchanged, although it is believed to produce biochemical changes which lead to greater stamina. On the other hand, exercises of high intensity such as weight lifting are known to be much more effective in producing muscle fibre hypertrophy.

One of the best examples of intensive exercise is weight lifting involving relatively heavy weights. The effect of this type of exercise has been studied using laboratory animals[73, 78]. In this work the animals were required to pull down a food basket which was attached to a pulley system. A known weight was attached at the other end of the pulley system and thus the amount of work performed by the animals in obtaining their food could be estimated. This type of exercise was found

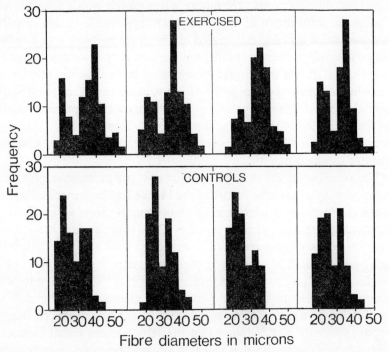

Fig. 3.5 Distribution histograms of fibre diameters in the biceps brachii muscles from half grown mice which have been exercised (top) and control mice of the same age. Each histogram is for a 100 fibre sample taken at random. (*From* Goldspink[73].)

to be very effective in producing fibre hypertrophy; again, the increase in mean fibre size was the result of a certain percentage of fibres undergoing the increase in size from the small 'phase' (20 μm diameter) to the large 'phase' (40 μm diameter) and not to a general increase in size of all fibres (see Fig. 3.5).

In mixed muscles it may be that certain types of exercise may cause the hypertrophy of one type of muscle fibre whilst a different type may stimulate the hypertrophy of another type of fibre. Further research is needed to clarify this point.

Other methods of increasing the work load on muscle fibre have been used. These methods usually involve the surgical removal of part of the muscle or the removal of a synergetic muscle[79-82]. Goldberg[79] carried out tenotomy of the gastrocnemius muscle of normal and hypophysectomized rats and showed that growth hormone was not required for work induced hypertrophy. Rowe and Goldspink[80] using essentially the same procedure found that the mouse soleus muscle which normally has a unimodal distribution of fibre sizes became bimodal. In this case the increased work load had not caused all the fibres to increase in size but caused a certain percentage of them to increase in cross-sectional area by 3 or 4 times. It was found that there was an increase in the number of fibres in the experimental muscles. Other workers have also recorded the formation of new fibres in greatly overloaded muscles, apparently resulting from splitting of the existing fibres[80, 82-84]. However, it is not known whether this is some form of regeneration of damaged fibres, or a normal adaptive mechanism.

The response of the muscle to different types of exercise poses the question as to what is the nature of the link or feedback mechanism between the mechanical event and the biochemical processes involved in the synthesis of more muscle proteins. Only very recently has it been possible to study cell growth at the molecular level and already several interesting findings have emerged concerning muscle fibre hypertrophy. Goldberg[85, 86] has shown that the incorporation of ^{14}C-leucine into both sarcoplasmic and myofibrillar protein is enhanced and the rate of degradation of these proteins is reduced, during work induced hypertrophy. Hamosh et al.[87] found that the RNA concentration increased during hypertrophy and that cell free systems prepared from muscles undergoing hypertrophy possess a greater ability to synthesize proteins. These workers found that L-phenylalanine was incorporated into protein at a faster rate by the microsomes fractions prepared from hypertrophied muscle in the absence and in the presence of artificial messenger RNA (poly U). They also found an increase in the RNA content of the microsome fractions. These two findings indicate that the enhanced ability to synthesize proteins during hypertrophy is due to an increase in ribosomes as well as messenger RNA.

Effect of nutrition

It is well known that reducing the food intake leads to a considerable reduction in muscle mass. In fact, excluding fat, muscle seems to be the tissue that is most susceptible to starvation. The decrease in muscle bulk has been shown to be associated with a decrease in mean fibre diameter in cattle[88], sheep[89], pigs[90], the laboratory mouse[74] and in humans[92]. The way in which the mean fibre diameter decreases has been studied in mouse muscle[81, 91]. The effect of starvation on muscles which are normally composed of large and small fibres, and therefore bimodal, was to reduce the number of large 'phase' fibres in the muscle so that the fibre size distribution plots tended to become unimodal. The decrease in fibre size is associated with a reduction in the number of myofibrils in the fibre and this accounts for the decrease in the contractile strength which accompanies starvation[82]. The way in which the myofibrils are removed from the fibres is not known.

Fibres of different types of muscles are known to be affected by starvation to different extents. The fibres of fast muscles of the laboratory mouse exhibited a decrease in size whilst those of the slow soleus muscle remain virtually unchanged by starvation[91] Cytospectrophotometric measurements of the succinic dehydrogenase concentrations in individual fibres have shown that a reduced food intake causes a reduction in the activity of this enzyme in the white fibres but not in the red fibres[93]. The explanation for the difference in response may lie in the kind of innervation of the fibre or the richness of the vascular supply of the muscle. However, it may be at a more biochemical level in that the reduced supply of fuel and raw material could be expected to affect the anabolic and catabolic pathways in different ways in fibres with different metabolisms.

The biochemical control of protein metabolism is not fully understood especially in relation to starvation. However, it appears that not only is rate of protein degradation increased during starvation but the rate of synthesis of new protein decreased. This may result in a shortage of one or more essential amino acids in the muscle fibres or it may be due to a reduction in the protein synthesizing capacity of the polyribosomes[95].

Type of innervation

Over the past 10 years or so it has become apparent that the nervous system exerts a very considerable influence on the way muscle fibres develop. It has long been recognized that denervation of muscles leads to atrophy and eventually the degeneration of the muscle fibres. However, it was not clear whether this was due to the change in the activity of the muscle or whether the nerve produces a substance or in some other way imparts stability to the muscle fibres. More recent work suggests that change in activity is important, but that the nerve may also exert a 'trophic influence'. The earlier work on the effect of denervation has been reviewed by Tower[96]. The structural changes that are associated with denervation are different from those resulting from simple disuse, in that many of the fibres, as well as undergoing atrophy, actually degenerate and this results in a reduction in the number of fibres in the muscle[97]. These changes in the muscle fibres are accompanied by a reduction in the size of the myofibrils[98] and an increase in the number of nuclei. Also the nuclei of the muscle fibres as well as increasing in number become more rounded in shape and their chromatin material appears to be more dispersed. Biochemical studies have shown that following denervation myofibrillar proteins are lost at a greater rate than the sarcoplasmic or connective tissue proteins[99].

Studies on incubated frog muscles have shown that denervation of the preparations resulted in a decreased incorporation of ^{14}C-valine into the myofibrillar proteins[100] but an increased incorporation into the sarcoplasmic proteins[101]. It seems, therefore, that the denervation selectively suppresses the synthesis of the myofibrillar proteins while it accelerates the synthesis of sarcoplasmic proteins. In addition to altering the rate of synthesis there is also evidence that the rate of breakdown of the proteins is increased. This seems to tie in with the fact that the size of the lysosomes is greater in denervated muscle and cathepsin activity levels are increased[98].

The rapidity of the weight loss following denervation varies between different muscles of the same animal and between the same anatomical muscles in different species. In general, the rate of weight loss can be related to the metabolic rate of the tissue[102]. Also the level at which the nerve supply is sectioned affects the rate of the decrease in muscle weight. Transection of the nerve near to the muscle produces a more rapid change[98], which suggests that some substance or substances pass down the axons.

In addition to stabilizing the highly differentiated state of the muscle fibres the nervous system also determines some of their characteristics. This was first demonstrated by the experiments of Buller, Eccles and Eccles[103] who crossed the nerve supply of a fast muscle and a slow muscle. Under these circumstances the contraction speed of the muscles changes to almost that of the muscles whose nerve supply they receive, thus demonstrating that the rate of contraction of the muscles is dictated by the type of innervation of the muscle fibres. The change in the speed of contraction following cross-innervation has been shown to be associated with a change in the myosin ATPase of the muscles[104]. Other studies have shown that cross-innervation of fast and slow muscles is also accompanied by changes in the levels of oxidative and glycolytic enzymes[105]. However, it is not known whether these other biochemical changes are the direct result of the cross-innervation or whether they are just part of the metabolic re-adjustments that would be necessary following the change in contraction speed and the change in level of activity of the fibres.

A study by McComas[106] on the number of functional motor units in dystrophic muscle has indicated the neurogenic nature of the disease. Therefore, from the medical standpoint, as well as from a fundamental point of view, it is very important that the influences of the nerve on muscle development should be more fully understood.

References

1. YOMADA, T. (1937), 'Der Determinationszustand des Rumpfmesoderms im Molchkeim nach der Gastrulation', *Roux Archir für Entwicklungsmechanik der Organismen*, **137** (2), 151–270.

2. MUCHMORE, W. B. (1958), 'The influence of embryonic neurol tissues on differentiation of striated muscle in Amblystoma', *Journal of Experimental Zoology*, **139**, 181–188.

3. HOLTZER, H. and DETWILER, S. R. (1953), 'Induction of skeletogenous cells', *Journal of Experimental Zoology*, **123**, 355–370.

4. MUCHMORE, W. B. (1968), 'The influence of neurol tissue on early development of somitic muscle in ventrolateral implants in Amblystoma', *Journal of Experimental Zoology*, **169**, 251–258.

5. YAFFE, D. (1968), 'Retention of differentiation potentialities during prolonged cultivation of myogenic cells', *Proceedings of the National Academy of Science of the United States*, **61**, 477–478.

6. MENDOFF, J. (1967), 'Enzymatic events during cartilage differentiation in the chick embryonic limb bud, *Developmental Biology*, **16**, 118–143.

7. ZWILLING, E. (1968), 'Morphogenetic phases in development', *The emergence of order in developing systems*. Ed. Locke, M., Academic Press: New York.

8. MARCHOK, A. C. and HERRMAN, H. (1967), 'Studies of muscle development 1. Changes in cell proliferation', *Developmental Biology*, **15**, 129–155.

9. HOLTZER, H. (1970), 'Mitosis and Myogenesis'. In *Physiology and Biochemistry of Muscle as a Food*. Eds. Briskey, E. J., Cassen, R. G. and Marsh, B. B., University of Wisconsin Press.

10. ISHIKAWA, H., BISCHOFF, R. and HOLTZER, H. (1968), 'Formation of arrowhead complexes with heavy meromyosin in a variety of cell types', *Journal of Cell Biology*, **43**, 312–328.

11. KONISBERG, I. R. (1963), 'Clonal analysis of myogenesis', *Science*, **140**, 1273.

*

12. MINTZ, B. and BAKER, W. W. (1967), 'Normal mammalian muscle differentiation and gene control of isocitrate dehydrogenase synthesis', *National Academy of Science of the United States*, **58**, 592–598.

13. MARK, G. A. and STRASSER, F. F. (1966), 'Pacemaker activity and mitosis in cultures of newborn rat heart ventricle cells', *Experimental Cell Research*, **44**, 217–233.

14. MOSKIN, E. and ASHFORD, T. P. (1968), 'Myocardial DNA synthesis in experimental cardiac hypertrophy', *American Journal of Physiology*, **215**, 1409–1413.

15. KELLY, A. M. and ZACKS, S. I. (1969), 'The fine structure of motor endplate morphogenesis', *Journal of Cell Biology*, **42**, 154–169.

16. ROBINS, N. and YONEZAWA, T. (1971), 'Developing neuromuscular junctions: First signs of chemical transmission during formation in tissue culture', *Science*, **172**, 395–397.

17. COLEMAN, J. R. and COLEMAN, A. W. (1968), 'Muscle differentiation and macromolecular synthesis', *Journal of Cellular and Comparative Physiology*, **72**, 19–23.

18. HEYWOOD, S. M. and RICH, A. (1968), '*In vitro* synthesis of native myosin, actin and tropomyosin from embryonic chick polyribosomes,' *Proceedings of the National Academy of Sciences U.S.*, **59**, 590–597.

19. HEYWOOD, S. M. and NWAGWU, M. (1969), 'Partial characterization of presumptive myosin messenger RNA', *Biochemistry*, **8**, 3839–3840.

20. SARKAR, S. and COOKE, P. H. (1970), '*In vitro* synthesis of light and heavy polypeptide chains of myosin', *Biochemical and Biophysical Research Communications*, **41**, 918–925.

21. FLORINI, J. R. and BREUER, C. B. (1965), 'Amino acid incorporation into protein by cell free preparations from rat skeletal muscle. III. Comparison of muscle and liver ribosomes', *Biochemistry*, **4**, 253–257.

22. BREUER, C. B. and FLORINI, J. B. (1965), 'Amino acid incorporation into protein by cell free systems from rat skeletal muscle. I. Properties of the muscle microsomal system', *Biochemistry*, **3**, 209–215.

23. SRIVASTAVA, U. (1969), 'Polyribosome concentration of mouse skeletal muscle as a function of age', *Archives of Biochemistry and Biophysics*, **130**, 129–139.

24. LARSON, P. F., HUDGSON, P. and WALTON, J. N. (1969), 'Morphological relationship of polyribosomes and myosin filaments in developing and regenerating skeletal muscle', *Nature London*, **222**, 1169.

25. WILLIAMS, P. E. and GOLDSPINK, G. (1971), 'Longitudinal growth of striated muscle fibres', *Journal of Cell Science*, **9**, 751–767.

26. SRIVASTAVA, U. and CHAUDHARY, K. D. (1969) 'Effect of age on protein and ribonucleic acid metabolism in mouse skeletal muscle', *Canadian Journal of Biochemistry*, **47**, 231–235.

27. CHEN, S. C. and YOUNG, V. R. (1968), 'Preparation and some properties of rat skeletal muscle polyribosomes', *Biochemical Journal*, **106**, 61–67.

28. BÁRÁNY, M., TUCCI, A. F., BÁRÁNY, K., VOLPE, A. and RECHARD, T. (1965), 'Myosin of newborn rabbits', *Archives Biochemistry and Biophysics*, **111**, 727.

29. TRAYER, I. P. and PERRY, S. V. (1967), 'Evidence for differences between foetal and adult myosins', *Biochemical Journal*, **97**, 36P.

30. TRAYER, I. P., HARMS, C. I. and PERRY, S. V. (1968), '3-Methyl histidine and adult and foetal forms of skeletal muscle myosin', *Nature London*, **217**, 452–453.

31. PERRY, S. V. (1970), 'Biochemical adaptations during development and growth in skeletal muscle', *Physiology and Biochemistry of Muscle as a Food*. Eds. Briskey, E. J., Cassens, R. G. and Marsh, B. B., University of Wisconsin Press.

32. GOLDSPINK, G. (1968), 'Sarcomere length during the post-natal growth of mammalian muscle fibres', *Journal of Cell Science*, **3**, 539–548.

33. ARONSON, J. (1961), 'Sarcomere size in developing muscles of a tarsonemid mite', *Journal of Biophysics and Biochemical Cytology*, **11**, 147–156.

34. SHAFIQ, S. A. (1963), 'Electron microscope studied on the indirect flight muscles of Drosophila melanogaster. II. Differentiation of the myofibrils', *Journal of Cell Biology*, **17**, 363–373.

35. AUBER, J. (1965), 'L'accroissement en longueur des myofibrilles et la formation de nouveaux sarcomeres on cours du developpement des muscles chez Calliphora erythrocephala', *Comptes Rendus des Seances de l'Academie des Sciences*, **261**, 4845.

36. RUSKA, R. H. and EDWARDS, G. A. (1957), 'A new cytoplasmic pattern in striated muscle fibre and its possible relation to growth', *Growth*, **21**, 73–88.

37. SCHMALBRUCH, H. Z. (1968), 'Noniusperioden und langen wachstum der quergestreiften Muskelfaser', *Zeitschrift für mikrobiologische-anatomishe Forschung*, **79**, 493–507.

38. HEIDENHAIN, M. (1913), 'Uber die Enstehung der quergestreiften Muskelsubstanz bei der Forelle', *Archiv für mikroskopishe Anatomie und Entwicklungsmechanik*, **83**, 427–522.

39. HOLTZER, H., MARSHALL, J. and FINCK, H. (1957), 'An analysis of myogenesis by the use of fluorescent antimyosin', *Journal of Biophysical and Biochemical Cytology*, **3**, 705–723.

40. ISHIKAWA, H. (1965), 'The fine structure of the myotendon junction in some mammalian skeletal muscles', *Archives of Histology of Japan*, **25**, 275.

41. MACKAY, B., HARROP, I. J. and MUIR, A. R. (1969), 'The fine structure of the muscle tendon junction in the rat', *Acta anatomica*, **73**, 588–604.

42. GRIFFIN, G. E., WILLIAMS, P. E. and GOLDSPINK, G. (1971), 'Region of longitudinal growth in striated muscle fibres', *Nature New Biology*, **232**, 28–29.

43. MOSS, F. P. (1968), 'The relationship between the dimensions of the fibres and the number of nuclei during normal growth of skeletal muscle in the domestic fowl', *Journal of Anatomy*, **122**, 555–564.

44. MAURO, A. (1961), 'Satellite cells of skeletal muscle fibres', *Journal of Biophysical and Biochemical Cytology*, **9**, 493–494.

45. MAURO, A., SHAFIQ, S. A. and MILHORAT, A. T. (1970), 'Regneration of striated muscle and myogenesis', *Excerpta Medica Internation Congress Series No. 218*.

46. ALLBROOK, D. B., HAN, M. F. and HELLMUTH, A. E. (1971), 'Populations of muscle satellite cells in relation to age and mitotic activity', *Pathology*, **3**, 233–243.

47. CRAWFORD, G. N. C. (1954), 'An experimental study of tendon growth on the rabbit', *Journal of Bone and Joint Surgery*, **36B**, 294–303.

48. CRAWFORD, G. N. C. (1961), 'Experimentally induced hypertrophy of growing voluntary muscle', *Proceedings of the Royal Society B*, **154**, 130–138.

49. ALDER, A. B., CRAWFORD, G. N. C. and EDWARDS, R. G. (1959), 'The effect of denervation on the longitudinal growth of voluntary muscle', *Proceedings of the Royal Society B*, **150**, 554–562.

50. TABARY, J. C., GOLDSPINK, G., TARDIEU, C., LOMBARD, M., TARDIEU, G. and CHIGOT, P. (1971), 'Nature de la retraction musculaire des I.M.C.', *Revue de Chirurgie orthopédique et réparatrice de l'Appareil Moteur Paris*, **51**, 463–470.

51. MORPURGO, B. (1897), 'Uber Activitats-hypertrophie der willkurtichen Muskeln', *Virchows Archiv für pathologische Anatomie und Physiologie und für klinische Medizin*, **150**, 522–554.

52. MacCALLUM (1898), 'Histogenesis of the striated muscle fibre and the growth of the human sartorius muscle', *Bulletin of Johns Hopkins Hospital*, **9**, 208–212.

53. GOLDSPINK, G. (1962), 'Studies on postembryonic growth and development of skeletal muscles', *Proceedings of the Royal Irish Academy B*, **62**, 135–150.

54. CHIAKULAS, J. J. and PAULY, J. E. (1965), 'A study of postnatal growth of skeletal muscle in the rat', *Anatomical Record*, **152**, 55.

55. ROWE, R. W. D. and GOLDSPINK, G. (1969), 'The growth of five different muscles in both sexes of mice. I. Normal mice', *Journal of Anatomy*, **104**, 519–530.

56. BRIDGE, D. T. and ALLBROOK, D. (1970), 'Growth of striated muscle in an Australian marsupial', *Journal of Anatomy*, **106**, 285–295.

57. BLACK-SCHAFFER, B., GRINSTEAD, C. E. and BRAUNSTEIN, J. N. (1965), 'Endocardial fibroelastosis in large mammals', *Circulation research*, **16**, 383–390.

58. STAUN, H. (1963), 'Various factors affecting number and size of muscle fibres in the pig', *Acta Agriculturae Scandinavica*, XIII, 293–322.

59. LUFF, A. R. and GOLDSPINK, G. (1967), 'Large and small muscles', *Life Sciences*, **6**, 1821–1826.

60. GOLDSPINK, G. (1970), 'The proliferation of myofibrils during muscle fibre growth', *Journal of Cell Science*, **6**, 593–603.

61. SHEAR, C. R. and GOLDSPINK, G. (1971), 'Structural and physiologicla changes associated with the growth of avian fast and slow muscle', *Journal of Morphology*, **135**, 351–360.

62. GOLDSPINK, G. (1971), 'Changes in striated muscle fibres during contraction and growth with particular reference to myofibril splitting', *Journal of Cell Science*, **9**, 123–138

63. HESS, A. (1961), 'Structural differences of fast and slow extrafusal muscle

fibres and their nerve endings in chickens', *Journal of Physiology (London)*, **157**, 221–231.

64. MORKIN, E. (1970), 'Postnatal muscle fibre assembly: localization of newly synthesized myofibrillar proteins', *Science*, **167**, 1499–1501.

65. HUXLEY, H. E. (1963), 'Electron microscope studies on the structure of natural and synthetic protein filaments from striated muscle', *Journal of Molecular Biology*, **7**, 281–308.

66. WALKER, S. M., SCHRODT, G. R. and BINGHAM, M. (1968), 'Electron microscope studies of the sarcoplasmic reticulum at the Z line level in skeletal muscle fibres of fetal and newborn rats', *Journal of Cell Biology*, **37**, 564–570.

67. SCHIAFFINO, S. and MARGRETH, A. (1969), 'Coordinate development of the sarcoplasmic reticulum and T system during post natal differentiation of rat skeletal muscle', *Journal of Cell Biology*, **41**, 855–875.

68. EDGE, M. B. (1970), 'Development of apposed sarcoplasmic reticulum at the T system and sarcolemma and the change in the orientation of Triads in rat skeletal muscle', *Developmental Biology*, **23**, 634–650.

69. SIEBERT, W. W. (1929), 'Untersuchungen über Hypertrophie des Skelettmuskels', *Zeitschrift für Klinische Medizin*, **109**, 350–359.

70. THORNER, S. H. (1934), 'Trainingsversuche an Hunden. 3 Histologische Besbachtungen an Herz- und Skelettmuskel', *Arbeitsphysiologie*, **8**, 359–370.

71. HOFFMAN, A. (1947), 'Weitere Untersuchungen über der Einfluss des Trainings auf die Skelettmuskulatur', *Anatomischer Anzeiger*, **96**, 191–203.

72. HELANDER, E. A. (1961), 'Influence of exercise and restricted activity on the protein composition of skeletal muscle', *Biochemical Journal*, **78**, 478–482.

73. GOLDSPINK, G. (1964), 'The combined effects of exercise and reduced food intake on skeletal muscle fibres', *Journal of Cellular and Comparative Physiology*, **63**, 209–216.

74. GOLDSPINK, G. (1965), 'Cytological basis of decrease in muscle strength during starvation', *American Journal of Physiology*, **209**, 100–114.

75. GOLDSPINK, G. (1970), 'The proliferation of myofibrils during muscle fibre growth', *Journal of Cell Science*, **6**, 593–604.

76. RICHTER, G. W. and KELLNER, A. (1963), 'Hypertrophy of the human heart at the level of fine structure', *Journal of Cellular Biology*, **18**, 195–206.

77. GORDON, E. E., KOWALSKI, K. and FRITTS, M. (1967), 'Adaptations of muscle to various exercises', *Journal of American Medical Association*, **199**, 103–104.

78. HOWELLS, K. and GOLDSPINK, G. (1973), 'Effects of exercise on normal and dystrophic hamster muscle'. (In preparation).

79. GOLDBERG, A. L. (1967), 'Work induced growth of skeletal muscle in normal and hypophysectomized rats', *American Journal of Physiology*, **213**, 1193–1198.

80. ROWE, R. W. D. and GOLDSPINK, G. (1968), 'Surgically induced muscle fibre hypertrophy', *Anatomical Record*, **161**, 69–76.

81. LESCH, M., PARMLEY, W. W., HAMMOSH, M., KAUFMAN, S. and SONNENBLICKE, E. H. (1968), 'The effects of acute hypertrophy on the contractile properties of skeletal muscle', *American Journal of Physiology*, **214**, 685–690.

82. VAN LINGE, B. (1962), 'The response of muscle to strenuous exercise', *Journal of Bone and Joint Surgery*, **44**, 711–721.

83. EDGERTON, V. R. (1970), 'Morphology and histochemistry of soleus muscle from normal and exercised rats', *American Journal of Anatomy*, **127**, 81–88.

84. HALL CRAGGS, E. C. B. (1970), 'The longitudinal division of fibres in overloaded rat skeletal muscle', *Journal of Anatomy*, **107**, 459–470.

85. GOLDBERG, A. L. (1968), 'Protein synthesis during work induced growth of skeletal muscle', *Journal of Cell Biology*, **36**, 653–657.

86. GOLDBERG, A. L. (1969), 'Protein turnover in skeletal muscle', *Journal of Biological Chemistry*, **244**, 3217–3229.

87. HAMOSH, M., LESCH, M., BARON, J. and KAUFMAN, S. (1967), 'Enhanced protein synthesis in a cell free system from hypertrophied skeletal muscle', *Science*, **157**, 935–937.

88. ROBERTSON, J. D. and BAKER, D. D. (1933), 'Histological differences in the muscles of full and rough fed steers', *University of Michigan Research Bulletin*, **200**.

89. JOUBERT, D. M. (1956), 'An analysis of factors influencing post natal growth and development of the muscle fibre', *Journal of Agricultural Science*, **47**, 59–102.

90. MCMEEKAN, C. P. (1940), 'Growth and development of the pig with special reference to carcass quality characteristics', *Journal of Agricultural Science*, **30**, 276–510.

91. ROWE, R. W. D. (1968), 'Effects of low nutrition on size of striated muscle fibres in the mouse', *Journal of Experimental Zoology*, **167**, 353–358.

92. MONTGOMERY, R. D. (1962), 'Muscle morphology in infantile protein metabolism', *Journal of Clinical Pathology*, **15**, 511–521.

93. GOLDSPINK, G. and WATERSON, S. E. (1971), 'The effect of growth and inanition on the total amount of nitroblue tetrazolium stain deposited in individual muscle fibres of fast and slow rat skeletal muscle', *Acta Histochemica*, **40**, p. 16.

94. BIRD, J. W. C., BERG, T. and LEATHERN, J. H. (1968), 'Activity of Liver and Muscle fractions of Adrenalectomized rats', *Proceedings of the Society of Experimental Biology and medicine*, **127**, 182–194.

95. YOUNG, V. R. (1970), 'The role of skeletal and cardiac muscle in the regulation of protein metabolism'. In *Mammalian protein metabolism*. Ed. Munro, H. N., Academic Press: New York and London.

96. TOWERS, S. S. (1939), 'The reaction of muscle to denervation', *Physiological Reviews*, **19**, 1–48.

97. GUTMAN, E. (1962), 'Denervated Muscle', *The effect of use and disuse on neuromuscular function*. Ed. Gutmann, E. and Hnik, P., Czech. Academy of Science.

98. PELEGRINO, C. and FRANZINI-ARMSTRONG, C. (1963), 'Recent contri-

butions of electron microscopy to the study of normal and pathological muscle', *Cytological Reviews*, **7**, p. 139.

99. FISHER, E. and RAMSEY, R. W. (1946), 'Changes in protein content and in some physicochemical properties of the protein during muscular atrophies of various types', *American Journal of Physiology*, **145**, 571–586.

100. MARGRETH, A., NOVELLO, F. and ALOISI, M. (1966), 'Unbalanced synthesis of contractile and sarcoplasmic proteins in denervated frog muscle', *Experimental Cell Research*, **41**, 666.

101. MUSCATELLA, V., MARGRETH, A. and ALOISI, M. (1965), 'On the different response of sarcoplasm and myoplasm to denervated in frog muscle', *Journal of Cellular Biology*, **27**, 1–24.

102. KNOWLTON, G. C. and HINES, H. M. (1936), 'Kinetics of muscle atrophy in different species', *Proceedings of the Society of experimental Biology and Medicine*, **35**, 394–420.

103. BULLER, A. J., ECCLES, J. C. and ECCLES, R. M. (1960), 'Differentiation of fast and slow muscles in the cat hind limb', *Journal of Physiology*, **150**, 399–416.

104. MOMMAERTS, W. F. H. M. (1970), 'The role of the innervation of the functional differentiation of muscle'. *Physiology and biochemistry of Muscle as a Food*, Ed. Briskey E. J., Cassens, R. G. and Marsh, B. B., Ch. 4, University of Wisconsin Press.

105. GUTH, L., WATSON, P. K. and BROWN, W. C. (1968), 'Effects of Cross reinnervation on some chemical properties of red and white muscles in the rat and cat', *Experimental Neurology*, **20**, 52–69.

106. McCOMAS, A. J., SICA, R. E. P. and CURRIES, S. (1970), 'Muscular Dystrophy: Evidence for a neural factor', *Nature,* **226**, 1263–1264.

4 Growth and Differentiation of Bone and Connective Tissue

J. J. PRITCHARD

Introduction

In this chapter 'bone growth' will be defined very broadly as the totality of the structural changes observable during the life of bones from the time of their first appearance, and emphasis will be placed on the cellular activities which underlie these changes.

In general, bones change in size, in shape, in internal architecture and in the kinds and proportions of the bony and other tissues of which they are composed. Hard bone matrix cannot stretch and so, unlike fibrous tissue or cartilage, a mass of bony tissue can only change in size or shape by surface addition of new bone, or surface removal of existing bone. Moreover, bone formation is not a primary, independent phenomenon, but only takes place at the expense of a pre-existing fibrous or cartilaginous tissue. Bone expansion is thus a peripheral, invasive process whose continuation depends upon preparatory growth changes in neighbouring fibrous tissue and cartilage. It follows therefore that bone growth, fibrous tissue growth and cartilage growth must, of necessity, be closely related and integrated processes. The growth of bony tissue at the expense of growing fibrous tissue is termed 'intramembranous ossification', and the bone so formed is called 'membrane bone'. The growth of bony tissue at the expense of growing cartilage is termed 'endochondral ossification' and the bone so formed is called 'cartilage bone'. Bones which begin their existence in a fibrous milieu are called 'membrane bones': those which are first modelled in cartilage are called 'cartilage bones'. However, during subsequent development, cartilage bone may be added to a membrane bone (Plate 4.1b), and membrane bone to a cartilage bone[3]; this apparent paradox arises because the word 'bone' refers to a particular kind of tissue, while 'a bone' designates an organ, composed of several kinds of tissues[37].

Bones change their internal architecture as a result of bone removal at certain sites and new bone formation at others. New bone formation within a bony mass is possible because bone is not solid, but is rather a labyrinth of hard material (*bone matrix*) riddled with channels containing blood vessels and soft perivascular connective tissue. The channels may be narrowed by accretion of new bone matrix on their walls at the expense of the space occupied by the perivascular connective tissue – a form of intramembranous ossification – or widened by bone resorption with replacement of bone matrix by loose vascular connective tissue. All bone changes, in fact, are associated with vascular changes. Growth in size of a mass of bone is essentially an extension of the hard labyrinth and the engulfment of surrounding blood vessels into its interstices; while surface resorption of bone involves the freeing of enclosed vessels.

All growth changes in bones are brought about by cells, and therefore it is of cardinal importance to try to understand the behaviour of the cells which live in and around a bone. Our understanding is still far from complete, but in recent years much insight has been gained by piecing together information of different kinds – morphological, histochemical and metabolic – from a variety of sources. It is now clear that we must envisage the cells in and around a bone as members of an interconvertible, modulating population. At a given moment some cells are relatively at rest, some are actively dividing, some are 'tooling-up' for a particular secretory task, some are actively secreting, while others are reverting to the resting state. Nowadays, also, we have a clearer picture as to how the cells of this population multiply, modulate and secrete materials from which intercellular matrices are constructed, or by means of which matrices are resorbed. We know much less, however, about the way these cellular activities are *controlled* quantitatively and qualitatively, temporally and spatially, in subservience to the overall programmes of development of individual bones and their integration into a harmoniously organized, functionally efficient skeleton.

It is difficult to discuss cellular behaviour without a reasonable classification of cell types, but unfortunately the dynamic nature of the bone cell population makes it virtually impossible to formulate a precise, unambiguous terminology. One would like to employ a consistent classification based on either morphology or function, but the inherited names in common use are derived in part from morphological appearances, in part from topographical relationships, and in part from supposed functional specializations.

Cells actively engaged in the manufacture of fibrous, cartilaginous and bony matrices are commonly called *fibroblasts, chondroblasts* and *osteoblasts* respectively; while the cells embedded in mature matrices are called *fibrocytes*, *chondrocytes* and *osteocytes*. But while the dividing line between osteoblasts and osteocytes is reasonably sharp because of their respective locations outside and inside hard bone matrix, the boundaries between chondroblasts and chondrocytes, and between fibroblasts and fibrocytes are not well defined. Further difficulties arise when the precursors of fibroblasts, osteoblasts and chondroblasts are under consideration. The embryonic ancestral cell of all three is reasonably called a *mesenchymal cell* but there is no consensus about either the status or the nomenclature of the cells from which the three cell types are recruited in later life. Some hold that mesenchymal cells with full embryonic potentiality persist into adult life: others believe that stem cells of more limited potentiality pervade the connective tissues throughout life: while others again postulate that the various connective tissue cells of adult life are inter-convertible, and reserve cells remaining over from embryonic life are quite unnecessary to account for cell recruitment during growth changes. The problem is that connective tissue cells other than chondroblasts and osteoblasts are all rather similar in appearance, and so it is probably simpler to regard them all as 'fibroblasts', using the term in a generic rather than a specific sense. In the field of cell classification more confusion is caused by over-splitting than by over-lumping where the criteria are imprecise. An exception should perhaps be made in the case of the immediate precursors of osteoblasts, for although they are of fibroblast type, they possess a very distinctive histochemistry which makes it possible to differentiate them from other fibroblasts[25]. Such cells were formerly called *osteogenic cells*[12], but are now more usually termed (*osteo-*) *progenitor cells*[39].

The multinucleated cells which remove bone matrix are *osteoclasts*: the very similar cells engaged in calcified cartilage matrix removal are generally called *chondroclasts*: but the cellular basis of fibrous tissue resorption is obscure. The origin of 'clast' cells is debated[13]: some undoubtedly arise through the fusion of progenitor cells, while others appear to come from monocytes and macrophages, and so are closely related to foreign-body giant cells.

Discussion of the broader aspects of cell behaviour during bone growth is conveniently introduced by describing the histological appearances, firstly during the early stages in the life history of a typical membrane bone, and secondly during the corresponding stages in the life of a cartilage bone. More detailed analysis can then be undertaken at certain sites where conditions for studying cellular behaviour are particularly favourable.

Early Development and Growth of Bones of the Face and Cranial Vault

At a site of impending bone development in the embryonic facial region, localized multiplication of mesenchymal cells gives rise to a compact 'blastema' of closely-packed progenitor cells supported within a meshwork of fine collagen fibres. The blastema is quickly broken up into a network of anastomosing cellular strands by invading wide, thin-walled, blood vessels. Coarse collagen fibres develop between the cells of the strands, augmenting the sparse fibre population carried over from the mesenchyme. Then beginning centrally and spreading centrifugally through the blastema, many of the progenitor cells in the strands transform into osteoblasts, in close association with which further collagen fibres are deposited. The *collagenous matrix* next to the osteoblasts becomes so dense that individual fibres are no longer recognizable in ordinary histological preparations, and it is now termed *osteoid*. Mineralization of the osteoid soon follows, and the resulting material is definitive, calcified *bone matrix*. Some osteoblasts are trapped in this matrix, and become osteocytes. Not all the progenitor cells of the blastema are converted into osteoblasts, however. Even in the centre, where osteoblasts first appear, the cells closest to the blood vessels remain in the progenitor state; while peripherally the blastema consists solely of progenitor cells.

The bone matrix framework develops as an expanding three-dimensional network, using the fibrous strands of the blastema as a scaffolding. The elements of the network are called bone *trabeculae*. In the most recently formed trabeculae the fibres vary in thickness and are irregularly arranged, and the osteocytes are large and closely but randomly associated. This early embryonic hard tissue is called *woven, non-lamellar* or *coarse-fibred bone*, in contradistinction to a more uniformly constructed *fine-fibred* type of bone, called *lamellar bone*, which appears later, and is laid down in successive sheets, or lamellae, rather than as a network.

Before long a layer of fibroblasts appears external to the blastema; and between these cells coarse collagen fibres are deposited which encapsulate the growing bone and delineate it from the surrounding general mesenchyme. It is convenient to call the whole fibroblast-wrapped region a *bone-territory*. At the centre of the bone territory there is a network of woven bone trabeculae interspersed with spaces containing blood vessels, progenitor cells and osteoblasts, the progenitor cells surrounding the vessels, the osteoblasts lying as a one-cell-thick, epithelium-like layer against the bone matrix. Peripherally, radially-projecting bone trabeculae are continuous with strands of osteoid, which in their turn continue into bundles of collagen fibres passing through the outer zone of progenitor cells into the fibroblastic layer. The part of the bone terri-tory external to the bony trabeculae is now called the *periosteum*, a structure which exhibits an outer stratum of densely packed, mostly parallel fibres and fibroblasts, and an inner looser stratum of fine fibres, progenitor cells, blood vessels and osteoblasts. These two strata are generally termed the *fibrous* and the *cambial* (or *osteogenic*) layers of the periosteum respectively. The fibrous layer is normally well defined in growing bone territories, but the cambial layer continues without any sharp boundary into the soft vascular connective tissue (*primary marrow*) lying between the bony trabeculae.

The bone territory as a whole, and the bony network within it, con-tinue to expand through the multiplication of progenitor cells in the cambial layer of the periosteum, and the conversion of some of them into osteoblasts. The vascular network in the deeper part of the cambial layer also expands as the parts of it closest to the growing bony tissue keep on being trapped in the bony network. The fibrous periosteum must also expand, but as it grows more in surface area than in thickness, the inten-sity of cell division and the rate of fibrous matrix production appear to be much less than in the cambial layer, and the histological appearances are correspondingly less dramatic.

Meanwhile, as the bony network expands peripherally, the interior of the network undergoes profound re-modelling. Osteoblasts are replaced by osteoclasts in many of the inter-trabecular spaces, and bone matrix is resorbed, giving rise to much larger bony cavities containing wide blood vessels and loose perivascular connective tissue. In time the connective tissue in the largest cavities will be colonized by haemopoietic cells, and become red bone marrow. In other situations osteoblasts continue to produce bone matrix, and the trabeculae become thicker, while the spaces between them become narrower. Continuation of these re-modelling activities eventually produces a bony structure with the com-pact cortex and the hollow, marrow-filled, medullary cavity (or cavities) typical of a mature bone. It should be noted that the compacting bone which thickens the original woven-bone trabeculae, is of the fine-fibred lamellar variety. The wider, more mature trabeculae therefore tend to retain a central core of woven bone.

The bone territories of the individual facial bones are at first isolated from one another by tracts of general mesenchyme, but as the territories expand they come into close relationship with one another, and their fibrous periostea partially fuse. An area of contact between two territories is called a *suture*, a type of fibrous joint. The cambial layers of the apposed periostea at a suture remain separate, however, and continue to support bone growth on each bone face for some time. Indeed the layers of progenitor cells and osteoblasts in the sutural cambial layers are usually thicker and more pronounced than in the cambium of non-sutural periosteum, indicating that the bones are growing faster at their sutural surfaces than elsewhere.

The early stages in the development of the bones of the cranial vault show some differences from those of the facial bones just described. In the cranial vault, centres of ossification appear within a fairly substantial pre-existing fibro-cellular membrane surrounding the brain. At such centres, accumulations of multiplying progenitor cells split the membrane into three layers, namely denser outer and inner fibrous layers and a looser middle zone, so that the fibrous periosteum of cranial bones, unlike facial bones, may be said to be present from the beginning. Because the approaching bones are separated by fibrous tissue rather than mesenchyme, the sutures in the cranial vault are established in a somewhat different way from those of the face. Apart from this however, facial and cranial bones are entirely comparable in their growth behaviour[29].

There is a matter of considerable theoretical importance concerning the behaviour of the periosteal cambium of facial and cranial bones in certain situations and circumstances, for the cambial layer locally, either wholly or in part, may be converted into *cartilage*. The progenitor cells become chondroblasts rather than osteoblasts and surround themselves with cartilage rather than osteoid matrix The cartilage then grows partly by internal (interstitial) expansion and partly by surface deposition. This type of cartilage is called *secondary*, or adventitious. It may have only a transitory existence[8], or it may continue to expand and contribute greatly to bone growth through endochondral ossification[33]. Progenitor cells are evidently bipotential at least. Two factors have been put forward to explain the switching from bone production to cartilage production, namely, (1) undue compression of the cambial cells[30], and (2) relative ischaemia from poor blood supply, growth of blood vessels having failed to keep up with cambial cell multiplication[12].

On the whole, however, expansile bone growth in the face and skull vault results from intramembranous ossification in the periosteal (including the sutural) cambium, the osteoprogenitor cells there first multiplying and producing an expanding pre-osseous fibrous scaffolding, and then some of them transform into osteoblasts and progressively ossify the scaffolding into a network of bone trabeculae. Deeper, older parts of the network meanwhile are being re-modelled. This method of growth is by no means peculiar to the face and skull vault, however. Except near the ends of cartilage bones, and where tendons and ligaments are attached, periosteal ossification everywhere exhibits similar phenomena. Bone growth in the exceptional cases mentioned presents some special features which will be referred to later.

Early Development and Growth of Cartilage Bones

A long cartilage bone like the humerus, or femur, makes its debut as a localized cellular blastema in the mesenchyme, but instead of being invaded at once by blood vessels, and starting bone formation, the greater part of the blastema is converted into cartilage – *primary* cartilage, to distinguish it from the secondary cartilage which forms at a relatively late stage at some surfaces of some membrane bones. The unchondrified blastema at the periphery of the cartilage is termed the *perichondrium*. The cartilage as a whole is termed the 'cartilage model' because, though minute in size, its shape is recognizably similar to that of the adult bone. Where cartilage models meet, an interzone of non-chondrified blastema indicates the site of a future joint. The perichondrium differentiates into an outer more fibrous layer, and an inner more cellular layer, rather like the periosteum.

The cartilage model grows in size partly by interstitial expansion, that is to say, by cartilage cell multiplication and production of more cartilage matrix internally, and partly by multiplication of cells in the deeper layer of the perichondrium and the conversion of some of them into cartilage cells.

Then, after a time, while the extremities of the model continue to grow in these ways, cartilage cell multiplication and matrix production cease in the middle segment of the model and the cells there first become greatly enlarged, or hypertrophic, the matrix between the cells undergoing compression and ultimately calcification, and then the cells die and disappear, leaving a honeycomb of calcified cartilage matrix. Simultaneously, the perichondrium around this belt of hypertrophying calcifying cartilage changes its character and becomes a periosteum; blood vessels invade it and the deeper cells transform into osteoprogenitor cells and osteoblasts. The perichondrial cells are evidently multipotential, and can switch readily from cartilage to bone production. The 'perichondrial' (now periosteal) osteoblasts deposit a thin collar of bone matrix around the zone of calcified cartilage, except at one site where blood vessels and progenitor cells erode their way from the periosteum into the honeycomb of calcified cartilage. Some of the invading progenitor cells fuse together and become chondroclasts which resorb many of the calcified partitions of the honeycomb, while others come to rest on surviving partitions, transform into osteoblasts, and bury the calcified cartilage beneath layers of bone matrix. Others again retain their progenitor status and accompany the blood vessels as they invade further and further into the cartilage model. Ossification thus begins perichondrally (producing a sheath of membrane bone) and continues endochondrally (producing a network of cartilage bone).

Cartilage cell hypertrophy, death, and matrix calcification, followed by vascular invasion, matrix destruction and new bone formation, progress steadily from the middle of the cartilage model towards the two ends; but the model is not used up, because there are bands of intense cartilage cell proliferation and new cartilage matrix production just beyond each of the hypertrophic 'fronts'. Before long a region of equilibrium between cartilage production and cartilage destruction is establishing a little way from each end of the model: these regions of the model constitute the *growth cartilages* or epiphyseal plates. Distal to them the ends of the model are typically expanded, forming the *cartilaginous epiphyses*. Their role is essentially an articular one, and they grow by radial expansion. Proximally the growth cartilages merge with zones of new bone formation, termed *metaphyses*. The growth cartilages are responsible for shaft lengthening, new cartilage being added distally and older cartilage being removed proximally and replaced by new bone, through the process of endochondral ossification.

Histologically a growth cartilage and its adjoining metaphysis show well defined bands, or zones, in each of which the cells are arranged in very orderly fashion and are engaged in a particular growth activity. Next to the cartilaginous epiphysis lies the *proliferative zone*, characterized by columns of flattened cells stacked like piles of coins. Mitotic figures are numerous. Next to this is a narrow *pre-hypertrophic zone* in which the cells are larger and plumper. Next to this again there is a wider *hypertrophic zone* in which the cells are very large indeed, and then there is a narrow *calcified zone* in which the cells are apparently degenerating and the matrix between them is mineralized. Mitotic figures are not seen in these last three zones. The succeeding zone belongs to the metaphysis, and it shows vascular loops and progenitor cells advancing into a honeycomb of residual calcified cartilage matrix, and occupying the cavities vacated by the hypertrophic cartilage cells. Next comes a zone in which osteoblasts have appeared and have deposited new bone matrix on the remains of the calcified cartilage matrix. Finally, in the oldest part of the metaphysis, remodelling is taking place, some of the calcified cartilage-cored bony trabeculae being resorbed by osteoclasts, while others are being thickened through continuing osteoblastic activity and new bone deposition.

It is clear that the appearances in a histological section of a growth cartilage are but a snapshot of the conveyor belt of events which constitute endochondral ossification. In orderly succession cartilage cells are multiplying, enlarging, dying and being replaced by advancing blood vessels and osteoprogenitor cells; while paralleling these cellular events the cartilage matrix is first increasing in volume, then being compressed, then undergoing calcification, and lastly being used as a scaffolding for bone matrix deposition.

While these events are taking place *inside* the cartilage model, and endochondral ossification is producing an ever-lengthening bony structure at its expense, bone formation is continuing on the *outside* of the model by intramembranous ossification under the fibrous periosteum. Bone formation here extends both radially and longitudinally, so that the jacket of periosteal membrane bone not only increases in width and adds to the girth of the bone as a whole, but also keeps pace lengthwise with endochondral ossification, matching the ever-increasing distance between the growth cartilages at the two ends of the bone. The advancing edges of periosteal new bone formation in lengthwise direction, in fact, always coincide with the pre-hypertrophic zones of the two growth cartilages[20].

The shaft of a long bone is evidently the product of both endochondral and intramembranous ossification. Much of the cartilage bone and some of the membrane bone, however, are subsequently resorbed in the process of formation of the marrow cavity: while at the same time the more superficial bone, initially cancellous, is compacted to form the bone cortex. Much later, independent centres of ossification appear within the cartilaginous epiphyses. These secondary centres expand and convert the cartilaginous epiphyses into bony epiphyses, which intervene between articular cartilage distally and growth cartilage proximally.

During adolescence the growth cartilages gradually cease their activities and are replaced by bone. The articular cartilages are then all that remain of the cartilage model.

The histology of epiphyseal bone formation is basically similar to that of metaphyseal bone formation, except that in the former the proliferative, hypertrophic and calcified zones of the cartilage are arranged concentrically rather than in linear sequence.

Cytology of Intramembranous Ossification

The morphology and behaviour of the cells engaged in membrane bone formation have been most often studied in fetal periosteum, but equally useful models are provided by the activated periosteum near a fracture, and the periosteum remaining after sub-periosteal bone resection.

In situations where new bone is being laid down beneath the periosteum, fibrous and cambial layers are clearly defined[25]. The fibrous periosteum with its densely packed collagen fibres is followed by a compact layer of progenitor cells embedded in a network of fine collagen fibres and permeated by a plexus of thin capillaries. The osteoblast layer, usually one cell thick, is closely applied to the underlying bone surface. Where the bone surface is scalloped the concavities are occupied by an oedematous-looking loose connective tissue containing wide sinusoidal blood vessels. When periosteum is stripped deliberately from a young bone, or when it is elevated by blood clot after an injury, the fibrous and progenitor layers tend to come away together, leaving the osteoblasts behind on the bone. The *progenitor cells* are by no means all alike, nor as a group are their morphological characteristics sharply delineated from the fibroblasts on the one hand and the osteoblasts on the other. In fact they form a graded series of cells which, under the electron microscope, as well as in ordinary histological preparations, appear more and more like fibroblasts as one passes outwards, and more and more like osteoblasts as one passes inwards. This is not to say that progenitor cells are simply modulating fibroblasts, but rather that their recognition as a distinctive group of cells is based on behavioural, genetic and chemical characteristics rather than on morphological ones.

As a group, progenitor cells are spindle-shaped but shorter and plumper than the fibroblasts of the fibrous layer of the periosteum, the inner progenitor cells being plumper than the outer ones. Progenitor cell cytoplasm is more basophilic than fibroblast cytoplasm, that of the inner cells being markedly so. The nucleus in the outer cells is centrally placed, but tends to lie eccentrically in the inner cells. The Golgi complex is larger, and the mitochondria more numerous, than in fibroblasts, and once again these features are accentuated as one passes inwards. These gradations in progenitor cytology are most clearly seen in oblique-tangential sections which exaggerate the thickness of the cell layers[25].

Autoradiography following flash labelling with tritiated thymidine demonstrates that many of the progenitor cells in active periosteum are in the S phase preceding mitotic division[24, 34], and indeed numerous mitotic figures can be seen in these cells in ordinary histological preparations, while after colchicine administration a very high proportion of the cells become arrested in metaphase. In tangential sections mitotic activity is seen to be particularly intense in the outer cells adjacent to the fibrous periosteum.

Histochemically, outer progenitor cells in fetal periosteum show a heavy accumulation of glycogen, unlike the inner cells in which glycogen is hardly demonstrable. The inner cells, on the other hand, show intense alkaline phosphatase activity, which reaches a peak as the osteoblast condition is approached, when the activity of this enzyme is probably greater than in any other cell in the body. Electron microscopy confirms this picture of increasing overall size, hypertrophy of the Golgi complex, multiplication of mitochondria, and displacement of the nucleus to one end as one passes inwards through the layers of progenitor cells[1, 31, 32]. It also accounts for the increasing cytoplasmic basophilia seen with the light microscope, for there is considerable hypertrophy of the ribosome-studded endoplasmic reticulum.

Definitive *osteoblasts*[26] characteristically form a single layer on the surface of the bone. With the light microscope they appear to be in lateral contact, and in EM preparations also, osteoblasts are often found with apposed plasma membranes, although rather more generally, tracts of collagen fibres intervene between the cells. Osteoblasts are big (15–30 μm) ovoid, pyriform or pyramidal cells, with a large, spherical, hypochromatic nucleus at one end of the cell containing 1–3 conspicuous nucleoli. In ordinary histological preparations the centre of the cell shows a round empty space called the juxta-nuclear vacuole, but with special silver staining it can be shown that the region is in reality occupied by an enormous Golgi complex as big as the nucleus, and the empty appearance results from the dissolution of Golgi material during histological preparation. The remaining third of the cell consists of intensely basophilic cytoplasm containing numerous filamentous mitochondria. Alkaline phosphatase activity in osteoblasts is intense, but perhaps not quite so marked as in mature progenitor cells. Pyrophosphatase activity is also high, but as alkaline phosphatase is able to dephosphorylate pyrophosphates very readily, this may simply be a reflection of the high content of the latter enzyme. Numerous esterases and oxidases have been reported in osteoblasts, but dehydrogeneses are less evident, and succinic dehydrogenase in particular is hard to demonstrate. Other workers have found phosphorylase, glycogen synthetase and collagenase in osteoblasts. Acid phosphatase activity can be shown, but it is not conspicuous, and certainly not when compared with osteoclasts.

Granules giving a strong PAS reaction, and presumably containing glycoproteins, have been described, and the presence of lipid material seems likely in view of the demonstration of highly osmiophilic droplets in the peripheral cytoplasm in EM preparations. The general appearance of osteoblasts under the electron microscope is, in fact, very much as one might expect from the results of light microscopy, but interesting additional features are brought out because of the higher resolution. The Golgi complex is truly enormous, and shows groups of flattened sacs and large numbers of large and small vesicles, many of the latter being continuous with the sacs. There are also coated vesicles, and some dense bodies which may be secretory granules, lysosomes or lipid drops. In the middle of the Golgi complex a pair of centrioles and associated microtubules can be seen.

In the general cytoplasm the rough endoplasmic reticulum is a most conspicuous feature. It is arranged as a collection of inter-communicating sacs studded with ribsomes. Some of the sacs are flattened, while others are irregularly dilated and contain fibrillary and particulate material in a moderately electron-dense amorphous matrix. The cytoplasm near the cell surface appears to be specialized, for it contains bundles of fine filaments, and coated vesicles. But perhaps the most striking feature of the osteoblast as seen with the EM is the presence of large numbers of long, fine, cytoplasmic processes projecting from the main body of the cell into the surrounding matrix, especially from the side of the cell next to which osteoid is being laid down.

Histochemically, osteoid stains intensely for acid mucopolysaccharides and glycoproteins, alkaline phosphatase and collagen. Phospholipids have also been reported. With the EM, osteoid exhibits densely packed collagen fibrils in an amorphous ground substance permeated with cytoplasmic processes from the osteoblasts. At the calcification front isolated clusters of hydroxyapatite crystals are seen, and then, a little deeper in, the mineralization becomes so dense that the fibrils are almost completely masked.

From these static morphological studies of periosteum overlying an actively growing bone surface, one can deduce a plausible dynamic picture of the events taking place. The progenitor cells next to the fibrous periosteum are evidently multiplying rapidly, while the more deeply-lying progenitor cells are transforming into osteoblasts. This transformation involves hypertrophy of the Golgi complex, multiplication of mitochondria and rough endoplasmic reticulum, and increases in enzyme activities of many kinds, especially alkaline phosphatase activity. It is clear that the osteoblast and its immediate precursors are very active cells and no one can doubt that they are responsible for the extracellular accumulation of collagen, mucopolysaccharides and glycoproteins, which constitutes osteoid formation.

The significance of the high alkaline phosphatase activity of the inner progenitor cells and osteoblasts is still not clear. Some workers believe that since pyrophosphates (present in most tissues) inhibit calcification, and since most alkaline phosphatases can act as powerful pyrophosphatases, the role of alkaline phosphatase in bone-forming tissues is to destroy the pyrophosphate inhibitor, and so allow spontaneous calcification to proceed.

Glycogen is very abundant in the progenitor cells of fetal periosteum, but is not so easily demonstrated in the similar cells of fracture callus. It may well be that glycogen accumulation is an adaptation to the relatively anaerobic conditions of fetal life, and is not an essential feature of the ossification process.

Recently it has been suggested that the long cytoplasmic processes of osteoblasts may be actively engaged in transporting calcium into the osteoid, the evidence for this being that calcification often seems to start near the tips of the processes.

Some of these general inferences about cell behaviour in the cambial layer of the periosteum have been confirmed in experiments using radioactive labels. Tritiated thymidine is at once taken up by progenitor cells about to divide, and the radioactive label persists in the nuclei of daughter cells right through to their transformation into osteocytes. In the active fetal periosteum of the rabbit femur the minimum time for such maturation is three days. Generally, however, the osteoblast stage alone lasts three days, and so the total maturation time is somewhat longer. Of course these figures should not be extrapolated to all bone surfaces in all animals at all ages.[35]

Collagen is very rich in glycine and (hydroxy-) proline, and therefore the rate of new matrix production on a bone surface can be correlated with cell maturation by simultaneous administration of ^{3}H-thymidine and ^{3}H-proline (or ^{14}C-glycine). It has been found by such methods that an osteoblast can produce its own volume of osteoid matrix daily[24].

Newly imprisoned osteocytes are very like osteoblasts, but as they mature and get more deeply buried they gradually lose their distinctive robust osteoblastic morphology and chemistry. They probably continue to produce a little matrix around themselves for a while, but whether or not they can later assume a clastic role and begin widening their lacunae again, is a question which is still undecided.

In adult life the cambial layer of the periosteum is rarely seen on normal bones, and the fibrous periosteum appears to extend right to the bone surface. However, very shortly after injury, infection or other disturbance, the cambial layer reappears in the vicinity of the damaged region and begins to behave like the cambium of a growing bone. The cambium is seemingly regenerated from the flattened 'resting' cells lying next to the bone surface. These cells enlarge to become active progenitor cells, multiply rapidly, differentiate into osteoblasts and begin to deposit bone matrix. The capillary bed grows *pari passu*. However, very close to the damaged region the progenitor cells in many cases do not transform into osteoblasts, but apparently modulate into cartilage cells instead, and a mass of secondary cartilage is formed which is only subsequently replaced by bone through endochondral ossification. In fracture repair it is very common, in fact, for the callus to contain a good deal of cartilage, and for cartilage union to precede bony union[28]. In the author's view the cells from which the cambium is regenerated after injury should be classified as resting progenitor cells. The evidence is consistent with the view that such cells are present throughout life on or near all normal bone surfaces, both external and internal, and that they have a very special potentiality for enlargement, rapid division and then modulation into either osteoblasts, chondroblasts or osteoclasts as and when an appropriate stimulus arises.

This view is strongly supported by the changes which occur in the residual periosteum after sub-periosteal resection of a length of rat rib[22]. Immediately after the operation only an inconspicuous single layer of alkaline phosphatase-positive cells could be seen on the denuded surface of the fibrous periosteum, but within a very few days these cells had multiplied and produced a thick blastema permeated with new blood vessels. A few days later numerous osteoblasts were present and bone formation was taking place rapidly. Some of the blastemal cells became cartilage cells, but the resulting secondary cartilage was short-lived. In about two weeks a fairly normal rib was regenerated, complete with cortex, marrow and periosteum. No such dramatic events took place in periosteum which had been scraped or cauterized, presumably because the resting progenitor cells had been destroyed, and the ordinary fibro-blasts of the repair tissue were unable to (or were not induced to) make a bone-forming blastema.

Nevertheless, it must not be supposed that cells with bone-forming potential are confined to the vicinity of bone surfaces. In *abnormal* circumstances bone formation may begin in almost any connective tissue in the body, and give rise to perfectly normal bones so far as their structural organization is concerned[4, 5]: such bones are simply in the wrong places! Neither the origin, nor the status, of the cells which give rise to such ectopic bones, are known. Nor is much known about the factors which trigger heterotopic ossification: there are certainly bone-inducing agents in transitional epithelium[4, 17], in hypertrophic cartilage[6], in de-calcified bone and dentine[38] and in damaged muscle[6], but their chemical nature has not been elucidated, though there is evidence that a protein is involved[36]. What is clear, however, is that the progenitor cells of bony tissues are committed towards bone formation in a way which other connective tissue cells are not. Thus progenitor cells can easily be persuaded to form bone in a matter of days, and with certainty: while other connective tissue cells, if they are to make bone, must be in a highly abnormal environment for weeks or months, and even then there is no certainty that they will do so.

So far we have been considering intramembranous ossification as it occurs in growing or activated periosteum, but such ossification may also begin in the middle of a tendon, or extend into a tendon or ligament from a bony surface. In such cases the new bone matrix is permeated by stout fibre bundles carried over from the pre-existing fibrous tissue to a much greater extent than in ordinary periosteal ossification, and the be-haviour of the cells is rather different.

At particular sites in turkey leg tendons, and at particular times, the tendon cells multiply rapidly and then transform into osteoblasts which secrete an amorphous material containing collagen precursors, muco-polysaccharides and phospholipids[23]. This secretion invests, and soaks into, the stout collagen fibres of the tendon. New collagen fibres then

Plate 4.1

(a) Developing cartilage bone. 6-month human fetus. Mallory. × 15.

(b) Secondary cartilage in clavicle. 6-month human fetus. H. & E. × 15.

(c) Cartilage and membrane bone in tibia. 6-month human fetus. Mallory. × 15.

(d) Periosteal intramembranous ossification. Chick radius. Oblique section. V.G. × 30.

(e) Intramembranous ossification. Newly-hatched turtle (Ridley) carapace. V.G. × 15.

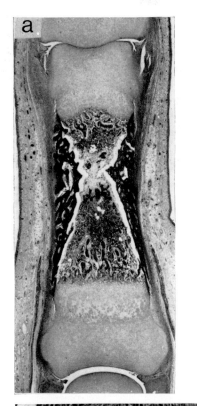
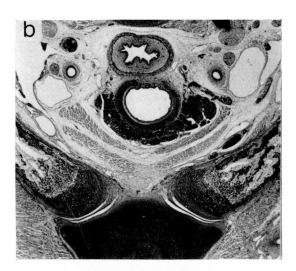
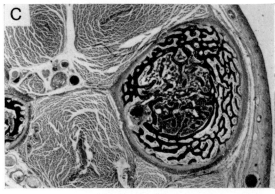
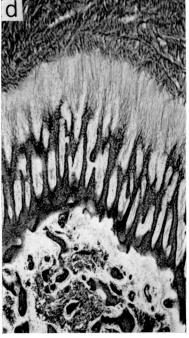
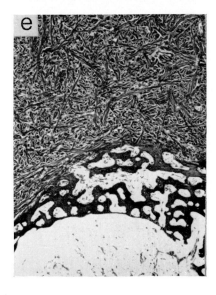

Plate 4.2

(a) Woven membrane bone formation. Periosteum of rat tibia. Oblique section. W. & V.G. × 95.

(b) Woven membrane bone formation. Mandible of fetal sheep. Wilder's Silver method. × 95.

(c) Compact lamellar bone. Rat femur. H. & E. × 95.

(d) Secondary cartilage formation. Periosteum of rat mandible. H. & E. × 585.

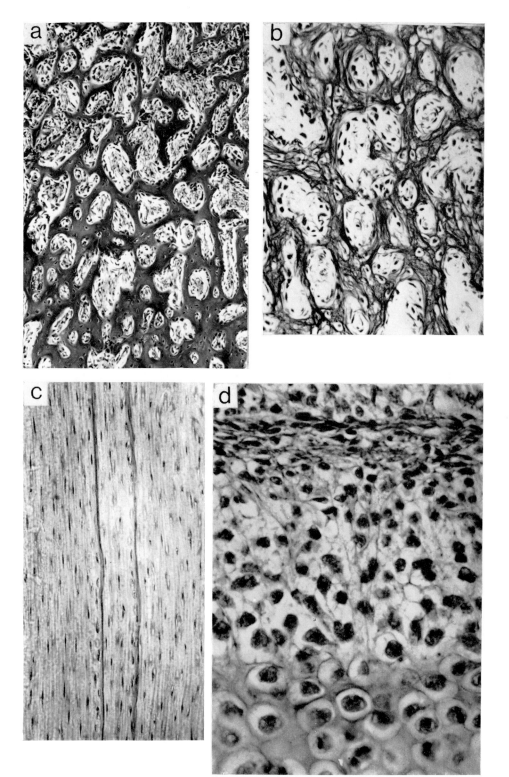

Plate 4.3

(a) Alkaline phosphatase activity. Mandible. 15-day rat embryo. Gomori. × 130.

(b) Centre of ossification. Pre-maxilla. 18-day rat embryo. Wilder. × 234.

(c) Centre of ossification. Mandible. 18-day rat embryo. Wilder. × 234.

(d) Centre of ossification. Pre-maxilla. 18-day rat embryo. Wilder. × 584.

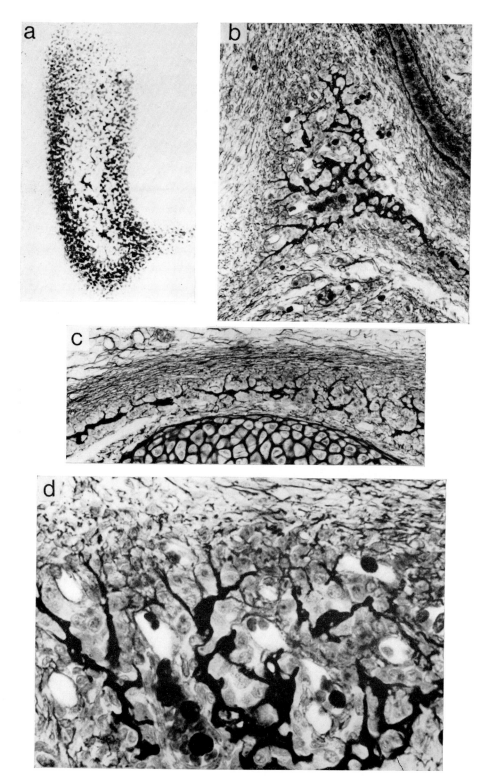

Plate 4.4

(a) Active periosteum. Tail vertebra of 1 week old rat. × 4000.

(b) Later progenitor cell. Active periosteum. Tail vertebra of 1 week old rat. × 12 600.

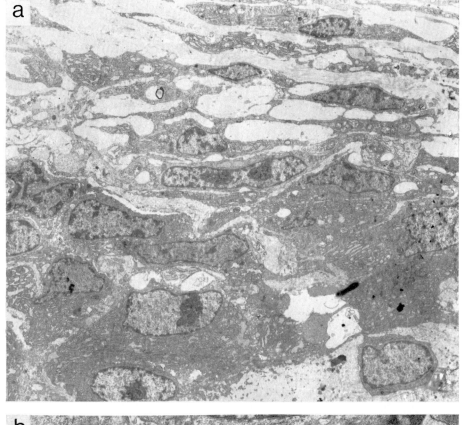

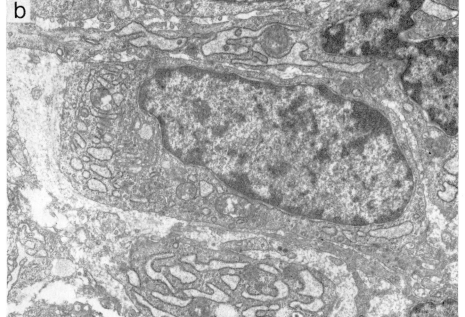

Plate 4.5

(a) Proliferative zone of growth cartilage. Young monkey. Scanning E.M. × 3200. (*Courtesy of* Dr A. Boyde.)

(b) Endochondral and intramembranous ossification. Upper end of tibia. 6-month human fetus. H. & E. × 15.

(c) A secondary centre. 1 week old rat tibia. H. & E. & Alcian Blue. × 30.

(d) An active growth cartilage. 1 week old rat tibia. H. & E. × 95.

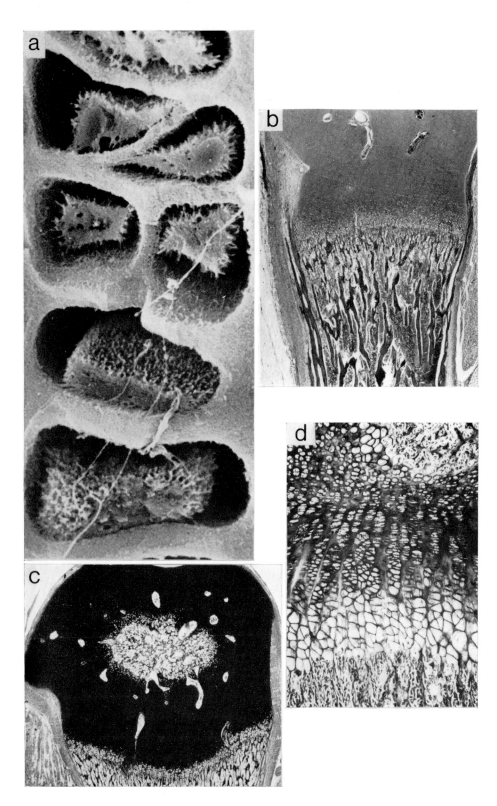

Plate 4.6

(a) Centre of ossification. Tail vertebra. Newborn rat. W. & V.G. × 100.

(b) Endochondral ossification. 2 week old chick radius. W. & V.G. × 15.

(c) Growth cartilage. 2 week old chick radius. W. & V.G. × 95.

(d) Growth cartilage. 2 week old chick radius. W. & V.G. × 585.

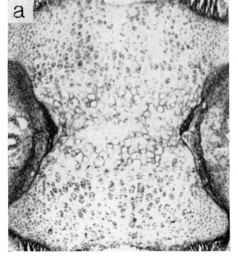

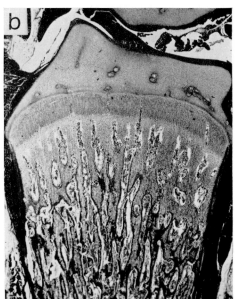

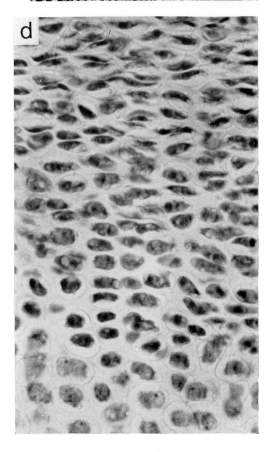

Plate 4.7

(a) Pre-hypertrophic cells. Lower end of radius. Newborn rat. H. & E. × 900.

(b) Early hypertrophic cells. Lower end of radius. Newborn rat. H. & E. × 900.

(c) Late hypertrophic cells. Lower end of radius. Newborn rat. H. & E. × 900.

(d) Metaphysis showing vascular-mesenchymal invasion of empty cartilage lacunae. Lower end of radius. Newborn rat. H. & E. × 900.

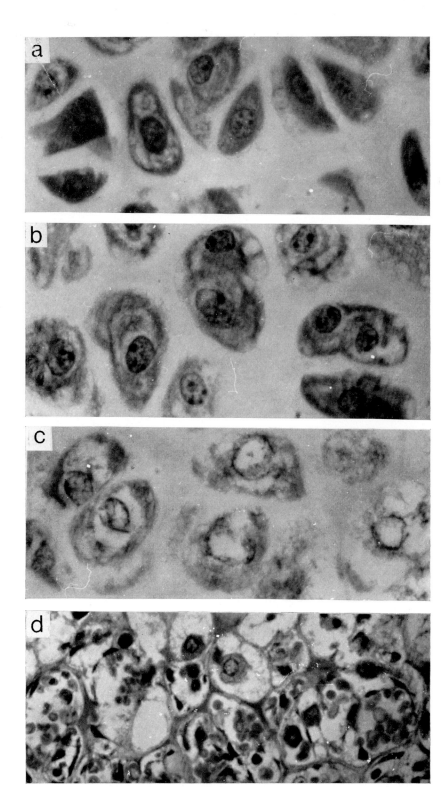

Plate 4.8

(a) Palatal suture. 90 mm C.R. Sheep fetus. W. & V.G. $\times$ 25.

(b) Membrane bone formation. Ramus of mandible. Human 6-month fetus. H. & E. $\times$ 95.

(c) Sagittal suture. 110 mm C.R. Sheep fetus. W. & V.G. $\times$ 25.

(d) Zygomatic suture. 6-month human fetus. H. & E. $\times$ 95.

(e) 'Resting' progenitor cells. Periosteum of adult rat tibia. H. & E. $\times$ 500.

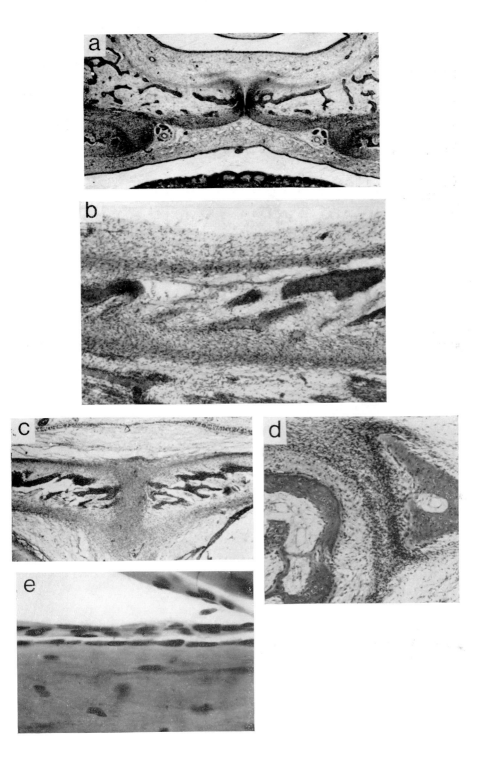

Plate 4.9

(a) Fibrous growth plate. Upper end of 1 week old rat tibia. H. & E. × 95.

(b) Fibrous growth plate. Upper end of 1 week old rat tibia. H. & E. × 585.

(c) Intramembranous ossification at expense of fibrous growth plate at upper end of 1 week old rat tibia. H. & E. × 585.

(d) Coarse woven (bundle) bone of tibial tubercle of 1 week old rat. H. & E. × 585.

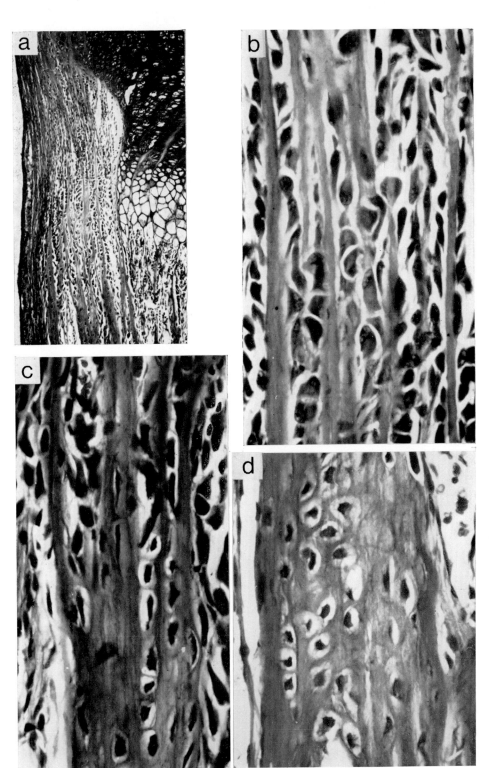

Plate 4.10

(a) Periosteum. Mandible of 18-day rat fetus. H. & E. × 584.

(b) Periosteum. Mandible of 6-month human fetus. H. & E. × 234.

(c) Periosteum. Mandible of 6-month human fetus. H. & E. × 234.

(d) Osteoblasts, osteoid and bone matrix. Young mouse. (*Courtesy of* Dr A. S. Dhawan.) × 10 000.

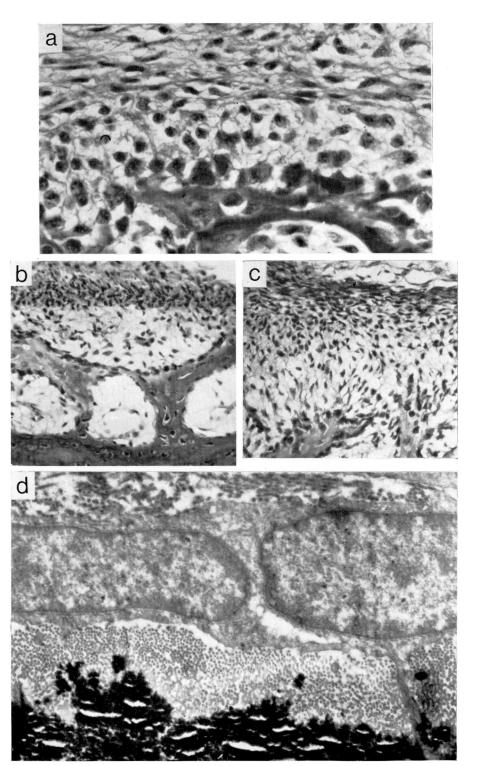

Plate 4.11

(a) Osteoblasts and a trabecula of new osteoid. Turtle embryo dermal bone. H. & E. $\times$ 584.

(b) Osteoblast showing Golgi apparatus. Aoyama. $\times$ 2600.

(c) Osteoblasts showing juxta-nuclear vacuole. Chick radius. $\times$ 730.

(d) Osteoblasts showing mitochondria. Rat embryo. Author's Silver method. $\times$ 850.

(e) Periosteal membrane bone showing osteoblasts and osteocytes. Rat tibia. W. & V.G. $\times$ 584.

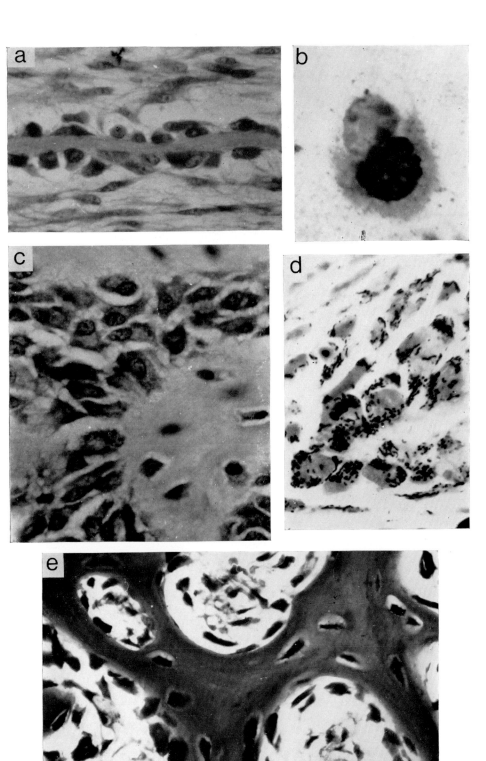

Plate 4.12

(a) Osteoblasts with basophilic cytoplasm. Fetal rat mandible. Pyronin-methyl green. × 584.

(b) Pre-hypertrophic chondrocyte. Chick embryo squamosal. × 9400.

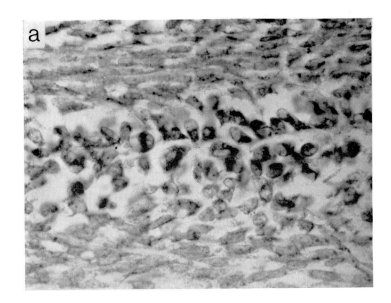

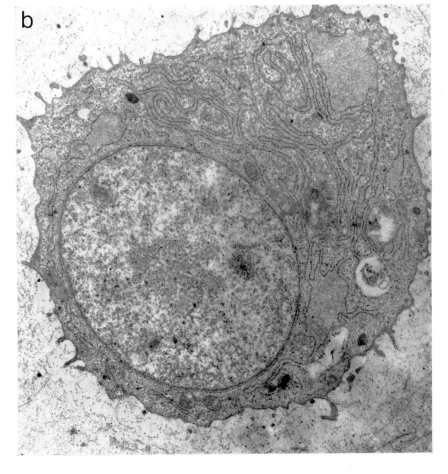

Plate 4.13

(a) Activated periosteum after fracture. Salamander humerus. H. & E. × 584.

(b) Activated periosteum 4 days after fracture. Rat tibia. H. & E. × 10.

(c) Activated periosteum. New bone and cartilage formation. 8-day fracture of very young chick radius. Mallory. × 15.

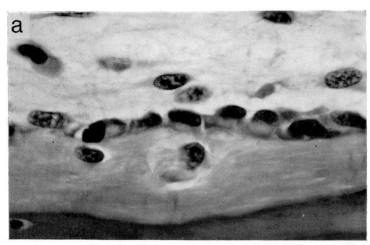

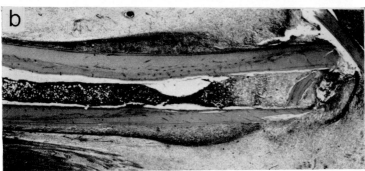

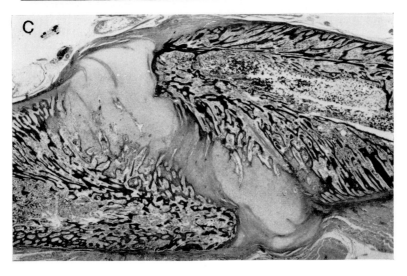

Plate 4.14

(a) Active periosteum after sub-periosteal removal of rat rib. H. & E. × 15.

(b) Regenerated rib from surviving periosteum 12 days after sub-periosteal resection of rat rib. H. & E. × 15.

(c) Bone overgrowth after removal of length of radius in young chick. × 1.

(d) Heterotopic ossification in alcohol-fixed transplant of hypertrophic cartilage. W. & V.G. × 100.

(e) Heterotopic ossification in rabbit muscle after injection of alcohol. Wilder. × 100.

(f) Heterotopic ossification in alcohol-killed callus transplant. W. & V.G. × 15.

(g) Rat skull. 58 days after damage to medial edge of left parietal bone *in utero*. × 2·5.

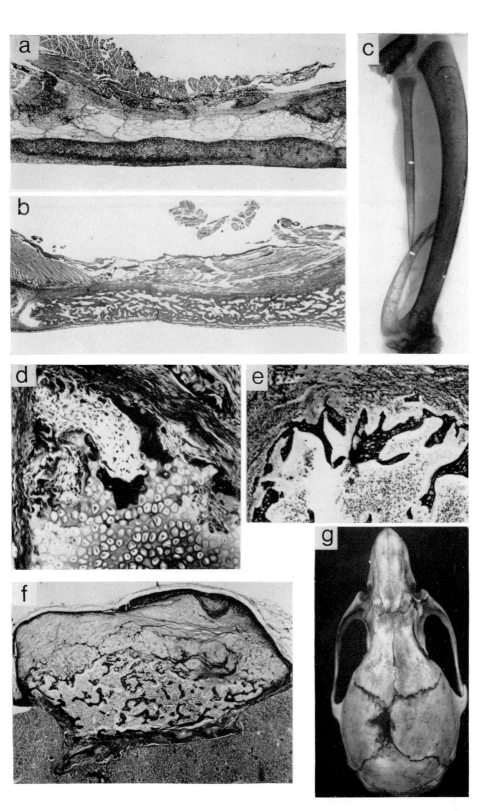

form in the osteoblast secretion, orientated more or less at right angles to the old tendon collagen. Next, hydroxyapatite crystals are deposited in large numbers, first in the new matrix, and then throughout the old tendon collagen as well. Osteoblasts surrounded by the calcified matrix become osteocytes. Blood vessels and immature connective tissue cells now invade the calcified 'tendon-bone' and tunnels are formed on whose walls a more mature type of bone is deposited.

A somewhat similar method of ossification occurs at sites of tendon attachment to bones. As the bone surface is approached the tendon cells in their parallel columns gradually enlarge, eventually assuming the morphology and histochemical characteristics of osteoblasts, adjacent to which the tendon collagen calcifies and becomes a primitive kind of bone matrix. The cell columns continue unbroken into the calcified matrix as lines of osteocytes, until the series is abruptly interrupted by tongues of vascularized connective tissue extending from the interior of the older bone. This connective tissue not only erodes and produces tunnels in the 'tendon bone', but also generates new osteoblasts which deposit new and more mature type of bone on the walls of the tunnels. In some situations, however, the tendon cells enlarge and become cartilage cells instead of bone cells, but the resulting calcified fibro-cartilage is eroded and replaced by new bone just like the tendon-bone.

This kind of bone growth has recently been investigated in some detail at the attachment of the patellar ligament to the tibial tuberosity in rats[2]. Here, about the time of birth, a portion of the ligament appears to become cut off and isolated between the cartilaginous tuberosity of the tibial epiphysis and the bony tuberosity of the tibial shaft, and since it occupies a similar position to the growth cartilage and must perforce grow at the same rate as the growth cartilage, it has been called the 'fibrous growth plate'.

The fibrous growth plate consists of parallel, longitudinal fibre bundles interspersed with columns of cells. The cells in the middle of the columns are fibroblast-like in general appearance; tritiated thymidine shows that they are mitotically very active, while tritiated proline indicates that there is active collagen deposition going on around them. Followed in the direction of the knee joint, the cells in the columns enlarge and become cartilage cells, which add to the bulk of the tibial epiphysis; but in the opposite direction the cells enlarge and become osteoblasts which contribute to the growth of the tibial shaft. The collagen fibres adjacent to the osteoblasts are calcified. Continuing downwards the calcified fibres become thickened by addition of new bone matrix, and at the same time cross partitions are formed, trapping some of the osteoblasts as osteocytes. In this region, collagen and mucopolysaccharide production, as demonstrated by ^{3}H-proline and ^{35}S-sulphate autoradiography, is intense; while the osteoblasts show the usual cytological and histochemical features associated with bone formation.

The new bone formed in this way at the tuberosity of the tibial shaft is not solid, however, nor does it last long, for regularly interspersed with the calcifying fibres and new bone trabeculae there are tongues of invading vascular tissue. Osteoblasts which are not trapped in the matrix, but find themselves on the walls of the vascular tunnels, continue to add new bone matrix until the tunnels are quite narrow. At the distal end of the tuberosity, osteoblasts give way to osteoclasts, and the bone is resorbed. And so the tuberosity as a whole migrates towards the knee, keeping pace with the elongation of the shaft.

In conclusion, it may be said that, despite some obvious histological differences, sites at which intramembranous ossification is taking place have certain basic features in common, namely, the presence of fibroblast-like (progenitor) cells which multiply and produce a fibrous matrix before transforming into osteoblasts and producing more and often a great deal more collagen; and at the same time, the matrix starts to calcify. A primitive kind of bone is thus formed which is normally short-lived, being soon invaded by vascularized connective tissue with bone-forming and bone-destroying potential which re-models it into a more mature type of bony tissue.

The bone formation which takes place within the vascular tunnels of *existing* bone is also classified as a type of intramembranous ossification. Such ossification may be primary or secondary. In the primary type, channels existing from the beginning of ossification are filled in through the continuing action of osteoblasts lining the walls of the channels. Concentric lamellae of bone matrix filling up such a primary vascular channel constitute a *primary osteon*. In the secondary type of ossification, the bone around an existing vascular channel is first eroded by osteoclasts, and then the cavity so produced is filled in again by osteoblasts, which lay down concentric lamellae to form a *secondary osteon*. Indeed the bone may be re-modelled over and over again in the course of a long life, tertiary osteons following secondary osteons and so on. However, such extensive remodelling is not obligatory, and at some sites primary osteones persist throughout life.

This 'osteonal' form of intramembranous ossification differs from that seen in the periosteum. It is characterized by a scarcity of collagen fibres in the pre-osseous tissue, and a paucity of progenitor cells, amongst which mitotic figures are few and far between. It would appear that the vascular channels in mature compact osteonal bone contain a rather scattered population of perivascular cells which can, however, modulate readily between the osteoblastic, progenitor and osteoclastic states, and so new cells do not often have to be recruited by means of progenitor cell multiplication. Because of their situation within rigid-walled tunnels, moreover, the cells of osteonal bone must first transform into osteoclasts and enlarge their tunnels before osteoblastic activity can be resumed. In the nature of things, therefore, osteonal ossification is a much more sedate affair than that which may be seen in active or activated periosteum.

Cytology of Endochondral Ossification

The general features of endochondral ossification are observable in any active growth cartilage, but it is convenient to take as our model the plate of cartilage between epiphysis and diaphysis at the end of a growing long bone.

With the light microscope a longitudinal section of such cartilage shows columns of cells separated by tracts of *inter*-columnar matrix. The *intra*-columnar matrix is in the form of thin partitions between the cell lacunae. In the distal part of a column (i.e. towards the epiphysis) the cells are flat and thin, like coins; but in the proximal part (i.e. towards the metaphysis) the cells are very large and spherical. Between the flat cells and the hypertrophic cells the columns show a narrow band of intermediate 'pre-hypertrophic' cells.

The cell types in adjacent columns tend to be in close register, so that the growth cartilage as a whole presents well defined transverse bands or zones of flat cells, pre-hypertrophic cells and hypertrophic cells. The last two or three hypertrophic cells in a column appear to be disintegrating (Plate 4.7c), though some observers doubt whether they actually break up and disappear[16].

The tracts of inter-columnar matrix are relatively wide in the flat cell zone, but are narrow and compressed-looking in the hypertrophic zone. Between the 'disintegrating' cells the inter-columnar matrix is quite heavily calcified, but the intra-columnar matrix partitions are much less densely mineralized, if at all.

The inter-columnar calcified cartilage matrix continues into the metaphysis, but the cartilage cell columns and intra-columnar matrix partitions cease abruptly at the metaphyseal vascular invasion front. The metaphysis is thus built on a scaffolding of calcified cartilage matrix in the form of a honeycomb whose interstices, vacated by hypertrophic cartilage cells, are colonized by blood vessels and progenitor cells. At first resorption of calcified cartilage is rife and the interstices are considerably widened, but deeper in the metaphysis osteoblasts become numerous, and the remaining calcified cartilage is buried beneath layers of new bone matrix. Deeper still, the new bone is also being re-modelled, and in such a way that the more mature metaphysis bone contains fewer, but thicker, bony trabeculae enclosing relatively wide inter-trabecular marrow spaces. Then, at the end of the metaphysis, as the medullary cavity is approached, large numbers of osteoclasts are seen, and bone resorption is rampant. The metaphysis is thus the site of intense osteoblastic and osteoclastic activity as solid cartilage is transformed into spongy bone and then into a marrow cavity.

In ordinary histological sections the nuclei and cytoplasm of the flat cells stain so heavily that little detail can be made out. The hypertrophic cells have nuclei which are large and spherical, and cytoplasm which is so vacuolated that it appears foamy. The matrix of the growth cartilage as a whole stains intensely and metachromatically with basic dyes, including alcian blue, but only the matrix associated with the hypertrophic cells is PAS positive. Presumably while all the matrix is rich in acid mucopolysaccharides, glycoproteins are chiefly confined to the hypertrophic zone.

As one passes through the flat cells into the zone of hypertrophic cells, glycogen storage increases to a maximum about half-way through the latter zone and then abruptly ceases. Alkaline phosphatase, on the other hand, is not demonstrable in the flat cells, but begins to appear in the hypertrophic cells and becomes very intense indeed as glycogen disappears. Even the matrix around the cells gives an intense reaction for the enzyme just before the zone of calcification is reached. This is very reminiscent of the situation in the progenitor and osteoblastic zones of fetal periosteum, where, just ahead of the calcification front, glycogen also disappears as alkaline phosphatase appears. Enzymatically and metabolically, in fact, hypertrophic cartilage cells seem very similar to the mature progenitor cells and osteoblasts.

Rather surprisingly, while published accounts of the ultrastructure of the bone-forming cells are in good general agreement, those dealing with the cells of the growth cartilages are confused and seemingly contradictory, possibly because of species differences, but more probably because of technical difficulties in preparing well-fixed material for electron microscopic examination. Moreover, electron micrographs show gradations in cytological characteristics, rather than abrupt changes, as one moves across the thickness of the growth cartilage; and, rather confusingly, some cells appear to be degenerating at all levels. The following account is an attempt to reconcile the observations of my colleagues on chick material[18, 21], with published accounts mostly based on small mammals[10, 19], but also on the chick[11].

The flat cells near the epiphyseal ends of the columns are greatly elongated in the transverse direction, and possess a mediocre complement of closely packed organelles in a highly condensed hyaloplasm containing many free ribosomes. The cells have been described as looking 'dehydrated' because of their overall electron density. Many of the flat cells are paired side to side, with their plasma membranes in contact, and rather often one of the partners is degenerating. The matrix nearby looks 'thin', with a loose arrangement of fine, non-banded collagen fibrils in an amorphous ground substance.

Deeper in the flat cell zone, as one approaches the pre-hypertrophic zone, the cells become plumper and the cytoplasm shows an extensive system of granular endoplasmic reticulum tubules, some of which expand into large, irregular cisterns. The reticulum as a whole, including the cisterns, contains granular, fibrillar and amorphous material. The Golgi complex is well developed, but rather dispersed, and it contains both large and small vesicles along with stacks of flattened, nongranular sacs. Moderate numbers of mitochondria are seen, and the cytoplasm also contains a miscellany of dense bodies, multivesicular bodies, granules, filaments and microtubules. The matrix around these cells is more voluminous, and its fibril density is higher, than that associated with the flat cells nearer the epiphysis; and in addition, dense bodies, and what look like sacs of extruded cytoplasmic material, are frequently seen in it.

These cytological features become even more marked in the pre-hypertrophic zone, where enormously dilated endoplasmic reticulum cisterns dominate the cytoplasm outside the voluminous Golgi zone with its high complement of very large vesicles. And then in the hypertrophic zone the appearances change rather abruptly. The cisterns disappear, and the whole endoplasmic reticulum collapses and becomes dispersed in a voluminous water-looking hyaloplasm. The Golgi complex also regresses, but the mitochondria become swollen and look oedematous, and many contain granular material. As the calcified zone is approached the cells look swollen to bursting point. Nuclei either shrink and become pyknotic, or else enlarge and rupture; while the watery cytoplasm is interspersed with fragments of the Golgi complex and endoplasmic reticulum. Rather surprisingly the mitochondria, though swollen, appear surprisingly normal otherwise. In the calcified zone, close to the vascular invasion front, only cell debris is to be seen – the cells themselves having apparently broken up completely.

The metaphysis begins with a profusion of capillary loops, chondroclasts and progenitor cells. The tips of the growing capillaries are sometimes in direct contact with the last intra-columnar partition of cartilage matrix, and their endothelial cells are plump and contain lysosome-like bodies. Chondroclasts are common on the inter-columnar calcified matrix, and are indistinguishable from osteoclasts in their fine structure. In addition, there are progenitor cells with large nuclei and rather scanty cytoplasm containing many free ribosomes. A little further into the metaphysics one finds typical osteoblasts, and associated with them there is a thin layer of bone matrix applied to the surface of the calcified cartilage.

Using tritiated thymidine it can be shown that many of the flat cells in growth cartilage are about to undergo cell division, particularly those nearer the epiphysis. None of the pre-hypertrophic or hypertrophic cells, however, show any signs of mitotic activity. Cell division reappears again in the progenitor cells and endothelial cells of the metaphysis.

With progressively longer intervals between administering tritiated thymidine and killing the animal it can be shown that the labelling apparently moves steadily from flat cells through to hypertrophic cells, though in reality the labelling remains at the same absolute level in the bone, and the apparent movement is due to hypertrophy of the labelled cells *in situ*, and the development on the epiphyseal side of new flat cells which are unlabelled. It is clear that the flat cell zone is a region of active and continuous cell proliferation, which would result in the piling up of such cells into an ever-lengthening column if it were not for the cessation of division in the middle of the column, and the transformation of flat cells into hypertrophic cells which calcify their matrix, die, and are replaced by invading capillaries and progenitor cells. The columns in fact are growing on the epiphyseal side while being destroyed on the metaphyseal side, and the constancy in the appearance of the columns over long periods of time in a series of histological sections is simply a reflection of the equilibrium which exists between column growth at one end and column destruction at the other. The inter-columnar matrix, of course, has to grow in the same way, and a steady supply of new matrix is essential, at least in the flat cell zone. Autoradiography after ^{35}S-sulphate administration bears this out, for the cartilage cells as a whole take up the labelled material and pass it on to the matrix, where presumably it is used in the sulphation of newly made chondroitin. Moreover, the cytology of the cartilage cells in the columns is indicative of intense synthetic activity, particularly in the older flat cells and the pre-hypertrophic cells, and this is borne out by the fact that ^{35}S uptake is highest in these cells.

Many questions remain to be answered regarding the quantitative aspects of growth cartilage dynamics. It is not yet clear, for example, whether the flat cells are self-perpetuating or whether they are progressively recruited from a zone of reserve cells adjoining the epiphysis. If the latter, one may envisage a single reserve cell dividing rapidly many times and producing a columnar 'clone' whose members then all cease to divide and begin their hypertrophic march to destruction; but it is difficult to imagine, if this be so, how the growth cartilage as a whole maintains such good cell registration in transverse direction. If the former is the true explanation, then position in the column must determine whether or not a cell will go on dividing, or cease dividing and begin to hypertrophy – a situation like that in the skin.

In the rabbit, the band of tritiated-thymidine-labelled (initially flat) cells reaches the metaphysis in a minimum of 3 days, which is therefore a measure of the life expectancy of a cell from the time it ceases dividing and begins to hypertrophy, to the time of its death. In man, the corresponding life expectancy is to be measured rather in weeks or months, because of the slowness of skeletal growth in our species.

Then there is the question of where the actual lengthening takes place in a growth cartilage. Cell hypertrophy, on the face of it, ought to contribute considerably to lengthening, but the contribution would be very much less if cell enlargement took place through resorption of the matrix partitions between the cells, as may well be the case. Matrix production, on the other hand, is clearly active in the flat cell zone, and as the zone as a whole does not increase very much in width, the new matrix must pile up in the longitudinal direction. On the whole, it seems likely that the flat cell zone is the principal contributor to lengthening, and this view is supported by the fact that the flat cell zone is never encased in a rigid bony corset like the hypertrophic cell zone invariably is.

Then there is the question of the fate of the hypertrophic cells. Despite the hopelessly degenerate appearance of these cells in the calcified zone, some authors are quite confident that they undergo rejuvenation and reappear in the metaphysis as progenitor cells.

Finally, there is the vexed question of the mode of resorption of the last cartilage matrix partition which separates the last hypertrophic cell from the tip of the advancing metaphyseal capillary. Some hold that resorption is undertaken by chondroclasts, others that capillary endothelium is invasive, while others again believe that the dying cartilage cells are responsible. At present the evidence is not sufficiently clear cut to enable one to choose between these alternatives.

Control of Bone Growth

The way in which genetic and environmental factors control the growth of fibrous tissue, cartilage and bone in the development of individual bones, and in the welding of bones into a harmoniously organized skeleton, is far from being understood. It seems likely, however, that cartilage is the pacemaker tissue in skeletal growth[27], except perhaps in the cranial vault where the expanding brain is a better candidate. Thus embryonic cartilage when transplanted, or cultivated *in vitro*, exhibit remarkable powers of autonomous growth and development – powers not possessed by fibrous and bony tissues, which if they grow at all, are forced to conform to their host tissues in respect of size and shape. Moreover, cartilage has the right physical characteristics for self-expansion and shape maintenance.

Furthermore, the development of bone in and around a cartilage model appears to be triggered by factors emanating from the cartilage, or at least from the calcified hypertrophic region of it. This conclusion is suggested by the very close spatial and temporal correspondence between cartilage cell hypertrophy and perichondrial ossification, and by the fact that bone forms regularly on the surfaces of transplanted growth cartilage[20]. Indeed, in rabbits even devitalized growth cartilage will induce bone[6].

It is apparent also that the periosteum, which is firmly attached to the epiphyseal cartilages (and their successors, the bony epiphyses) at the ends of a bone, must be subjected to considerable tension during growth as a result of the activities of the growth cartilages. Moreover, the subperiosteal bony trabeculae near the bone ends, which run very obliquely, are continued as, and are anchored to, the fibrous periosteum by means of tense, obliquely-running fibrous strands which traverse the cambial layer. The direction of these strands suggests that their attachments to the fibrous periosteum are being pulled towards one or other epiphysis. A unifying hypothesis is that the separating growth cartilages tense the fibrous periosteum and in so doing induce it to lengthen by interstitial formation of new collagen fibres; at the same time, the oblique fibrous strands between bone and fibrous periosteum are of necessity tensed also, and the tense strands in their turn stimulate and direct the growth of the bony trabeculae.

This hypothesis has been tested by Crilly[7], who found that if the periosteum of a growing chicken radius is severed completely the two segments of the periosteal tube spring apart, and the sub-periosteal bony trabeculae formed subsequently have a different orientation from those formed before the operation. He also found, and this is an observation of great theoretical importance, that the growth cartilages grew considerably faster than normal after severing the periosteum in this way, and normal growth rate only returned with the repair of the periosteum. The implication of this is that not only do the growth cartilages tense the periosteum and stimulate periosteal growth, but the tense growing periosteum in its turn has a braking influence on the growth cartilages. All this suggests that the harmony and co-operation between cartilage, fibrous tissue and bone growth in the development of a bone depend basically on a feedback relationship between cartilage and fibrous tissue, with cartilage taking the lead. The initial deposition of bony tissue seems wholly subservient and secondary to the events taking place in growing cartilage and fibrous tissue. Crilly's results, incidentally, explain the phenomenon of bone overgrowth after fracture in young animals.

A feedback between cartilage and fibrous tissue growth also seems to be of importance in the correlation of the growth of the separate bones of the skeleton. Harrison[14, 15] has demonstrated that maintenance of the shape of the growing pelvis involves a sliding of the ilium relative to the sacrum at the sacro-iliac joint, and that this is brought about by tension in the sacro-iliac ligaments resulting from cartilage growth at the iliac crest. Growth correlation in the fibrous and cartilaginous parts of the upper tibial growth plate, referred to earlier, can be explained on similar lines, that is to say, growth in the cartilaginous portion of the plate stretches the fibrous plate and stimulates it to grow at a similar rate.

In the development of the cranial vault, there is no doubt that the principal stimulant to bone growth is brain expansion, acting primarily on the fibrous 'brain capsule' in which the cranial bones are embedded. Thus if part of the brain is removed, the growth of the fibrous capsule is retarded, and the bones are smaller. The cranium, in effect, is made to fit the brain! That the cranial bones themselves are not prime movers in the growth process is shown by the experiment of lightly cauterizing the growing edges of the bones in a young animal: the fibrous brain capsule continues to grow, and forms a skull of normal shape and size, even though the individual bones have an abnormal shape, and the sutural patterns are bizarre[9].

If the foregoing analysis is correct, then prime responsibility for skeletal growth outside the skull vault and face resides with the cartilage models and the cartilaginous growth plates which succeed them: and it is therefore to the cartilage which we should first direct our attention when we are considering the effects of genetic, hormonal, mechanical, nutritional and other factors on the growth and development of the skeleton. Of course, not all bone growth phenomena are subordinate or related to cartilage activity: for example, internal re-modelling appears to be influenced *directly* by mechanical, hormonal, vascular and meta-bolic factors, unlike changes in size and shape which are controlled in the main via cartilage. Thus the bones of a paralysed limb, through disuse, tend to be grossly porous internally, yet approximate to the normal in overall size.

Finally, the great question which must be faced, is how all the various factors affecting bone growth are transduced into stimuli which evoke an answering response in osteoprogenitor cells, so that their proliferative activity, and the matrix-forming and -destroying activities of their progeny, are mobilized, controlled and harmonized in space and time, in accordance with the grand strategy of skeletal development and adult homeostasis programmed in the genes. The question is being tackled at cell, tissue and organ level, *in vivo* and *in vitro*. The accumulated data is overwhelming. The cells in bone are evidently the final target for many kinds of influence – nutritional, gaseous, hormonal, mechanical and electrical – but the means by which cellular co-operation is achieved so as to produce a skeleton which is both elegant and efficient, self-maintaining, self-repairing and self-adapting, are still far from clear: though if one were to hazard a guess in the present state of knowledge, it would be to suggest that the transducing of the mechanical field into an electrical one is at the heart of the matter.

References

1. ASCENZI, A. and BENEDETTI, E. L. (1959), 'An electron microscopic study of foetal membranous ossification', *Acta. anatomica*, **37**, 370–385.
2. BADI, M. H. (1972), 'Ossification in the fibrous growth plate at the proximal end of the tibia in the rat', *Journal of Anatomy, London*, **111**, 201–209.
3. DE BEER, G. R. (1937), *The development of the vertebrate skull*, Oxford University Press, London.
4. BRIDGES, J. B. (1958), 'Heterotopic ossification in the ischaemic kidney of the rabbit, rat and guinea pig', *Journal of Urology*, **79**, 903–908.
5. BRIDGES, J. B. (1959), 'Experimental heterotopic ossification', *International Review of Cytology*, **8**, 253–278.
6. BRIDGES, J. B. and PRITCHARD, J. J. (1958), 'Bone and cartilage induction in the rabbit', *Journal of Anatomy, London*, **92**, 28–38.
7. CRILLY, R. G. (1972), 'Longitudinal overgrowth of chicken radius', *Journal of Anatomy, London*, **112**, 11–18.
8. DIXON, A. D. (1953), 'The early development of the maxilla', *Dental Practitioner*, **3**, 331–336.
9. GIRGIS, F. G. and PRITCHARD, J. J. (1958), 'Effects of skull damage on the development of suture patterns in the rat', *Journal of Anatomy, London*, **92**, 40–57.
10. GODMAN, G. C. and PORTER, K. R. (1960), 'Chondrogenesis studied with the electron microscope', *Journal of Biophysical and Biochemical Cytology*, **8**, 719–760.
11. HALL, B. K. and SHOREY, C. D. (1968), 'Ultrastructural aspects of cartilage and membrane bone differentiation from common germinal cells', *Australian Journal of Zoology*, **16**, 821–840.
12. HAM, A. W. (1930), 'A histological study of the early phases of bone repair', *Journal of Bone and Joint Surgery*, **12**, 827–844.
13. HANCOX, N. M. (1972), 'The osteoclast', In *The Biochemistry and Physiology of Bone*, Vol. 1, 2nd Edn. Ed. Bourne, G. H., Academic Press: New York and London.
14. HARRISON, T. J. (1958a), 'The growth of the pelvis in the rat – a mensural and morphological study', *Journal of Anatomy, London*, **92**, 236–260.
15. HARRISON, T. J. (1958b), 'An experimental study of pelvic growth in the rat', *Journal of Anatomy, London*, **92**, 483–488.
16. HOLTROP, M. E. (1966), 'The origin of bone cells in endochondral ossification' *Calcified Tissue. Proceedings of European Symposium*, **3**, 32–36.
17. HUGGINS, C. B. (1931), 'The formation of bone under the influence of the urinary tract', *Archives of Surgery*, **22**, 377.
18. KARASHANI, J. T. (1972), 'Intramembranous ossification and secondary chondrogenesis', Ph.D. Thesis. The Queen's University of Belfast.
19. KEMBER, N. F. (1960), 'Cell division in endochondral ossification', *Journal of Bone and Joint Surgery*, **42B**, 824–839.
20. LACROIX, P. (1951), *The organization of bones*. Translated by S. GILDER. J. A. Churchill: London.

21. LEVAI, G. (1972), 'Fine structure of chick radius', *Journal of Anatomy, London*, **113**, 293–294.
22. MULHOLLAND, H. C. and PRITCHARD, J. J. (1959), 'The fracture gap', *Journal of Anatomy, London*, **93**, 590.
23. NYLEN, M. V., SCOTT, D. and MOSLEY, V. M. (1960). In *Calcification in biological systems*, American Association for the Advancement of Science, Washington, D.C.
24. OWEN, M. (1963), 'Cell population kinetics of an osteogenic tissue', *Journal of Cellular Biology*, **19**, 19–32.
25. PRITCHARD, J. J. (1952), 'A cytological and histochemical study of bone and cartilage formation in the rat', *Journal of Anatomy, London*, **86**, 259–277.
26. PRITCHARD, J. J. (1972a), 'The osteoblast'. In *The Biochemistry and Physiology of Bone*, Vol. 1, 2nd Edn. Ed. G. H. Bourne. Academic Press New York and London.
27. PRITCHARD, J. J. (1972b), 'The control or trigger mechanism induced by mechanical forces which causes responses of mesenchymal cells in general and bone opposition and resorption in particular', *Acta Morphologica Neerlando-Scandinavica*, **10**, 63–69.
28. PRITCHARD, J. J. and RUZICKA, A. J. (1950), 'Comparison of fracture repair in the frog, lizard and rat', *Journal of Anatomy, London*, **84**, 236–261.
29. PRITCHARD, J. J., SCOTT, J. H. and GIRGIS, F. G. (1956), 'The structure and development of cranial and facial sutures', *Journal of Anatomy, London*, **90**, 73–86.
30. ROUX, W. (1895), *Gesammelte Abhandlungen über Entwicklungsmechanik.* Engelmann: Leipzig.
31. SCOTT, B. L. and PEASE, D. C. (1956), 'Electron microscopy of the epiphyseal apparatus', *Anatomical Record*, **126**, 465–496.
32. SHELDON, H. and ROBINSON, R. A. (1957), 'Electron microscope studies of crystal collagen relationships in bone', *Journal of Biophysical and Biochemical Cytology*, **3**, 1011–1016.
33. SYMONS, N. B. B. (1952), 'The development of the human mandibular joint', *Journal of Anatomy, London*, **86**, 326–332.
34. TONNA, E. A. and CRONKITE, E. P. (1961), 'Cellular response to fracture studied with tritiated thymidine', *Journal of Bone and Joint Surgery*, **43A**, 352–362.
35. TONNA, E. A. (1966), 'A study of osteocyte formation and distribution in ageing mice complemented with H^3 proline autoradiography', *Journal of Gerontology*, **21**, 124–130.
36. URIST, M. R. and STRATES, B. S. (1971), 'Bone morphogenetic protein', *Journal of Dental Research*, **50**, 1392–1406.
37. WEINMAN, J. P. and SICHER, H. (1955), 'Bone and bones'. *Fundamentals of bone biology*, 2nd Edn., H. Kimpton, London.
38. YEOMANS, Y. D. and URIST, M. R. (1967), 'Bone induction by decalcified dentine implanted into oral, osseous and muscle tissues', *Archives of Oral Biology*, **12**, 999–1008.
39. YOUNG, R. W. (1964), 'Specialization of bone cells'. In *Bone Dynamics.* p. 117. Ed. Frost, H. M. Little, Brown & Co.: Boston.

5 Differentiation and Growth of Cells of the Skin

F. J. EBLING

Evolution of the Vertebrate Skin

The skin of all vertebrates consists of a cellular epidermis which overlies a dermis of connective tissue elements[1-5]. It arises by the juxtaposition of two major embryological contributions. During gastrulation prospective mesoderm is brought into contact with the inner surface of the prospective epidermis, which is the entire surface ectoderm from the union of the neural folds in the median dorsal line to the median ventral line of the embryo. That this prospective epidermis originates from a surface area of the early gastrula has been shown in both amphibia[6, 7] and birds[8] by vital staining.

The mesoderm not only provides the source of the dermis, but its presence is essential for the differentiation of the epidermal structures from the overlying ectoderm. In classical embryology the dermis was believed to arise only from the dermatome, one of the three arbitrary divisions of the somite, but this view is contradicted by the evidence that isolates of unsegmented lateral plate mesoderm as well as early limb buds of both chick and mouse, will produce normal feathers and hair, respectively, when grafted into the coelom of embryonic chick hosts[1]. While, therefore, there is no doubt that the dermatome of the somite contributes to the formation of the dermis in the dorsal and dorsolateral regions of the body, it seems that the dermis of the limbs, flank and venter is derived from the somatopleure.

As the source of the pigment cells, the neural crest also makes an important contribution – albeit small in bulk – to the skin. These cells variously described as melanophores or chromatophores are common to all vertebrates. They are found in both dermis and epidermis, including the epidermal derivatives feathers and hairs, and, especially in lower vertebrates, also in some other sites such as the perivascular layers, meninges and peritoneum. Proof that they originate from the neural crest has been established by experiments on many vertebrate groups[9]. In amphibia, extirpation of the neural folds of neurulae results in total absence of pigment cells from the operated region, and isolated neural folds produce pigment cells *in vitro* or when transplanted to the flanks of other embryos. In birds, explants of neural crest produce typical melanophores when cultured *in vitro* or grafted into embryonic coelom[1]. In mammals, only tissues containing cells from the neural crest can produce pigment in grafts[10].

The epidermis[2, 4, 5] of *Amphioxus* is a simple columnar epithelium; in the young animal the cells are ciliated but a 'cuticular' border with a thin layer of mucus develops in the adult. In Cyclostomata the epithelium becomes stratified. The very abundant mucus characteristic of lampreys and hagfish is formed by mucous cells, but two other types – club and granular cells – are present; all arise from specific precursors in the basal layer. Fish have a stratified epidermis which is similarly characterized by many unicellular glands or goblet cells. The skin of elasmobranchs is pierced by placoid scales, similar to teeth, with a dentine core of dermal origin coated by enamel secreted from epidermal cells; the skin of bony fishes contains hard scales of various kinds within the dermis.

Amphibia show two important trends related to emergence from water: the cornification of the outermost layer of the epidermal surface and the development of multicellular glands. The evolutionary background is reflected in changes during development. Embryonic epidermis contains two layers of cells, many of which have cilia, and a cuticular border. The cilia disappear sometime after hatching; the cuticular border is then essentially similar in structure to that of cyclostomes. At the end of larval life the amphibian epidermis becomes markedly thicker, is without unicellular glands and develops a superficial stratum corneum[11]. This horny layer is periodically sloughed.

Within the land vertebrates the epidermis performs some remarkable feats. Reptiles are characterized by their highly keratinized scales[12], birds and mammals by feathers[13, 14] and hair[3, 15, 16], respectively. The evolutionary relationships of these epidermal specializations are debatable. The feather is usually regarded as a homologue of the reptilian scale whereas the hair follicle is considered to be a new structure arising from the fold of epidermis between the scales[17]. One suggestion is that hairs were evolved from sensory structures which were sparsely distributed in the skin of reptilian ancestors[18]. In development, the earliest stages of scale, feather and hair are indistinguishable, but the formation of feathers[19] and hairs[20] later involves the aggregation of groups of dermal cells, a process which is not evident in the organization of scales[12]. The feather follicle differs from the hair in that it first arises as a dome of epidermis and dermis and only later sinks into the body of the skin, whereas the hair follicle starts by an inward growth of epidermis which then invests the dermal papilla.

In snakes and lizards the entire horny layer is from time to time moulted; in birds and mammals epidermal keratinocytes are continually shed and replaced, but the feathers[13, 14] and hairs are – like reptile scales – moulted at intervals. Thus in birds and mammals part of the developmental process is periodically re-enacted throughout adult life; indeed, the adult hair follicle undergoes a cycle of change which is remarkably similar to a section of its embryonic history[15, 16].

Reptiles are often believed to have only a few skin glands, which are variously located in different species, holocrine, and probably concerned with producing scent for sexual communication. However, a variety of holocrine glandular specializations within the epidermis have been described for lizards[21–23]. Birds have a single cutaneous gland, the uropygial or preen gland at the base of the tail[14]. It functions by holocrine secretion to produce an oily material which the bird spreads over its feathers with its beak. Mammals, on the other hand have a variety of sebaceous[24] and sweat[25] glands, as well as mammary glands.

In each species the nature of the skin varies between the regions of the body. Striking differences occur in the nature of the epidermal outgrowths, feathers and hair, in birds and mammals. But there are also less obvious differences in skin thickness, pigmentation and distribution of glands. The origin of this regional specificity in ontogeny and the way it is produced are clearly matters of importance.

Skin of adult vertebrates retains its regional specificity when transplanted to other sites on the body. For example, hair follicles do not change the period of their active phase, and continue to produce hairs of the original length, even though the timing of their cycles may slowly change[26]. Regional specificity becomes established early in ontogeny[1]. In a variety of larval amphibians skin transplanted to other animals has been shown to retain its individual characteristics. Moreover, after metamorphosis the transplant produces a spotting pattern typical of the region of the animal from which it came.

Exactly when specificity arises is not clear, but, in *Amblystoma*, the type of skin seems already determined by the young tail-bud stage, since flank ectoderm transplanted to the mid-dorsum produces no dorsal fin, whereas mid-dorsal epidermis transplanted to flank does so[27].

The question remains how regional differences of the skin are produced. Do they become established in the ectoderm itself or do they depend on reaction between ectoderm and underlying mesoderm? Experiments on the feather germ and the hair follicle have demonstrated the importance of interaction between the components. In essence, prospective dermis does not become organized in the absence of ectoderm nor does ectoderm produce its appendages without the necessary underlying mesoderm. It appears that the mesoderm functions as an inductor of the epidermal structure, but that the specificity of the response depends to a major extent upon the overlying ectoderm. The evidence will be further considered below.

The Reptilian Scale

Development

The integument of reptiles[12] is divided into scales. In crocodiles the dermis undergoes ossification, and the carapace of the turtles and tortoises is composed of bony plates which form in the dermis. In all these forms de-squamation of the epidermis is slow and inconspicuous. But in snakes and lizards, as well as in *Sphenodon*, the entire horny layer is moulted at intervals; thus the epidermis shows more than one generation of cells and a duplicated set of layers.

In snakes and lizards the scales overlap in most regions of the body, though all the dermal and epidermal layers of each scale are continuous with those of the next. Thus each scale has an outer and an inner surface and a flexible hinge region.

In early embryos the epidermis is represented by a single epithelial layer of cuboidal cells lying upon a basement membrane. By division of the cells, the stratum germinativum first differentiates an outer periderm and later an embryonic stratum corneum.

In the next stage of development the symmetrical scale rudiments become apparent. There are no pronounced proliferations of cells in either the epidermis or dermis which could account for the elevation of the rudiments; the progress from symmetry to the asymmetrical scale is gradual (Fig. 5.1). The embryonic scale has a continuous periderm over its surface. Immediately beneath this is a lightly staining granular layer, which is the presumptive 'Oberhautchen'. The remainder of the embryonic stratum corneum consists of five to six layers of flattened cells which are granular and darkly staining. Keratinization begins in the

Oberhautchen and spreads down through the embryonic stratum corneum; the outer layers give rise to the β-layer of the first formed epidermal generation and deeper-lying layers give rise to the α-layer. Thus, excepting for the periderm which is shed, the epidermis now has the

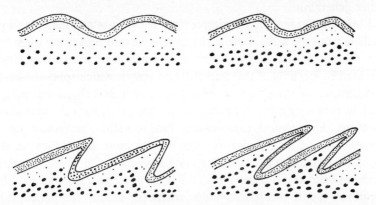

Fig. 5.1 Change in shape of early symmetrical scale anlage to perfect embryonic scale with outer and inner scale surfaces and hinge region. Note growth of deep dermal strand. Medium stippling, epidermis; light stippling, superficial dermis; dark stippling, deep dermis. (*After* Maderson[12].)

structure of the resting condition in the adult. In lizards only one epidermal generation is present at hatching, but in snakes, where the first slough occurs in the first few days of embryonic life, there are two.

Moulting

The structure of the adult epidermis and the details of the moult have been described in the colubrid snake, *Elaphe taeniura*[12, 28], the rat snake *Ptyas korros*[29] and in lizards[30-34]. The process is essentially similar in all Squamata. During the moult, a whole 'epidermal generation' is lost. It comprises six distinct cell layers: the Oberhautchen, the β-layer of compact keratinized material, the mesos layer, the α-layer of loose keratinized material and the underlying lacunar and clear layers (Fig. 5.2C). After shedding (Fig. 5.2D) the functional body surface is formed by the erstwhile inner epidermal generation which is immature and as yet incomplete. During the resting phase or 'intermoult period', which is of variable length, little proliferative activity is seen (Fig. 5.2A). At the end of the resting phase, proliferative activity in the germinal layer recommences to complete the remaining components of this generation, namely the lacunar tissue and the clear layer (Fig. 5.2B). The outer generation is now complete. A renewal phase, prior to the next moult, now follows. The stratum germinativum undergoes rapid proliferation to form the Oberhautchen, β- and mesos layers, and part of an α-layer of a new epidermal generation. When shedding occurs, the mature clear layer of the outer generation separates from the Oberhautchen of the inner generation.

Eosinophil granulocytes migrate from the dermis and may play a part in dispersing the living cells of the clear layer and separating them from the Oberhautchen of the inner epidermal generation. Shedding is initiated by swelling of the dermal blood vessels which splits the keratin and removal of the skin may be aided by specific behavioural patterns such as rubbing against obstacles. Holocrine glandular cells may also be involved. Histological examination has revealed that scales on the posterior abdominal region of lizards may show concavities on their outer surfaces which house 'generation glands', i.e. holocrine glands deriving from and maturing with the basic epidermal generation. These are of two types, 'escutcheon scales' where the cellular component is laid down immediately above the clear layer and 'β-glands', in which the glandular material derives either from the Oberhautchen or from the underlying β-cells or from both. The material of the β-glands is exposed when the outer generation is moulted; that of the escutcheon scales appears to be left behind by an outward shift of the splitting zone.

Feathers

Development of feathers

Feathers are not uniformly distributed over the skin but are confined to definite tracts known as pterylae, which are separated by naked areas or apteria[5, 13, 14, 19]. One such tract extends over the dorsum from head to tail and there are others on each side of the venter and on the thighs, wings and head. The tracts become evident in embryonic development as they become populated with feather germs, an event which starts at about the fifth day of incubation in the chick. Each feather germ at first consists of a dome composed of an aggregation of mesodermal cells covered by a two-layered ectoderm. Feather germ formation starts in the centre of the tract and moves peripherally until a characteristic number of germs is laid down; by the tenth or eleventh day of incubation the tracts are complete and well defined.

The feather papillae each produce a succession of types of plumage throughout life. The first to be formed are the fluffy down feathers, of

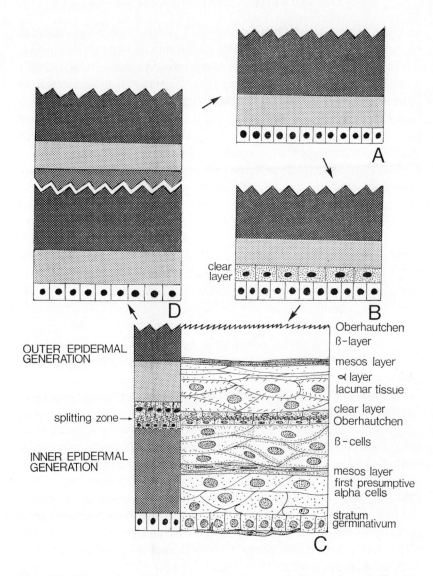

Fig. 5.2 Changes in epidermal structure in association with periodic skin shedding in lizards. A. Immediately after beginning of resting phase. B. End of resting phase. C. Renewal phase, shown with histological details and providing a key to the diagrams. D. 'Pre-shedding' condition; the outer epidermal generation will be lost and the previous inner epidermal generation will become the functional body surface. (*After* Maderson, P. F. A. (1970), 'Lizard glands and Lizard hands: models for evolutionary study', *Forma et functio*, **3**, 1792-1804.)

simple structure, consisting of a short quill or calamus bearing a circle of filamentous barbs having finer barbules without any interlocking hooks. The soft undercoat of the adult is composed of plumules which closely resemble the juvenile down feathers. But juveniles and adults develop an outermost covering made up of definitive or contour feathers, characterized by a long shaft or rachis bearing on each side a vane made up of parallel barbs held together by interlocking barbules, a system particularly well developed in the flight and tail feathers. A third type of feather, the small filoplume, has a thin rachis with barbs only at the tip.

In the formation of a down feather (Fig. 5.3), the dome of the germ first elongates into a tapering epidermal cylinder enclosing the mesodermal cone. The epidermis has three layers, the outer forms a protective sheath, the middle gives rise to the barbs, barbules and calamus of the feather, and the inner surrounds the mesodermal pulp. The barbs and barbules arise in sequence as longitudinal epidermal ridges, and during this process the base of the feather cylinder sinks below the surface into a tube-like follicle (Fig. 5.3D).

The protective sheath does not rupture to release the barbs until the chick dries after hatching. The dermal core covered by a thin layer of epidermis remains as a permanent structure – the feather papilla – to produce all succeeding feathers.

The first set of contour feathers of the juvenile plumage begins to form before the down feather is shed. Prior to the formation of the contour feather (Figs. 5.4, 5.5), the epidermal cells covering the surface of the dermal core of the persisting papilla multiply to form a thick ring or 'collar'. Proliferation of these collar cells produces an epidermal cylinder, and all the parts of the feather develop from an intermediate layer of this cylinder. The rachis forms by rapid growth at one side of the collar; the barb ridges (primordia of the barbs, barbules and barbicels) arise opposite to the shaft, parallel to each other and perpendicular to the collar. As the barb ridges increase in length they move towards and fuse with the elongating rachis. The feather cylinder gradually emerges from the mouth of the follicle, the pulp is gradually withdrawn towards the base and the sheath is shed allowing the vane to unfold.

Dermo-epidermal interactions in birds

The feather germ has provided a classical material for studying the interactions of dermis and epidermis in the production of cutaneous appendages. The dermal papilla has been shown to be essential to the formation of a new feather; if it is destroyed or removed, the epidermal cells alone cannot regenerate a feather.

By transplantation experiments, Lillie and Wang[35-37] were the first to demonstrate that the dermal papilla acts as an inductor of feather

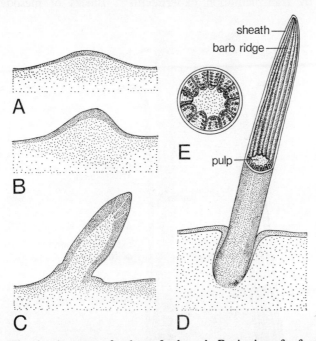

Fig. 5.3 The development of a down feather. A. Beginning of a feather-germ formation. Note aggregation of mesodermal cells (primordium of dermal papilla and pulp) and slight thickening of overlying ectoderm. B. Feather-germ protruding from the surface of skin. Note increase in thickness of both tissue layers. C. Elongation of feather germ into a tapering epidermal cylinder enclosing a mesodermal core. Epidermal cells of the cylinder wall are beginning to undergo rearrangement, longitudinally, into barb-ridges. D. Feather near time of hatching; basal portion sunk beneath skin surface in tubular follicle lined with epidermis. Dermal papilla now located permanently in floor of follicle. Wall of epidermal cylinder is divided into a series of longitudinal barb-ridges – primordia of barbs and barbules – surrounding central pulp and protected by external sheath. E. Transverse section through feather showing 11 barb-ridges, central pulp and external sheath. (*After* Rawles[13, 19].)

formation, but that the type of feather produced depends upon the tract origin of the epidermis. Thus a saddle follicle, stripped of its epidermis, induced a breast type of feather when transplanted to the floor of an empty breast follicle. Conversely, a breast dermal papilla produced a saddle feather when placed in a saddle follicle from which the papilla had been removed.

That feather tract specificity of the epidermis is acquired early in ontogeny by interaction with the underlying mesoderm was similary shown by transplantation experiments[38]. Blocks of mesoderm and

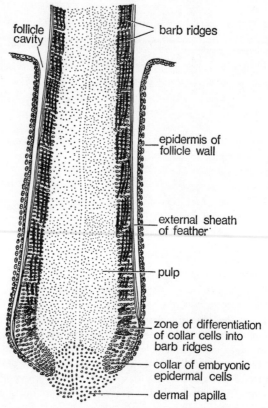

follicle cavity

barb ridges

epidermis of follicle wall

external sheath of feather

pulp

zone of differentiation of collar cells into barb ridges

collar of embryonic epidermal cells

dermal papilla

Fig. 5.4 Developing contour feather in longitudinal section. (*After* Rawles[13, 19].)

adhering ectoderm were excised from the dorsal side of the wing bud in $3\frac{1}{2}$–4 day old embryos and replaced by pure mesoderm from the prospective thigh. The implanted mesoderm became covered by ectodermal cells migrating from the surrounding wing but at a stage before the formation of the feather primordia. The contour feathers which subsequently developed had all the characteristics of the femoral tract from which the transplanted mesoderm had come, and not of the wing, from which the ectodermal cells had migrated.

That the special nature of the epidermis – whether, for example, it

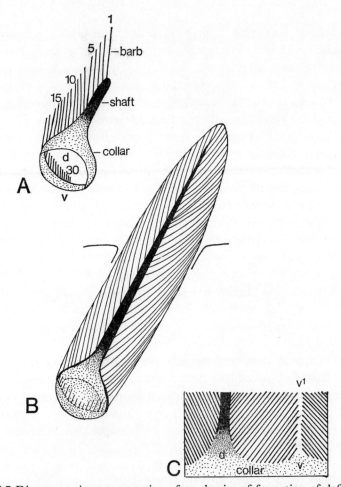

Fig. 5.5 Diagrammatic representation of mechanics of formation of definitive feather vane from embryonic cells of the collar. Dermal papilla and feather sheath are not shown. A. Early stage in formation of epidermal cylinder. Shaft (rachis) develops along dorsal wall, d, of cylinder parallel to its long axis. Barb-ridges forming one vane-half originate on each side of the ventral locus, v, opposite shaft. As each barb-ridge lengthens it shifts dorsally to its junction with the shaft where growth terminates. At any given time during feather development a series of stages in barb-ridge formation ranging from those completely formed at tip of shaft (nos. 1–10) to rudiments at ventral locus, v (nos. 29, 30). B. Later stage showing feather cylinder ('pin-feather') emerging from mouth of follicle (only a fraction of the numerous barb-ridges is shown). After emergence, the pulp is gradually withdrawn towards the base and the surrounding sheath shed allowing the feather parts to unfold to form the characteristic vane. C. Interior view of base of developing feather cylinder opened and spread flat to show relationships of shaft, barb-ridges and collar. Feather cylinder would normally split along mid-ventral line v^1–v where apices of barbs of the two vane halves meet. (*After* Rawles[13, 19].)

forms feather, scale, claw, spur, beak or uropygial gland – is determined during embryogenesis by a specific dermal stimulus has also been experimentally demonstrated in the chick[19, 39]. By treatment with trypsin, dermis and epidermis were separated in skin excised from a prospective feathered region (mid-dorsum) and a prospective scaled region (foot) before and during the formation of the respective primordia. The tissues were recombined and cultured on the chorio-allantoic membranes.

Epidermis taken from the prospective feathered region between 5 and $7\frac{1}{2}$ days of incubation invariably gave rise to feathers when grown with dermis taken from the foot between 9 and 11 days of incubation. However, when grown with foot dermis taken at 12 days, scales were occasionally induced, and when the dermis was taken between 13 and 15 days they were regularly developed. Thus this epidermis between 5 and $7\frac{1}{2}$ days is capable of producing either feather or scale, but the dermis of the foot does not become capable of inducing it to form scale before 13 days of incubation. Epidermis taken from the prospective feathered region of 8–$8\frac{1}{2}$ day embryos failed to produce scales when grown with foot dermis from 13–15-day embryos; thus the fate of the epidermis becomes irrevocable after 8 days.

The reciprocal combinations gave even more clear cut results. Epidermis from foot placed with dermis from feather tract always produced feathers irrespective of the age of either component. Moreover, even at 12–13 days of incubation, when it had already begun to form scales, these could be changed to feathers by juxtaposition of feather tract dermis.

The avian moult cycle and its control
When the feather is fully formed, the feather follicle enters a resting phase and the completed feather remains attached to the base of the resting follicle by its calamus. Each follicle continues, however, to undergo cyclical periods of activity (anagen) alternating with periods of rest (telogen)[5, 14]. The growth phase occurs prior to the moulting of the old plumage. The germinal cells of the collar surrounding the dermal papilla proliferate to form a new feather, which gradually pushes the old one out of the follicle canal (Fig. 5.6). The old feather remains attached to the tip of the keratinized sheath of the new feather; when the sheath splits to reveal the new feather, the old is lost. As between pterylae the moult occurs in sequence and is symmetrical, starting in the wings and spreading in turn to the body, head and neck. In most birds there is an ordered sequence within each pteryla.

The control of moulting appears to involve several levels and is matched in complexity only by the situation in the hair follicle. Each feather follicle appears to have some intrinsic control, since resting follicles can be induced to enter anagen by plucking the feather. At the

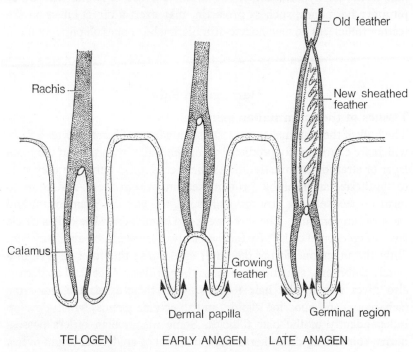

Fig. 5.6 Stages of feather growth and moult. (*After* Spearman[5].)

other end of events, the overriding factor which controls both repro-duction and feather development is the seasonal change in the photo-period as recorded by daylight falling on the retina. In passerine birds, for example, a spring moult occurs during a lengthening photoperiod and an autumn moult during a shortening one.

Environmental changes may be presumed to influence the feather follicle cycle through the endocrine system. Though the first down feathers develop without hormonal influence, most feathers in juveniles and adults require thyroid hormone. Thyroidectomy prevents formation of all except wing feathers, and large doses of thyroxine will induce moulting by speeding up the anagen phase. Conversely, feather forma-tion is retarded by a high level of oestrogenic hormones, which explains why moulting does not occur during egg laying and why gonadectomy of fowls results in continuous renewal of plumage instead of an annual moult. Thus it seems that both thyroid and gonads influence activity of the feather follicle. These organs are, in turn, controlled by the thyrotrophic and gonadotrophic hormones of the pituitary, completing the link from the hypothalamus, the brain and the environment. Other pituitary hormones, such as prolactin, may exert a direct effect on the feather follicle, as do melanocyte-stimulating hormones on pigmentation.

Mammalian Skin

Tissues of the mammalian skin[2, 3, 5, 40]

The bulk of the mammalian epidermis is made up of keratinocytes[5, 40, 41], which are continuously produced throughout life by mitosis in the basal layer or stratum germinativum. As these cells move outwards they start to synthesize an insoluble protein, keratin, within them, and come to form respectively the stratum spinosum, the stratum granulosum and the stratum corneum. The three week old human fetus has only a single layer of epidermal cells[42]. By four weeks two layers can be distinguished. Only the inner one forms the adult epidermis; the outer cells form a purely embryonic structure known as the periderm. The fetal epidermis also gives rise to the hair follicles, with their associated holocrine (sebaceous) and apocrine glands, and to eccrine glands, which develop independently of the hair follicles. Some mammalian orders possess horns, comprising dense keratinized material of epidermal origin over a bony core[3, 5]. At the distal ends of the digits, claws are produced in most mammals, but they are modified into hooves in ungulates and into nails in the more advanced primates.

Sebaceous glands occur in all mammals except whales[43]. Apocrine glands associated with hair follicles are found in all mammalian groups except whales, elephants, sea cows and scaly anteaters. Free eccrine glands occur in the foot-pads of all mammals with the exception of the above named orders and bats. They are present in hairy skin, however, only in a few primates, including tree shrews, Old World monkeys, apes and man; they appear in the course of primate evolution to replace the apocrine glands. Whether or not primate eccrine glands should be regarded as homologues of foot-pad glands is debatable; it is of interest that in the human embryo those on hairy skin develop later than those on the palmar and plantar surfaces.

Specialized aggregations of epidermal glands[43] are widespread amongst species and occur in many regions of the body, for example, in the nasal (goat, sheep), facial (bats), temporal (elephant), occipital (camel) and chin (rabbit, cat, bats) areas, as well as in the angle of the jaw (rodents), neck (bats), flanks (shrews) and in the inguinal and anal regions of many forms. Some of these glands are purely apocrine, others, such as the side glands of shrews, the inguinal glands of rabbits and the muzzle glands of bats, have both sebaceous and apocrine components. Others again, such as the supracaudal glands of the guinea-pig and pre-putial glands of rodents, are purely sebaceous. Mammary glands are present in all mammals and are considered to be homologues of either the apocrine or possibly the sebaceous glands.

Three other types of cell also come to reside in the adult epidermis. Specialized cells found in apposition to intraepidermal neurites with expanded tips, and characterized under the electron microscope by possession of membrane-limited spherical granules, probably represent the Tastzellen of Merkel[44], and may be sensory elements (Plate 5.1). Pigment-forming melanocytes (Plate 5.2) are present in the basal layers of the epidermis and in the matrices of the hair follicles. They are secretory cells, each of which injects pigment granules by way of dendritic processes into a group of epidermal or hair cells with which it is associated[45-48]. Melanocytes are formed from the neural crest, migrate to their epidermal sites and appear to retain substantial powers of movement even in the adult. Pigment free dendritic cells[49-52] which are negative DOPA-staining but ATP-ase positive have been described in the epidermis of man, rhesus monkeys, guinea-pigs and hairless mice, and are considered to be identical with gold chloride staining cells originally described by Langerhans[53]. The cells, as viewed by the electron microscope (Plate 5.3), are characterized by a granule having a 'racquet' profile[54]. Since fully differentiated Langerhans cells can be demonstrated in human fetal skin before melanocytes become functional[55], and skin deprived of melanocytes through exclusion of the neural crest derivatives has a full complement of Langerhans cells[56], it is now considered that Langerhans cells are not related to melanocytes. The occurrence of cells in every respect identical to Langerhans cells in the dermis in the skin condition histiocytosis-X suggests that epidermal Langerhans cells may arise from wandering histiocytes of mesenchymal origin[57].

The dermis[58] contains a variety of cells. Fibroblasts or fibrocytes, as the resident inactive cells are usually designated, are responsible for the secretion of the fibrous components of the dermis, namely collagen and elastin. They also probably secrete the mucopolysaccharides of the ground substance, though the mast cell has also been implicated. Fibroblasts are developed from primitive wandering cells of the mesenchyme, which also gives rise to several other cell types occurring in the dermis.

Mast cells[59] are especially found around small blood vessels and in serous membranes. Human mast cells contain – though it is not certain that they manufacture – heparin and histamine. In some species, e.g. rats, they have also been shown to contain serotonin. Under the electron microscope they are seen to contain granules of two types, one with a lamellated appearance.

Tissue macrophages, wandering cells or histiocytes, are also present. In addition, various blood cells: polymorphonuclear granulocytes, lymphocytes, monocytes and even plasma cells may also be found. Some of these dermal cells vary morphologically, and the distinctions between them may not be clear cut.

Development of the superficial epidermis

The most studied mammalian epidermis from the developmental viewpoint is that of man, which may be identified in the third week of fetal life as a single layer of cells[42]. Their plasma membranes which face the amniotic fluid have occasional cilia and small microvilli[60]. Within the cells, in addition to the common organelles, are extensive glycogen deposits and, although tonofilaments are not present, there is a skein of fine filamentous material parallel to the basement membrane.

By 4 to 6 weeks two layers (Plate 5.4 and Fig. 5.7) can be seen in most areas of the body, namely a periderm and a stratum germinativum[61]. These layers are somewhat loosely coherent and contact between the cells is effected not only by apposition of the plasma membranes, but by interlocking villous processes and by desmosomes which appear at intervals. Both periderm and stratum germinativum cells contain similar organelles – mitochondria, rough membranes, Golgi bodies – as well as glycogen granules, and they are not distinguishable on cytological grounds. No tonofilaments are present. A well-defined continuous plasma membrane, about 250–300 nm thick, underlies the basal plasma membranes of the germinative cells.

Between 8 and 11 weeks a middle layer starts to form. At this stage glycogen is abundant in all layers and microvillous projections occur at the surface of the periderm.

By 12 to 16 weeks (Plate 5.5) there are one or more intermediate layers[62, 63]. The cells contain abundant glycogen both within and between them; mitochondria, Golgi complexes and a few tonofilaments are now identifiable.

Microvilli now become much more numerous and globular structures appear on the surface of the periderm (Plate 5.6), remaining attached to the cells by pedicles of cytoplasm. Between 16 and 26 weeks the intermediate layers increase in number, and by 21 weeks keratohyalin granules appear in the uppermost layer. The globular elevations of the periderm break off into the amniotic fluid; by 26 weeks of age the whole of the periderm starts to separate from the embryo[64, 65].

The periderm is not simply a protective cover for the underlying epidermis, but appears to be an important functional embryonic organ. The development of microvilli and globular elevations increases the surface area exposed to amniotic fluid and, coupled with development of vesicles within the cytoplasm of the cells, suggests that the system is concerned with fluid exchange across the peridermal plasma membrane. The peak of this development of the periderm occurs between 12–16 weeks, after which it transforms into flattened squames. It has been suggested that the periderm plays a major role in production of amniotic fluid, a rôle later performed by the umbilical cord epithelium[60].

The significance of the glycogen in embryonic epidermis is uncertain. It is undoubtedly formed *in situ* from glucose, which occurs in a concentration of 30 mg% in the amniotic fluid. Glycogen is also found in the developing hair follicle and amylophosphorylase, an enzyme concerned with glycogen formation, has been identified at an early stage in differentiation[66]. Possibly the transitory appearance of glycogen indicates, as does its accumulation in adult muscle, a temporary dependence of the developing embryonic tissue on anaerobic glycolysis.

The origin of the cornified material is still a subject for discussion and embryology does not solve the problem. As possible precursors of the epidermal keratin; two materials, namely the heavily staining granules of keratohyalin in the granular layer and the tonofilaments in the basal layer, have been recognized[41]. Some authors, while not denying some contribution of interfilamentous material, believe tonofilaments to be the most important component, recognizing a continuous gradual maturation of fibrillar material from basal layer to stratum corneum[67, 68]. Others have proposed a two-component origin[69, 70]; some lay greater stress on the rôle of the non-fibrous keratohyalin and suggest that the contribution of the fibrils is minimal[71]. At the extreme is the view that keratinization does not commence before the granular layer and that the tonofibrils, which also can be demonstrated in non-keratinizing

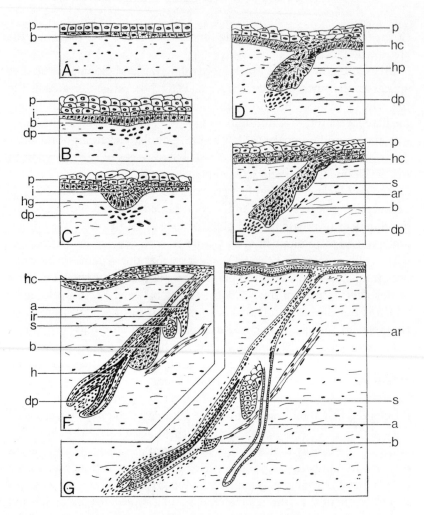

Fig. 5.7 Development of epidermis, hair follicles and associated structures in human skin.

A. Section of skin of fetus at about 4 weeks. The periderm is clearly seen (p) and a basal germinative layer (b) appears in occasional areas (see Plate 5.5).

B. Skin at about 11 weeks. The epidermis is made up of basal cells, cuboidal in shape (b), and cells of the stratum intermedium (i) are beginning to appear above them. The periderm (p) consists of two or three cell layers. Mesenchyme cells (dp) are beginning to aggregate below a presumptive hair follicle. This is the 'primitive hair germ' or 'pre-germ stage'.

C. Hair germ (hg) stage. Basal cells are now columnar and starting to grow downwards.

D. Hair peg (hp) stage. Cells of the so-called 'hair canal' (hc) form a solid strand. Compare with Plate 5.7.

cells such as leucocytes and fish epidermis, play no part in the process[72].

Studies of fetal material[60] show that keratohyalin invariably appears within the bundles of tonofilaments, but do not provide decisive information on its source. The fact that the disappearance of nuclei in the periderm is not accompanied by the appearance of keratohyalin and the finding that early granular cells appear to have healthy nuclei both suggest that the granules are not a product of nuclear degeneration.

The hair follicle
Development of the hair follicle
The development of the hair follicle has been described for a number of mammals, for example the opossum[73], rats and mice[74, 75], cattle[76], sheep[77] and man[42, 78, 79]. The rhesus monkey[80] and man[81–83] have received detailed attention at the ultrastructural level.

In man, the first rudiments of the hair follicles appear in the regions of the eyebrows, upper lip and chin. These are the sites on which, in other mammals, vibrissae are borne and such specialized hairs develop much earlier than the general pelage. General development of body hair in man does not begin until the fourth fetal month. The development of the follicle is best described by stages (Fig. 5.7).

(1) The pre-germ stage (Fig. 5.7B). The first sign of a hair follicle is a crowding of nuclei in the basal layer of the epidermis, the so-called primitive hair germ or pre-germ stage. These alterations in the epidermis probably precede the visible accumulation of underlying mesodermal nuclei.

(2) The hair germ stage (Fig. 5.7C). The pre-germ passes rapidly into the hair germ stage, in which the basal cells become high, the nuclei become elongated and the whole structure starts to grow downward into the dermis. At the same time mesenchymal cells and fibroblasts increase

---

E. Bulbous hair peg. Note solid 'hair canal' (hc), sebaceous gland rudiment (s), bulge (b) for attachment of developing arrector muscle (ar).

F. Later stage showing apocrine rudiment (a), sebaceous gland (s) now partially differentiated, and bulge (b). The dermal papilla (dp) has been enclosed and a hair (h) is starting to form, with an inner root sheath (ir). Compare with Plate 5.8.

G. Complete pilosebaceous unit of axillary skin from a 26 week old fetus. The sebaceous gland (s) is well differentiated and the apocrine gland (a) is canalized. (*From* Rook, Wilkinson and Ebling[42] by permission of Blackwell Scientific Publications.)

in number to form the rudiment of the hair papilla beneath the hair germ.

(3) Hair peg stage (Fig. 5.7D and Plate 5.7). The hair germ starts to grow obliquely downwards into the mesenchyme in the shape of a solid column of epithelial cells that seems to push the mesodermal nuclei before it. The outer cells of the hair peg become arranged radially to the long axis and are columnar in shape, those at the advancing matrix end being conspicuously tall and narrow. The column becomes enveloped in a sheath of mesodermal cells.

(4) Bulbous peg stage (Fig. 5.7E and Plate 5.8). The follicle continues to elongate. The advancing border becomes bulbous and envelops part of the underlying mesoderm to form the dermal papilla. At the same time two epithelial swellings appear on the posterior wall of the follicle. The lower one is the bulge to which the arrector muscle becomes attached, and the other is the rudiment of the sebaceous gland. In many follicles a third bud later appears above the sebaceous gland; this is the rudiment of the apocrine gland.

The bulbous hair peg continues its downward growth and differentiates. The first cells of the inner root sheath begin to form above the region of the matrix, and above the root sheath the inner cells of the follicle grow upwards into the epidermis to form the hair canal. The central cells of this follicular plug undergo some form of degeneration or partial keratinization, so that as the hair grows it appears gradually to push out the plug from the end of the canal.

Within the hair follicle there is more than one period of glycogen accumulation. The early hair germ cells lose glycogen very quickly, but the outer root sheath and bulge to which the arrector muscle becomes inserted soon re-acquire glycogen.

In the adult, hair follicles are arranged in patterns of which the group of three is the most characteristic. It appears that the first follicles develop over the surface at fixed intervals of between 274 and 350 μm. As the skin grows these first germs become separated, and new rudiments develop between them when a critical distance, dependent on the region of the body, has been reached. Thus in any region of developing skin there may be follicles in different stages of development. There is no large scale destruction of follicles during post-natal development, only a decrease in actual density as the body surfaces increase, nor do any new follicles develop in adult skin.

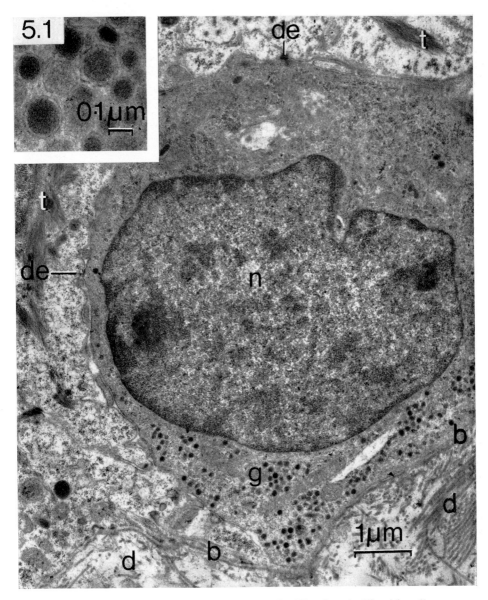

Plate 5.1 A Merkel cell in human epidermis. The dermis (d) with collagen fibres is seen in the lower part of the picture. b, basement membrane; n, nucleus of Merkel cell; g, spherical granules (see inset); de —, desmosomes making connections with adjacent basal keratinocyte; t, tonofilaments. (Electron micrograph by A. S. Breathnach, reproduced from Rook, Wilkinson and Ebling[43], by permission of Blackwell Scientific Publications.)

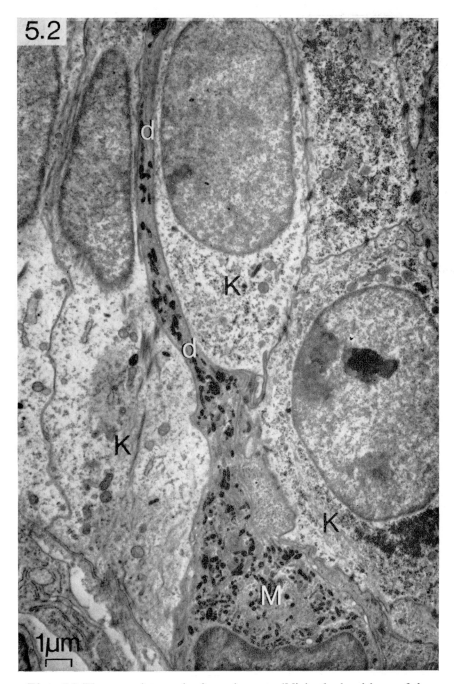

Plate 5.2 Electron micrograph of a melanocyte (M) in the basal layer of the epidermis of a human fetus. Note the slender dendritic process (d) extending between the surrounding keratinocytes (K). The melanocyte has its cytoplasm packed with dense pigmented melanosomes, which it has hardly, as yet, begun to transfer to neighbouring keratinocytes. (Electron micrograph reproduced from Breathnach, A. S., 1971 'Melanin pigmentation of the skin', by permission of Oxford University Press.)

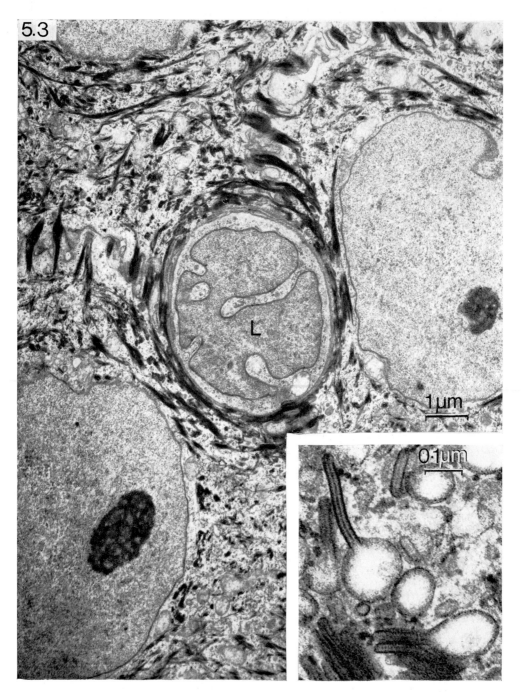

Plate 5.3 A Langerhans cell (L), with its characteristically indented nucleus, situated between cells of the stratum spinosum of the human epidermis. Inset: Langerhans granules showing racquet-shaped profiles. (Electron micrograph by A. S. Breathnach, reproduced from Rook, Wilkinson and Ebling[43] by permission of Blackwell Scientific Publications.)

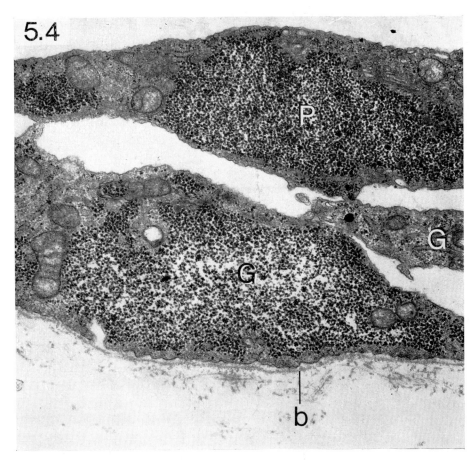

Plate 5.4 Electron micrograph of bi-laminar epidermis from the crown of the head of a 6 week old human embryo. A germinative cell (G) is shown lying on the basement membrane (b), with part of another germinative cell peripherally overlapping it. Lying over them is a periderm cell (P). As regards general cytological features, periderm and germinative cells are practically identical and both contain abundant glycogen granules. × 23 500. (*From* Breathnach and Robins[61], by permission of the editor of the British Journal of Dermatology.)

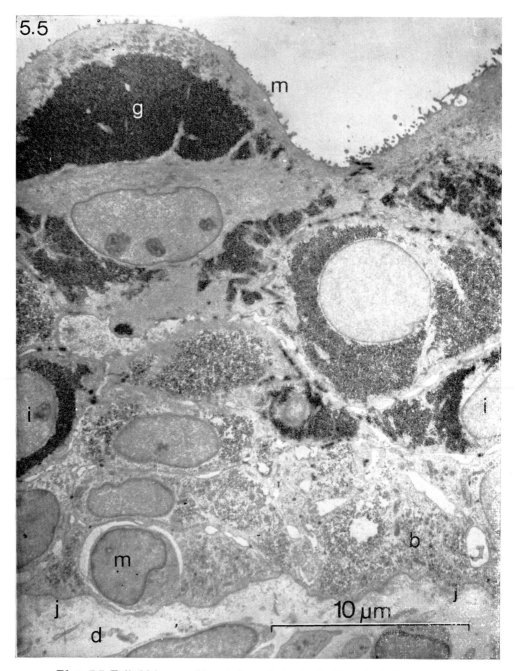

Plate 5.5 Full-thickness epidermis from the back of a 14 week old human fetus. Osmium fixation and lead staining. The periderm cells are full of glycogen (g) and have microvilli (m) at their amniotic border. Cells of the intermediate layer (i) also contain glycogen. Basal layer cells (b) have lost glycogen at this stage. Just above the dermo-epidermal junction (j), is seen a melanocyte (m); the surrounding space indicates that it is a recent immigrant from the dermis (d). (Electron micrograph by A. S. Breathnach, reproduced from Rook, Wilkinson and Ebling[43], by permission of Blackwell Scientific Publications.)

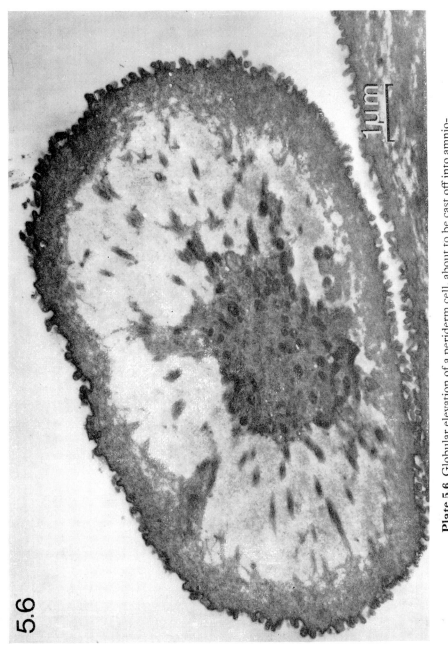

Plate 5.6 Globular elevation of a periderm cell, about to be cast off into amniotic fluid in a human embryo about 16 weeks old. PTA stain; unstained area is glycogen. (Electron micrograph by A. S. Breathnach, reproduced from Rook, Wilkinson and Ebling[43] by permission of Blackwell Scientific Publications.)

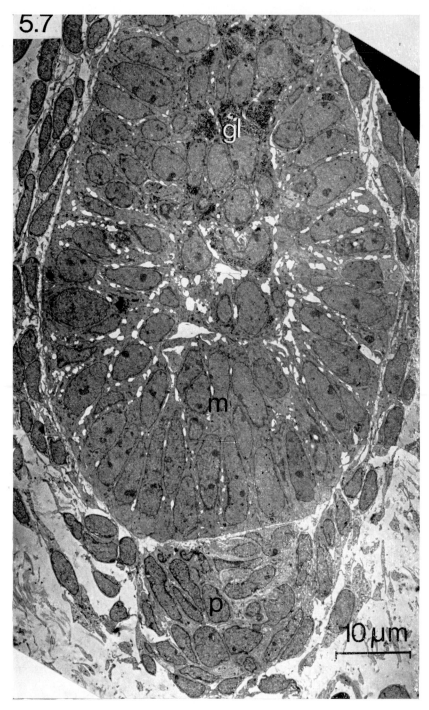

Plate 5.7 Montage of central longitudinal section through lower two-thirds of hair peg from developing scalp follicle at 15–16 weeks. gl, glycogen in central cells; m, matrix cells; p, dermal papilla. (Electron micrograph from Robins and Breathnach [82] by permission of the Journal of Anatomy.)

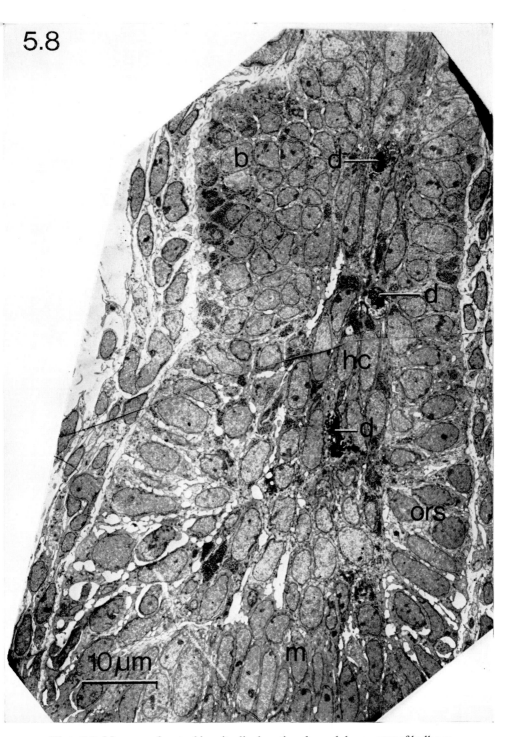

5.8

Plate 5.8 Montage of central longitudinal section through lower part of bulbous peg. The hair-cone (hc) extends proximally from the matrix region of the bulb (m) to the level of the bulge (b), and its cells are readily distinguishable from those of the prospective outer root sheath (ors). (d) electron-dense elements in hair cone. (*From* Robins and Breathnach [82] by permission of the Journal of Anatomy.)

Plate 5.9 Pre-melanosomes and melanosomes from hair-bulb melanocyte of a human fetus. × 81 375. (*From* Breathnach, A. S. (1971), *Melanin Pigmentation of the Skin*, by permission of Oxford University Press.)

Dermo-epidermal interactions in mammals
The underlying mesoderm is essential for the development of hair follicles. If epidermis is isolated from mouse embryos between the eleventh day of gestation (before any hair germs can be seen) and the fifteenth day (by which time vibrissae are already keratinizing), further development of the follicles cannot take place *in vitro*. Hair follicles will, however, form if epidermis is recombined with dermis from the same or different sites, and it has even proved possible to obtain organized vibrissa follicles by grafting co-mingled but dissociated embryonic dermal and epidermal cells on to the chorio-allantoic membrane of chick embryos[84, 85].

Is the regional specificity of the hair type determined by the dermis, or does it reside in the epidermis? That the specificity resides in the epidermis, in mammals as in birds, is reinforced by the finding that epidermis from the mouse snout will form vibrissa follicles when combined with dermis from the mid-dorsum[84]. However, the supposition that, because hair of local type is always produced irrespective of the source of the inducing dermal papilla, type specificity is proved to reside in the epidermis has been challenged. When the large dermal papilla of a vibrissa follicle is transplanted to the ear, it shrinks to the appropriate local size in the process of inducing a hair of local type[86, 87]. Thus although the vibrissa papilla has initiated hair formation, the hair actually forms under the influence of local ear dermis; regional specificity lies neither in the local epidermis nor in the transplanted papilla, but in the local dermis. The classical picture of dermal and epidermal interactions may, indeed, be much too simple.

A further complication is that dermal papillae can themselves be regenerated from the lower third of the vibrissa follicle. Isolated lengths of the follicle wall, composed of inner root sheath, outer root sheath and mesenchyme, regenerated dermal papillae at their proximal ends and even produced short whiskers when they were transplanted into ear skin. It seems, therefore, that either or both the outer root sheath and the mesenchymal layer are essential for papilla regeneration[88, 89].

The mammalian moult cycle and its control
The adult hair follicle consists of a deep pocket of epidermis known as the outer root sheath which is surrounded in turn by a non-cellular vitreous membrane and a sheath of connective tissue[15, 16]. The active (anagen) follicle (Fig. 5.8A) has at its base a proliferating epidermal matrix which surrounds a central papilla of the underlying dermis. From

this matrix arise all the layers of the hair, namely medulla, cortex and cuticle, and the inner root sheath which is keratinized and is lost as the hair emerges from the skin surface.

Hairs grow to a maximum length which depends upon the region of the body. The follicles then pass through a transitional phase (catagen) to a resting phase (telogen)[15, 16]. In catagen (Fig. 5.8B, C) the melano-cytes cease activity, resorb their dendrites and become indistinguishable from the matrix cells. The middle region of the bulb becomes constricted to a thin column of cells; above this the base of the hair becomes expanded and keratinized as a 'club', and at its lower end the dermal papilla becomes released from the epidermal investment. From the onset of catagen the vitreous membrane thickens enormously and causes a corrugation of the epithelial strand. The club hair then moves towards the skin surface, lengthening the epithelial column. Finally, as the fol-licle passes into telogen, the epithelial strand shortens from below until it is reduced to a small nipple, the secondary germ (Fig. 5.8D).

The next cycle of activity starts by cell division in the secondary germ and the downgrowth of an epithelial column which engulfs the dermal papilla and forms a new bulb (Fig. 5.8E). Subsequently, the keratinized dome of the newly forming hair, encased in its inner root sheath, emerges by the side of the old club hair, which is sooner or later shed.

Activity of hair follicles and moulting of the club hairs follows a pattern in many animals; for example, in the rat, bands of activity start in the venter and pass symmetrically over the flanks to the mid-dorsum and finally to the head and tail[90, 91]. In some animals the patterns are less clear. In the scalp of man, for example, follicles appear to moult at random, so that at any one time an average of about 13% of the follicles are in telogen[16]. The guinea pig was believed to show a similar mosaic pattern of moulting. But in newborn guinea pigs all follicles are active, and in the young animal it seems probably that there is a measure of synchrony within each of the three or four follicular types, even if follicles are out of phase with other types[92].

It has been shown by experiments on the rat, that if the site of a group of follicles is changed by skin grafting[26], or if the follicles are trans-planted to another animal in a different phase of the moult[93, 94], they continue for a long time in the rhythm of their donor sites and out of phase with their new neighbours. These facts demonstrate that the follicles have an intrinsic rhythm. Follicular cycles can be re-phased by plucking the resting hair from the follicle or by wounding[95, 96]. Either of these procedures can induce a follicle to become active sooner than it would have done spontaneously. It has been shown that in rats, oestradiol testosterone or adrenal steroids delay the initiation of follicular activity, and oestradiol also delays the shedding of club hairs, so that the moult is accelerated by gonadectomy or adrenalectomy; conversely, thyroid hormone accelerates follicular activity and thyroidectomy or

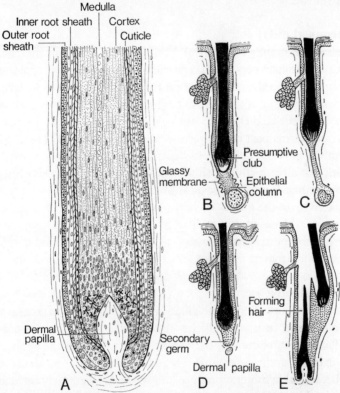

Fig. 5.8 The human hair cycle. (A) An active follicle. (B) Early catagen. (C) Late catagen, showing ascent of the presumptive club. (D) Telogen. (E) Early anagen. (*From* Ebling, F. J. (1964), 'The Hair Follicle'. In *Progress in the Biological Sciences in Relation to Dermatology*. Vol. 2. Eds. Rook, A. and Champion, R. H. By permission of Cambridge University Press.)

inhibition of the thyroid delays the moult[94, 96]. Oestradiol has also been shown to delay the onset of follicular activity in the guinea pig[92].

In temperate zones many adult wild mammals have moults in the spring and the autumn, so that the physical nature and colour of the pelage becomes adapted to changes in temperature and background[90, 91]. The skin cycle seems to resemble the sexual cycle in that it is influenced by changes in the environment, in particular the photoperiod, with possibly a small modifying effect of temperature. These environmental influences, as demonstrated for the gonadal cycle, probably act through the hypothalamus, the hypophysis and the endocrine organs already mentioned. The pineal may also play a rôle in such mediation. Thus in its control system the hair follicle bears a remarkable similarity to the feather follicle.

Sebaceous glands

The sebaceous glands arise as solid bulges (Fig. 5.7F) on the posterior surfaces of the hair pegs[66]. At first they contain moderate amounts of glycogen, but soon the cells in the centre lose this and become large and foamy as they accumulate drops of lipid. The human sebaceous glands become differentiated at 13–15 weeks and are then large and functional. The sebum forms part of the vernix caseosa. At the end of fetal life the sebaceous glands are well developed and large over the entire surface of the skin, but after birth the size is rapidly reduced and the glands enlarge to become functional again only after puberty.

The activity of the glands is controlled by hormones[97–100]. Sebaceous secretion is increased by androgens and by thyroid hormone and diminished by oestrogens. In rats, at least, there is evidence that the response to androgens may be mediated by a pituitary hormone.

Apocrine glands

In man, rudiments of apocrine glands develop as buds peripheral to the sebaceous gland in many follicles (Fig. 5.7F), including some on the scalp, face, chest, abdomen, back and legs as well as in the axilla, mons pubis, external auditory meatus, eyelids, circumanal area, areola region of the breast, labia minora, prepuce and scrotum where they canalize and survive in the adult.

In the human axillary organ[101] the apocrine primordia become apparent late in the fifth fetal month (Fig. 5.7G), by which time the eccrine glands are already well advanced in development. In the sixth month the apocrine glands are all elongated and coiled at the base and some show the beginning of a lumen. During the ninth month they become fully formed. They remain small in early childhood, but become well differentiated and on morphological evidence appear functional by eight years of age.

Specialized apocrine glands concerned with scent production occur in many mammals. In the rabbit, at least, these are hormonally influenced[102].

The mammary glands[103]

The first structures which are unquestionably concerned with the development of mammary glands are the milk lines or milk crests which appear as narrow ridges of ectoderm on either side of the midline, within a previously existing structure known as the mammary band. The mammary band gradually disappears and the milk line diminishes in length, becoming interrupted to form a series of thickenings of ectodermal cells which sink into the dermis to become the mammary buds. After a pause in development, the bud elongates and the distal part penetrates into the mesenchyme to form the primary mammary cord. In some species this may branch. As the mammary bud grows, the mesoderm differentiates into four distinct layers which give rise, respectively, to the smooth muscle of the nipple, the stroma of the nipple, the connective tissue between the mammary lobules and the interlobular septa.

The mammary crest appears in the human embryo of 7 mm in length. By 20–30 mm the mammary buds have assumed a circular shape and lie below the surface of the surrounding epidermis. There is little further development until the fifth month when the surface of each bud spreads out with a depression forming in its centre. At the same time the deeper layer of the bud proliferates to produce 15–25 secondary buds; these gradually lengthen to form solid cellular cords, which expand at their extremities and become surrounded with mesenchyme. Later, the cellular cords bifurcate, develop a lumen and become ducts; meanwhile the superficial ectoderm of the bud de-squamates and keratinizes to leave an inverted nipple area. At term, canalization of the cords is almost complete. The nipple depression becomes flattened and then somewhat everted, as the underlying dermis thickens, so that ducts now open to the exterior.

Information on the rôle of hormonal factors in the growth of fetal mammary glands is mainly based on rodents. The formation of mammary primordia in the mouse appears to be independent of the influence of hormones from the fetal gonads up to the fourteenth day of gestation. After this time, however, the male differs from the female in that the mammary buds fail to connect to the exterior and nipples are not formed. In the rat the mammary bud appears to retain its connection with the skin, although nipples are not formed. Androgens injected into the mother or into the fetus before day 14 cause the mammary primordia of the female fetus to follow the male pattern of growth.

Mammary growth remains quiescent between birth and puberty. At the onset of regular ovarian cycles, however, there is a marked acceleration in growth. In rodents, the rapidly growing glands form club-shaped terminal end buds at their periphery. In man, more extensive branching occurs and the foundations of lobules which will be borne on the branches are laid down.

During pregnancy further extension and branching of the duct system occurs, alveoli develop on the ducts and lobules are developed in the alveoli. The development of the mammary gland at this stage is hormonally controlled. Most information is based on experiments with rodents, and the exact hormonal requirements may vary in different species. In rodents, full mammary growth can be induced by anterior pituitary hormones – prolactin and growth hormone – acting in combination with ovarian and adrenal steroids. Oestrogen promotes duct growth, and oestrogen and progesterone together promote alveolar development. But both these effects require the presence of adrenocorticosteroids and of either or both the mentioned pituitary hormones.

Eccrine glands

No development of human eccrine glands takes place before 12 weeks. In embryos of 12 weeks (crown-rump length 6–7 cm) the rudiments of eccrine sweat glands are first identifiable as regularly-spaced undulations of the stratum germinativum. Cells which form the anlagen are oblong, palisading and lie closely together, but otherwise they do not differ from the rest of the stratum germinativum. By 14–15 weeks (crown-rump length of embryo 8–10 cm) the tips of the eccrine sweat gland rudiments have penetrated deeply into the dermis and have begun to form the coils. In the overlying epidermis, columns of cells which are destined to form the intra-epidermal sweat ducts are recognizable. Each column is composed of two distinct cylindrical layers, comprising two inner cells within which the formation of the lumen takes place and two outer cells which are elongated and curved so that they embrace the inner cylinder.

The development of the intra-epidermal[104] sweat duct is of some interest, because it forms in a different way from the intradermal sweat duct[105]. In lumen formation within the epidermis, intracytoplasmic cavities are formed within the inner cells (Fig. 5.9A). These cavities enlarge, coalesce within each cell, and then break through the cell membranes of the opposite cells and unite. The formation of the intracellular cavities is preceded by the appearance of dense bodies full of vesicles which probably represent lysosomes.

In the intradermal segment, on the other hand, the lumen forms intercellularly by dissolution of the desmosomal attachment plaques between

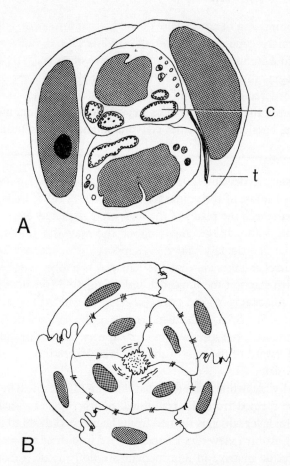

Fig. 5.9 Lumen formation in eccrine glands. (A) Intra-epidermal lumen formation; 16 week-old human embryo. Cavities (c), each containing small vesicles of
pinched-off cytoplasm, are forming within two opposing inner cells. The plasma
membranes between the inner cells are intact. Two outer cells, with prominent
tonofibrils (t) along their inner sides, flank the inner cells. *After* Hashimoto,
Gross and Lever[104]. (B) Intradermal lumen in 15 week-old embryo. The lumen
is formed in the centre of four opposing inner cells; it is bordered by microvilli
and contains small vesicles. There is a filamentous zone around the lumen. The
lateral borders of the basal cells at this stage become convoluted and interdigitating. (*After* Hashimoto, Gross and Lever[105].)

the cells comprising the inner core of the eccrine duct germ (Fig. 5.9B). Electron microscope examination of a 14 week-old embryo reveals groups of four luminal cells surrounded by five outer cells. By 15 weeks, a lumen begins to form in the centre of four opposing inner cells. This lumen is bordered by microvilli and contains small vesicles. As development proceeds the lumen increases in size, and the plasma membrane, between the basal cells, becomes convoluted and interdigitating.

Nails

The nails start to develop in the 9-week human embryo from invaginations, or nail folds, of the epidermis on the dorsal tip of each digit. By the twentieth week the plates of the nails are completely formed. There is, however, some disagreement about the way the fold forms the plate[106]. The traditionally held view, supported by experimental studies in the squirrel monkey, is that the nail plate is formed solely from a matrix which starts at the proximal limit of the fold (possibly involving part of the adjacent roof) and extends below the nail to the distal margin of the lunula (Fig. 5.9A). The distal region of the nail bed does not contribute to the nail, but merely provides support for it. The cuticle is an extension forward of the roof of the fold which keratinizes and seals on to the nail plate.

A second view, while accepting that dorsal and intermediate layers of the nail are formed from the traditional matrix, suggests that there is also a ventral layer which is formed from the nail bed distal to the lunula (Fig. 5.9B). A third view divides the nail bed into three zones, namely a proximal fertile matrix, an intermediate sterile bed and a distal fertile matrix which contributes a small quantity of material to the underside of the nail (Fig. 5.9C). An increased contribution from this distal matrix may account for some pathological conditions of the nail.

Electron microscope studies of the developing nail folds in 16–18 week human embryos[107] reveal two horizontal layers, the ventral and dorsal matrices (Fig. 5.9D). Cells at the apex of the fold show no signs of keratinization, but at about a quarter of the distance towards the cuticle keratinization starts. Both in the dorsal and ventral matrices cells first develop tonofibrils and later keratohyalin granules, and eventually transform into horny cells densely packed with keratin. The dorsal matrix plays a definite productive rôle in the formation of the embryonic nail plate. But beyond the level of the cuticle, the thickening ventral matrix becomes transformed into the nail bed which alone continues to contribute cells to the nail plate.

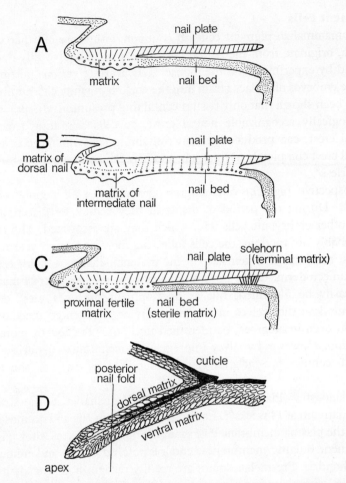

Fig. 5.10 A, B, C: diagrams illustrating possible methods of nail formation. *After* Samman[106].

A. Traditional view: the nail plate is formed solely from a single proximal matrix.

B. Hypothesis that ventral layer of nail plate is formed from the nail bed.

C. Third view that distal terminal matrix, which is separated from proximal matrix by intermediate sterile matrix, contributes some material to underside of nail.

D. Schematic depiction of nail in 18 week-old human embryo. The vertical line indicates the beginning of keratinization. (*After* Hashimoto, Gross, Nelson and Lever[107].)

*

Pigment cells

That mammalian pigment cells, in common with those of other verte-brates, originate from the neural crest has been unequivocally demon-strated by experiments on the mouse[10]. By grafting various portions of mouse embryos of a black strain into the coelom of white chick embryos, it has been shown that only tissues containing presumptive neural crest, histologically recognizable neural crest, or cells migrating from the neural crest, can produce melanin-containing cells; in the absence of neural crest completely normal skin and hair can be formed, but they are colourless[9].

Prospective pigment cells migrate gradually to all regions of the body[46]. During the period of dispersal they cannot be distinguished from other embryonic cells with which they are associated. Movement is probably not random, the cells following more or less predetermined routes, moving dorso-ventrally in the mesenchymal tissues adjacent to the skin ectoderm. In the early human embryo the neural crest material can easily be identified, but as development proceeds most of the elements lose themselves in the mesenchyme. Pigmented melanocytes cannot, even in Negroes, be identified until after the fourth month of gestation. However by silver impregnation techniques, dendritic cells said to contain pre-melanin can be identified as early as the tenth week.

Melanocytes (Plate 5.2) are readily distinguishable in the stratum germinativum at 14 weeks[55]. Lacking the tonofilaments and desmosomes along the plasma membrane characteristic of keratinocytes, they possess prominent bulging membranes, endoplasmic reticulum and numerous mitochondria. Pre-melanosomes are readily recognizable as rod-shaped granules with characteristic striations (Plate 5.9).

Mature melanocytes lose much of their power to move. If melano-cytes in the pigmented region are destroyed, the region lacks pigment for the remainder of the animal's life. However some powers of migra-tion are retained. The pigmented areas of piebald animals increase in size during life, and if a piece of pigmented skin is grafted into a colour-less area the phenomenon of pigment spread can be experimentally demonstrated. Various hypotheses to explain this have been put for-ward, including the proposal that pigment spread involves the infective colour transformation of Langerhans or other colourless dendritic cells into pigment-forming cells by contact with adjacent melanocytes. How-ever, all the known facts can be explained on the hypothesis that there is a differential migration of pigment cells and, at least for the spread of black pigmentation into white skin, this is the currently accepted explanation[108].

Langerhans cells

Langerhans cells[49–50] are dendritic cells of a form similar to melanocytes but they contain no pigment and they do not darken when incubated with dihydroxyphenylalanine. Langerhans[53] demonstrated their existence in epidermis by gold chloride staining and they can be shown in non-pigmented skin from guinea pigs by methylene blue. DOPA-negative dendritic cells which are ATPase positive occur in human epidermis and have been shown in a number of other mammals. Some authors believe that ATPase activity is not specific for Langerhans cells but is shared by effete melanocytes. Langerhans cells can, however, be identified under the electron microscope, where they differ from melanocytes in possessing granules with rod-like profiles bearing a striated line and often blown out at one end to give a racquet shape[54].

Some authors have regarded Langerhans cells as neural elements, Schwann cells or interstitial cells which have migrated from the dermis to the epidermis; others have proposed they are effete melanocytes. Evidence on their origin is provided by two pieces of embryological evidence.

Firstly, Langerhans cells have been shown to be present in the basal and lower intermediate layers of the fetal epidermis as early as 14 weeks[55]. They were readily distinguished from keratinocytes by the absence of tonofilaments and desmosomes and by the characteristic granules which were similar to those of the adults, except that there was no blowing out of the boundary membranes. Thus Langerhans cells occur in the human embryo at a time when melanocytes are still immature and functionless.

Secondly, Langerhans cells have been demonstrated in grafts of embryonic mouse skin without any neural crest component[56]. Putative neural crest-free grafts from limb buds of 10 day old mouse embryos and control neural crest-containing material derived from somite regions were transplanted to spleens of histocompatible recipient adult male mice, and removed after 3 or 5 weeks.

The neural crest containing grafts contained both identifiable melanogenic melanocytes and other non-keratinocytes, that is Langerhans cells or possibly amelanotic melanocytes. The neural crest-free grafts, on the one hand, lacked not only melanocytes, but other neural crest derivatives, neuroaxons and Schwann cells. On the other hand, they contained non-keratinocytes in which typical Langerhans granules could be seen in electron micrographs.

Merkel cells

Merkel gave the name Tastzellen to cells near the base of the rete peg in the snout skin of the mole[44]. After fixation in osmium tetroxide, such cells appeared large and vesicular and, since intra-epidermal neurites with expanded tips (Merkel's discs) were often found adjacent to them, Merkel believed the cells to be transducers of physical stimuli. Similar cells have been described in the snout of the opossum[109]. They also occur in the tactile corpuscles, or Haarscheiben, which are specialized epidermal pads that have been described in man[110, 111], sheep and a number of other mammals[112–114]. These epidermal pads generally occupy a fixed position in relation to certain hair follicles, and an atypical guard hair follicle with which they are associated, known as the tylotrich follicle, is possibly common to most mammals.

Merkel cells have been revealed by electron microscopy in the basal layer of the embryonic epidermis of both the sheep[115] and man[116]. They are characterized by their content of many small membrane-limited granules. In the human embryo they have been identified at 16 weeks in the primitive mesenchyme, where they are surrounded by Schwann sheaths, in epidermis where the sheath becomes retracted, and in nail matrices.

On the grounds that it has desmosomes, some authors have considered the Merkel cell to be a modified keratinocyte. In the sheep, for example, the Merkel cells appear initially as isolated structures not associated with neurites, though they may become innervated in older fetuses. However intra-epidermal Merkel cells differ significantly from adjacent keratino-cytes at all stages of development, and there are no transitional or inter-mediate forms. There is general agreement that it is unlikely that the cell is of mesenchymal origin. It seems more likely that the cells originate from neuro-ectoderm and migrate into the epidermis.

Dermal constituents

Mesenchymal cells from somatopleure as well as somitic mesoderm give rise to a range of blood and connective tissue cells, including the fibro-blasts and mast cells of the dermis and the fat cells of the subcutis.

The human embryonic dermis is at first very cellular, and in the second month the dermis and subcutis are not distinguishable from each other. Fibrillar components shortly make their appearance, and regular bundles of reticulin and collagen fibres are evident by the end of the third month. Later, the papillary and reticular layers become distinct and at the fifth month, the connective tissue sheaths are formed around the hair follicles. Elastic fibres probably develop somewhat later. Beneath the dermis is a looser tissue characterized by fat islands which begin to form in definite places.

At first, the undersurface of the epidermis is smooth, but during the fourth month, at the same time as the hair follicle starts to develop, it becomes irregular.

Touch pads become recognizable on the hands and fingers, and on the feet and toes, by the sixth week, and reach their greatest development at the fifteenth week. After this they flatten and become indistinct. It is these areas, however, which determine the pattern of dermatoglyphs – the systems of papillary ridges – which take their place.

References

1. RAWLES, M. E. (1955), 'Skin and its derivatives'. In *Analysis of Development*. Ed. Willier, B. H., Weiss, P. A. and Hamburger, V., 499–519, Saunders: Philadelphia and London.
2. WARREN, A. (1959), *Textbook of Comparative Histology*, Oxford University Press: New York.
3. GABE, M. (1967), 'Le tégument et ses annexes'. In *Traité de Zoologie*. Ed. Grasse, P., Vol. XVI: 1 Mammiféres, 1–233, Masson et Cie: Paris.
4. MADERSON, P. F. A. (1972), 'The vertebrate integument', *American Zoologist*, **12**, 13–171.
5. SPEARMAN, R. I. C. (1973), '*The Integument*', *A Textbook of Skin Biology*. Cambridge University Press.
6. VOGT, W. (1925), 'Gestaltungsanalyse am Amphibienkeim mit örtlicher Vitalfärbung. I. Methodik', *Roux Archiv für Entwicklungsmechanik der Organismen*, **106**, 542–610.
7. VOGT, W. (1929), 'Gestaltungsanalyse am Amphibienkeim mit örtlicher Vitalfärbung. II. Gastrulation und Mesodermbildung bei Urodelen und Anuren', *Roux Archiv für Entwicklungsmechanik der Organismen*, **120**, 384–706.
8. PASTEELS, J. (1937), 'Études sur la gastrulation des vertébrés méroblastiques. III. Oiseaux', *Archives de biologie*, **48**, 381–488.
9. RAWLES, MARY E. (1948), 'Origin of melanophores and their role in development of color patterns in vertebrates', *Physiological Reviews*, **28**, 383–408.
10. RAWLES, MARY E. (1947), 'Origin of pigment cells from the neural crest in the mouse embryo', *Physiological Zoology*, **20**, 248–266.
11. SPEARMAN, R. I. C. (1968), 'Epidermal keratinization in the salamander

and a comparison with other amphibia', *Journal of Morphology*, **125**, 129–144.

12. MADERSON, P. F. A. (1965), 'The structure and development of the squamate epidermis'. In *Biology of the Skin and Hair Growth*. Eds. Lyne, A. G. and Short, B. F., 129–153. Angus and Robertson: Sydney.

13. RAWLES, MARY E. (1960), 'The integumentary system'. In *Biology and Comparative Physiology of Birds*. Ed. Marshall, A. J., Vol. I, 189–240, Academic Press: New York.

14. SPEARMAN, R. I. C. (1971), 'Integumentary System'. In *Physiology and Biochemistry of the Domestic Fowl*. Eds. Freeman, B. H. and Bell, D. Vol. 2, 603–620, Academic Press: London.

15. MONTAGNA, W. and ELLIS, R. A. (1958), (Eds.). *The Biology of Hair Growth*, Academic Press: New York.

16. EBLING, F. J. (1964), 'The hair follicle'. In *Progress in the Biological Sciences in Relation to Dermatology 2*. Eds. Rook, A. and Champion, R. H., 303–323, Cambridge University Press.

17. SPEARMAN, R. I. C. (1964), 'The evolution of mammalian keratinized structures'. In *The Mammalian Epidermis and its Derivatives*, No. 12, Symposia of the Zoological Society of London. Ed. Ebling, F. J., 67–81, Academic Press: London.

18. MADERSON, P. F. A. (1972), 'When? Why? and How?: Some speculations on the evolution of the vertebrate integument', *American Zoologist*, **12**, 159–171.

19. RAWLES, MARY E. (1965), 'Tissue interactions in the morphogenesis of the feather'. In *Biology of the Skin and Hair Growth*. Eds. Lyne, A. G. and Short, B. F., 105–128. Angus and Robertson: Sydney.

20. PINKUS, H. (1958), 'Embryology of hair'. In *The Biology of Hair Growth*. Eds. Montagna, W. and Ellis, R. A., 1–32, Academic Press: New York.

21. QUAY, W. B. (1972), 'Integument and the environment: glandular composition, function and evolution', *American Zoologist*, **12**, 95–108.

22. MADERSON, P. F. A. (1968), 'The epidermal glands of *Lygodactylus* (Gekkonidae, Lacertilia)', *Breviora*, **288**, 1–35.

23. MADERSON, P. F. A. and CHIU, K. W. (1970), 'Epidermal glands in gekkonid lizards: evolution and phylogeny', *Herpetologica*, **26**, No. 2, 233–238.

24. MONTAGNA, W. (1963), 'Comparative aspects of sebaceous glands'. In *Advances in Biology of Skin*, Vol. 4, *The Sebaceous Glands*. Eds. Montagna, W., Ellis, R. A. and Silver, A. F., 32–45. Pergamon Press, Oxford.

25. WEINER, J. S. and HELLMANN, K. (1960), 'The sweat glands', *Biological Reviews*, **35**, 141–186.

26. EBLING, F. J. and JOHNSON, E. (1959), 'Hair growth and its relation to vascular supply in rotated skin grafts and transposed flaps in the albino rat', *Journal of Embryology and Experimental Morphology*, **7**, 417–430.

27. BODENSTEIN, D. (1952), 'Studies on the development of the dorsal fin in amphibians', *Journal of Experimental Zoology*, **120**, 213–245.

28. MADERSON, P. F. A. (1965), 'Histological changes in the epidermis of snakes during the sloughing cycle', *Journal of Zoology*, **146**, 98–113.

29. MADERSON, P. F. A., CHIU, K. W. and PHILLIPS, J. G. (1970), 'Changes in the epidermal histology during the sloughing cycle in the rat snake *Ptyas korros* Schlegel, with correlated observations on the thyroid gland', *Biological Bulletin Marine Biological Laboratory, Woods Hole, Mass.*, **139**, 304–312.

30. MADERSON, P. F. A. (1967), 'The histology of the escutcheon scales of *Gonatodes* (Gekkonidae) with a comment on the squamate sloughing cycle', *Copeia*, **4**, 743–752.

31. MADERSON, P. F. A. and LICHT, P. (1967), 'Epidermal morphology and sloughing frequency in normal and prolactin treated *Anolis carolinensis*', *Journal of Morphology*, **123**, 157–171.

32. FLAXMAN, B. A., MADERSON, P. F. A., ROTH, S. I. and SZABO, G. (1968), 'Ultrastructural data on the mature and immature epidermal generations of lizards', *American Zoologist*, **8**, 787.

33. LILLYWHITE, H. B. and MADERSON, P. F. A. (1968), 'Histological changes in the epidermis of the subdigital lamellae of *Anolis carolinensis* during the shedding cycle', *Journal of Morphology*, **125**, 379–401.

34. MADERSON, P. F. A., FLAXMAN, B. A., ROTH, S. I. and SZABO, G. (1972), 'Ultrastructural contributions to the identification of cell types in the lizard epidermal generation', *Journal of Morphology*, **136**, 191–210.

35. LILLIE, F. R. and WANG, H. (1941), 'Physiology of development of the feather. V. Experimental morphogenesis', *Physiological Zoölogy*, **14**, 103–133.

36. WANG, H. (1943), 'The morphogenetic functions of the epidermal and dermal components of the papilla in feather regeneration', *Physiological Zoölogy*, **16**, 325–350.

37. LILLIE, F. R. and WANG, H. (1944), 'Physiology of development of the feather. VII. An experimental study of induction', *Physiological Zoölogy*, **17**, 1–31.

38. CAIRNS, J. M. and SAUNDERS, J. W., Jr. (1954), 'The influence of embryonic mesoderm on the regional specification of epidermal derivatives in the chick', *Journal of Experimental Zoology*, **127**, 221–248.

39. RAWLES, MARY E. (1963), 'Tissue interactions in scale and feather development as studied in dermal-epidermal recombinations', *Journal of Embryology and Experimental Morphology*, **11**, 765–789.

40. MONTAGNA, W. (1962), *The Structure and Function of Skin*, 2nd Edn., Academic Press: New York.

41. MONTAGNA, W. and LOBITZ, W. C. (Eds.) (1964), *The Epidermis*, Academic Press, New York and London.

42. PINKUS, F. (1910), 'The development of the integument'. In *Manual of Human Embryology*. Eds. Keibel F. and Mali F. P. Vol. I, 243–291 Lippincott: Philadelphia.

43. ROOK, A., WILKINSON, D. S. and EBLING, F. J. G. (1972), *Textbook of Dermatology*, 2nd Edn. Ch. 53 and 54, Blackwell Scientific Publications: Oxford.

44. MERKEL, F. (1875), 'Tastzellen und Tastkörperchen bei den Hausthieren

und beim Menschen', *Archiv für mikroskopische Anatomie und Entwicklungs-mechanik*, **11**, 636–652.

45. BILLINGHAM, R. E. and SILVERS, W. K. (1960), 'The melanocytes of mammals', *Quarterly Review of Biology*, **35**, 1–40.

46. BOYD, J. D. (1960), 'The embryology and comparative anatomy of the melanocyte', In *Progress in the Biological Sciences in Relation to Dermatology*. Ed. Rook, A., 3–14, Cambridge University Press.

47. RILEY, V. and FORTNER, J. G. (Eds.) (1963), 'The pigment cell: molecular, biological, and clinical aspects', *Annals of the New York Academy of Sciences*, **100**, 1–1123.

48. MONTAGNA, W. and HU, F. (Eds.) (1966), *Advances in Biology of Skin*, Vol. VIII, *The pigmentary system*, Publication No. 183 from the Oregon Regional Primate Research Center. Pergamon Press: Oxford and New York.

49. ZELICKSON, A. S. (1965), 'The Langerhans cell', *Journal of Investigative Dermatology*, **44**, 201–212.

50. BREATHNACH, A. S. and WYLLIE, L. M. A. (1967), 'The problem of the Langerhans cells'. In *Advances in Biology of Skin*, Vol. VIII, *The Pigmentary System*. Eds. Montagna, W. and Hu, F., 97–113, Pergamon Press: Oxford.

51. BREATHNACH, A. S. (1968), 'The epidermal Langerhans cell', *British Journal of Dermatology*, **80**, 688–689.

52. NIEBAUER, G. (1968), *Dendritic cells of human skin*, Karger: Basel.

53. LANGERHANS, P. (1868), 'Über die Nerven der menschlichen Haut', *Virchows Archiv für pathologische Anatomie und Physiologie und für klinische Medizin*, **44**, 325–337.

54. WOLFF, K. (1967), 'The fine structure of the Langerhans cell granule', *Journal of Cell Biology*, **35**, 468–473.

55. BREATHNACH, A. S. and WYLLIE, L. M. (1965), 'Electron microscopy of melanocytes and Langerhans cells in human fetal epidermis at fourteen weeks', *Journal of Investigative Dermatology*, **44**, 51–60.

56. BREATHNACH, A. S., SILVERS, W. K., SMITH, J. and HEYNER, S. (1968), 'Langerhans cells in mouse skin experimentally deprived of its neural crest component', *Journal of Investigative Dermatology*, **50**, 147–160.

57. TARNOWSKI, W. M. and HASHIMOTO, K. (1967), 'Langerhans' cell granules in histiocytosis X. The epidermal Langerhans' cell as a macrophage', *Archives of Dermatology and Syphilology*, **96**, 298–304.

58. MONTAGNA, W., BENTLEY, J. P. and DOBSON, R. L. (1968), *Advances in Biology of Skin*, Vol. X, *The dermis*, Publication No. 396, from the Oregon Regional Primate Research Center. Appleton-Century-Crofts: New York.

59. SMITH, D. E. (1963), 'Electron microscopy of normal mast cells under experimental conditions', *Annals of the New York Academy of Sciences*, **103**, 40–52.

60. BREATHNACH, A. S. (1971), 'Embryology of human skin. A review of ultrastructural studies', *Journal of Investigative Dermatology*, **57**, 133–143.

61. BREATHNACH, A. S. and ROBINS, J. (1969), 'Ultrastructural features of epidermis of a 14 mm. (6 weeks) human embryo', *British Journal of Dermatology*, **81**, 504–516.

62. BREATHNACH, A. S. and WYLLIE, L. M. (1965), 'Fine structure of cells forming the surface layer of the epidermis in human fetuses at fourteen and twelve weeks', *Journal of Investigative Dermatology*, **45**, 179–189.

63. HASHIMOTO, K., GROSS, B. G., DiBELLA, R. J. and LEVER, W. F. (1966), 'The ultrastructure of the skin of human embryos. IV. The epidermis', *Journal of Investigative Dermatology*, **47**, 317–335.

64. HOYES, A. D. (1968), 'Electron microscopy of the surface layer (periderm) of human foetal skin', *Journal of Anatomy*, **103**, 321–336.

65. HOYES, A. D. (1968), 'Ultrastructure of the cells of the amniotic fluid', *Journal of Obstetrics and Gynaecology of the British Commonwealth*, **75**, 164–171.

66. SERRI, F. and HUBER, W. M. (1963), 'The development of sebaceous glands in man'. In *Advances in Biology of Skin*, Vol. 4, *Sebaceous Glands*. Eds. Montagna, W., Ellis, R. A. and Silver, A. F., 1–18, Pergamon Press: Oxford.

67. BRODY, I. (1964), 'Cytoplasmic components in the psoriatic horny layers with special reference to electron-microscopic findings'. In *The Epidermis*. Eds. Montagna, W. and Lobitz, W. C., Jnr., 551–572, Academic Press: New York.

68. BRODY, I. (1964), 'Observations on the fine structure of the horny layer in the normal human epidermis', *Journal of Investigative Dermatology*, **42**, 27–31.

69. ROTH, S. I. and CLARK, W. H. (1964), 'Ultrastructural evidence related to the mechanism of keratin synthesis'. In *The Epidermis*. Eds. Montagna, W. and Lobitz, W. C., Jnr., Academic Press: New York.

70. ODLAND, G. F. (1964), 'Tonofilaments and keratohyalin'. In *The Epidermis*. Eds. Montagna, W. and Lobitz, W. C., Jnr., 237–249, Academic Press: New York.

71. MERCER, E. H. (1961), *Keratin and keratinisation*, Pergamon Press: Oxford.

72. JARRETT, A. (1973), '*The physiology and pathophysiology of the skin*', Vol. 1. *The epidermis*, Academic Press: London and New York.

73. GIBBS, H. F. (1938), 'A study of the development of the skin and hair of the Australian oppossum, *Trichosurus vulpecula*', *Proceedings of the Zoological Society of London*, **108**, 611–648.

74. FRASER, D. A. (1928), 'The development of the skin of the back of the albino rat until the eruption of the first hairs', *Anatomical Record*, **38**, 203–223.

75. HARDY, M. H. (1969), 'The differentiation of hair follicles and hairs in organ culture'. In *Advances in Biology of Skin*, Vol. IX, *Hair Growth*. Eds. Montagna, W. and Dobson, R. L., 35–60, Pergamon Press: Oxford.

76. LYNE, A. G. and HEIDEMAN, M. J. (1959), 'The pre-natal development of skin and hair in cattle (*Bos taurus* L.)', *Australian Journal of Biological Sciences*, **12**, 72–95.

77. RYDER, M. L. and STEPHENSON, S. K. (1968), *Wool Growth*, Academic Press: London and New York.
78. STÖHR, P. (1903–1904), 'Entwicklungsgeschichte des menschlichen Wollhaares', *Anatomische Hefte Abt. I.* **23**, 1–66.
79. PINKUS, H. (1958), 'Embryology of hair'. In *The Biology of Hair Growth*. Eds. Montagna, W. and Ellis, R. E., 1–32, Academic Press: New York and London.
80. BELL, M. (1969), 'The ultrastructure of differentiating hair follicles in fetal rhesus monkeys'. In *Advances in Biology of Skin*, Vol. IX, *Hair Growth*. Eds. Montagna, W. and Dobson, R. L., 61–81, Pergamon Press: Oxford.
81. BREATHNACH, A. S. and SMITH, J. (1968), 'Fine structure of the early hair germ and dermal papilla in the human foetus', *Journal of Anatomy*, **102**, 511–526.
82. ROBINS, E. J. and BREATHNACH, A. S. (1969), 'Fine structure of the human foetal hair follicle at hair-peg and early bulbous-peg stages of development', *Journal of Anatomy*, **104**, 553–569.
83. HASHIMOTO, K. (1970), 'The ultrastructure of the skin of human embryos. V. The hair germ and perifollicular mesenchymal cells. Hair germ-mesenchymal interaction', *British Journal of Dermatology*, **83**, 167–175.
84. KOLLAR, E. J. (1966), 'An *in vitro* study of hair and vibrissae development in embryonic mouse skin', *Journal of Investigative Dermatology*, **46**, 254–262.
85. KOLLAR, E. J. (1972), 'The development of the integument: Spatial, temporal and phylogenetic Factors', *American Zoologist*, **12**, 125–135.
86. COHEN, J. (1969), 'Interactions in the skin', *British Journal of Dermatology*, **81**, Supplement 3, 45–56.
87. COHEN, J. (1969), 'Dermis, epidermis and dermal papillae interacting'. In *Advances in Biology of Skin*, Vol. IX, *Hair Growth*. Eds. Montagna, W. and Dobson, R. L., 1–18. Pergamon Press: Oxford.
88. OLIVER, R. F. (1969), 'The vibrissa dermal papilla and its influence on epidermal tissues', *British Journal of Dermatology*, **81**, Supplement 3, 55–65.
89. OLIVER, R. F. (1969), 'Regeneration of the dermal papilla and its influence on whisker growth'. In *Advances in Biology of Skin*, Vol. IX, *Hair Growth*. Eds. Montagna, W. and Dobson, R. L., 19–33, Pergamon Press: Oxford.
90. EBLING, F. J. (1965), 'Comparative and evolutionary aspects of hair replacement'. In *The Comparative Physiology and Pathology of The Skin*. Eds. Rook, A. J. and Walton, G. S., 87–102, Blackwell Scientific Publications: Oxford.
91. LING, J. K. (1972), 'Adaptive functions of vertebrate molting cycles', *American Zoologist*, **12**, 77–93.
92. JACKSON, D. and EBLING, F. J. (1972), 'The activity of hair follicles and their response to oestradiol in the guinea-pig *Cavia porcellus* L.', *Journal of Anatomy*, **111**, 2, 303–316.
93. EBLING, F. J. and JOHNSON, E. (1961), 'Systemic influence on activity of hair follicles in skin homografts', *Journal of Embryology and Experimental Morphology*, **9**, 285–293.

94. EBLING, F. J. and JOHNSON, E. (1964), 'The control of hair growth', *Symposia of the Zoological Society of London*, **12**, 97–130.

95. JOHNSON, E. and EBLING, F. J. (1964), 'The effect of plucking hairs during different phases of the follicular cycle', *Journal of Embryology and Experimental Morphology*, **12**, 465–474.

96. EBLING, F. J. and HALE, P. A. (1970), 'The control of the mammalian moult', *Memoirs of the Society for Endocrinology*, **18**, 215–237.

97. STRAUSS, J. S. and EBLING, F. J. (1970), 'Control and function of skin glands in mammals,' *Memoirs of the Society for Endocrinology*, **18**, 341–371.

98. EBLING, F. J. (1970), 'Steroid hormones and sebaceous secretion'. In *Advances in Steroids*, Vol. 2. Ed. Briggs, M. H., 1–39, Academic Press: London.

99. EBLING, F. J. (1972), 'The response of the cutaneous glands to steroids', *General and Comparative Endocrinology*, Supplement 3, 228–237.

100. EBLING, F. J. (1974), 'Hormonal control and methods of measuring sebaceous gland activity'. In *Advance, in Biology of Skins* Vol. XIV, *Sebaceous glands and acne vulgaris*. Eds. Montagna, W., Bell, M. and Strauss, J. S. *Journal of Investigative Dermatology*, **62**, 161–171.

101. MONTAGNA, W. (1959), 'Histology and cytochemistry of human skin. XIX. The development and fate of the axillary organ', *Journal of Investigative Dermatology*, **33**, 151–161.

102. WALES, N. A. M. and EBLING, F. J. (1971), 'The control of apocrine glands of the rabbit by steroid hormones', *Journal of Endocrinology*, **51**, 763–771.

103. COWIE, A. T. and TINDAL, J. S. (1971), *The physiology of lactation*, Edward Arnold: London.

104. HASHIMOTO, K., GROSS, B. G. and LEVER, W. F. (1966), 'The ultrastructure of the skin of human embryos. I. The intraepidermal eccrine sweat duct', *Journal of Investigative Dermatology*, **45**, 139–151.

105. HASHIMOTO, K., GROSS, B. G. and LEVER, W. F. (1965), 'The ultrastructure of human embryo skin. II. The formation of the intradermal portion of the eccrine sweat duct and the secretory segment during the first half of embryonic life', *Journal of Investigative Dermatology*, **46**, 513–529.

106. SAMMAN, P. D. (1972), 'The nails'. In *Textbook of Dermatology*. Eds. Rook, A., Wilkinson, D. S. and Ebling, F. J. G., Vol. 2, 1642–1671, Blackwell Scientific Publications: Oxford.

107. HASHIMOTO, K., GROSS, B. G., NELSON, R. and LEVER, W. F. (1966), 'The ultrastructure of the skin of human embryos. III. The formation of the nail in 16–18 weeks old embryos', *Journal of Investigative Dermatology*, **47**, 205–217.

108. BILLINGHAM, R. E. and SILVERS, W. J. (1963), 'Further studies on the phenomenon of pigment spread in guinea pigs' skin', *Annals of the New York Academy of Sciences*, **100**, 348–363.

109. MUNGER, B. L. (1965), 'The intraepidermal innervation of the snout skin of the opossum', *Journal of Cell Biology*, **26**, 79–97.

110. MUSTAKALLIO, K. K. and KIISTALA, U. (1967), 'Electron microscopy of Merkel's Tastzelle', *Acta dermato-venereologica Stockh.*, **47**, 323–326.

111. SMITH, K. R. (1970), 'The ultrastructure of the human Haarscheibe and Merkel cell', *Journal of Investigative Dermatology*, **54**, 150–159.

112. STRAILE, W. E. (1960), 'Sensory hair follicles in mammalian skin: the tylotrich follicle', *American Journal of Anatomy*, **106**, 133–147.

113. PATRIZI, G. and MUNGER, B. L. (1966), 'The ultrastructure and innervation of rat vibrissae', *Journal of Comparative Neurology*, **126**, 423–436.

114. MANN, S. J. (1969), 'The tylotrich follicles as a marker system in skin'. In *Advances in Biology of Skin*, Vol. IX, *Hair Growth*. Eds. Montagna, W. and Dobson, R. L., 399–418, Pergamon Press: Oxford.

115. LYNE, A. G. and HOLLIS, D. E. (1971), 'Merkel cells in sheep epidermis during fetal development', *Journal of Ultrastructure Research*, **34**, 464–472.

116. HASHIMOTO, K. (1972), 'The ultrastructure of the skin of human embryos X. Merkel tactile cells in the finger and nail', *Journal of Anatomy*, **111**, 99–120.

6

Differentiation and Growth of Cells in the Gonads

BERNARD GONDOS

Introduction

The gonads, ovary and testis alike, contain three principal cell types: *germ cells*, involved in the formation of gametes; *supporting cells*, important in maintaining germ cell nutrition and maturation; and *stromal cells*, capable of differentiating into hormone-secreting cells. The three types of tissues undergo specialized patterns of differentiation and growth, beginning in the embryo and continuing into adulthood, with complex changes affecting the gonadal cells at many different stages of pre-natal and post-natal development. Understanding of the cellular changes requires separate consideration of the different cell types; at the same time, the relationship of each cell type to the overall gonadal structure and function must be kept in mind.

Development of ovary and testis begin in a similar manner, with formation of an indifferent gonadal primordium. After a short time, sexual differentiation occurs, bringing about specific patterns of cellular differentiation. The patterns of differentiation have fundamental similarities in the ovary and testis, but there are also sharp divergences related to the differing patterns of gamete formation and hormonal activity in the two sexes.

The present chapter is divided into five parts: the opening sections describe the indifferent stage of development (p. 170) and the stage of sexual differentiation (p. 172); the main sections deal with the development of the testis (p. 173) and the development of the ovary (p. 185) a closing section summarizes the principal similarities and differences in ovarian and testicular development (p. 195). The material presented generally relates to mammals, but in most major aspects applies to other vertebrates as well. Those interested in special aspects of gonadal development in non-mammalian vertebrates and invertebrates should consult additional references indicated at the end of the chapter.

The Indifferent Gonad

The primordia of the gonads arise as bilateral thickenings of mesodermal tissue, the *genital ridges*, on the mesial surface of the mesonephros. Gonadal outgrowths, composed of an outer layer of epithelium and underlying mesenchyme, gradually enlarge, projecting into the coelomic cavity. During this time, germ cells migrate to the genital ridges from the region of the yolk sac[1]. The germ cells become situated in the

Table 6.1 *Origin and characteristics of gonadal cell precursors*

Cell type	Origin	Precursor cell	Characteristics
Germ cells	extragonadal	primitive germ cell	Large, round to oval, amoeboid, single, alkaline phosphatase and PAS-positive, eccentric nucleus, prominent nucleolus, spherical mitochondria
Supporting cells	coelomic epithelium	epithelial cell	Elongated, irregular nucleus, many ribosomes, rod-shaped mitochondria, closely associated with germ cells
Stromal cells	gonadal mesenchyme	mesenchymal cell	Elongated, irregular nucleus, rough endoplasmic reticulum, rod-shaped mitochondria, loosely arranged.

epithelial layer on the surface of the gonadal outgrowths. Early investigators assumed that germ cells were formed within the epithelium, and therefore named it the *germinal epithelium*. Since it is now clear that germ cells have an extragonadal origin, alternate terms such as *surface epithelium*, *coelomic epithelium*, and *covering epithelium* have come into use to describe the outer gonadal layer. There is no objection to the term germinal epithelium, if it is understood that germ cells present are there as a result of migration rather than *de novo* origin. The indifferent gonad, then, contains three cellular elements – *germ cells* derived from an extragonadal site, *epithelial cells* present on the gonadal surface, and *mesenchymal cells* located in the gonadal stroma (Table 6.1).

Germ cells

The germ cells of the indifferent gonad are large oval cells which can be selectively stained for alkaline phosphatase[2]. After migrating to the gonad, they are at first located within the surface epithelium, then subsequently move into the gonadal parenchyma. They are generally arranged singly surrounded by closely attached epithelial cells which move with them into the underlying parenchyma. Increase in the number of germ cells occurs during this period due to continuing migration and mitotic proliferation[3]. Histologic sections reveal large numbers of dividing germ cells, particularly within the surface epithelium. The number of germ cells increases from less than a hundred to many thousand by the end of this period[4]. The early germ cells, designated *primitive germ cells* (Plate 6.1), have an eccentric nucleus and oval shape with irregular cytoplasmic projections suggesting amoeboid activity[5]. The nucleus contains finely distributed chromatin and a prominent nucleolus. Cytoplasm includes abundant glycogen producing a positive PAS stain[6], but relatively few other organelles. The latter consist of scattered ribosomes, occasional lipid droplets and small numbers of spherical mitochondria. The spherical shape of mitochondria is a distinguishing feature of germ cells throughout their development. Arrangement of germ cells in pairs is occasionally seen toward the end of this stage of development.

Epithelial cells

The portion of the coelomic epithelium covering the developing gonad is of the usual mesothelial type. The only difference from the coelomic lining elsewhere is the presence of germ cells interspersed among the epithelial cells. The germ cells are easily distinguished from the epithelial cells by their large size, round to oval shape, pale staining cytoplasm and spherical mitochondria[7]. The epithelial cells are elongated, with irregular nuclei, dark staining cytoplasm and numerous ribosomes and rod-shaped mitochondria (Plate 6.2). The epithelial cells located adjacent to germ cells bend around the germ cells to which they are closely attached. No transitional cells between germ cells and epithelial cells have been observed.

Mesenchymal cells

Contained within the primitive gonadal stroma are loosely arranged elongated cells representing undifferentiated mesenchymal elements. Large spherical germ cells which have moved into the mesenchymal region are interspersed among the undifferentiated mesenchyme. Surrounding the germ cells are elongated cells derived from the surface epithelium[8]. At this stage, the mesenchymal cells and the epithelial cells are difficult to distinguish on the basis of cytologic characteristics (Fig. 6.1). The epithelial cells can be identified by the manner in which they curve around the surface of the germ cells, with adjacent cell membranes in close apposition. The mesenchymal cells remain undifferentiated, lacking cytoplasmic structures associated with hormonal activity at this stage of development.

Sexual Differentiation

Sex differentiation is genetically determined at the time of fertilization. In the presence of a Y chromosome (XY), testes will develop; in the absence of a Y chromosome (XX), ovaries will develop. The nature of the sex-determining genes and the manner in which they control gonadal differentiation are not known[9, 10]. Other than the presence of a peripheral nuclear chromatin mass (Barr body) in the cells of females[11], the first morphologic evidence of sexual differentiation is gonadal sex differentiation, which precedes sex-specific differentiation of other structures in the reproductive system.

The initial changes distinguishing male and female gonads are architectural rather than cellular. The testis, which differentiates earlier than the ovary, is identified by the formation of a clear separation between the surface epithelium and the underlying parenchyma by a layer of loose connective tissue, the *tunica albuginea* (Plate 6.3). At the same time, groups of germ cells and supporting cells that have moved into the gonadal parenchyma from the surface epithelium become demarcated from the surrounding mesenchyme by a basal lamina, thus forming *sex cords* which gradually take on the appearance of tubules[12]. In the female gonad during the corresponding time period, clusters of germ cells and supporting cells extend into the gonadal parenchyma (Plate 6.4), but demarcation from the surface epithelium and surrounding mesenchyme is delayed, the tunica albuginea and mesenchyme remaining poorly developed[13].

Whether gonadal sex differentiation depends on antagonistic inductive substances responsible for male and female differentiation[14] or on hormonal regulators present only in males[15] is not clear. Evidence for the former is the predominant development of the cortex in the ovary and the medulla in the testis. Evidence for the latter includes the following observations: (1) testicular differentiation consistently precedes ovarian differentiation; (2) in cattle in which twin fetuses have a common placental circulation, the female develops a rudimentary testis (free-martin), while the male develops normally[16]; (3) sexual differentiation of the reproductive tracts is determined by the presence or absence of substances produced by the testis[17]. To date, neither male hormones nor other inductive substances responsible for gonadal sex differentiation have been identified.

Sexual differentiation results in the division of the gonad into two main compartments: one contains groups of germ cells and supporting cells (sex cords), and the other consists of mesenchymal elements, or stromal cells, distributed around and between the germ cell-supporting cell groups. In the testis, the sex cords will eventually develop into *seminiferous tubules*, while in the ovary the cell groups are broken up into *follicles*. The seminiferous tubule and the follicle are the basic units of gonadal structure involved in gamete formation. The germ cells ultimately give rise to mature gametes, and the supporting cells develop into Sertoli cells in the testis and granulosa cells in the ovary (Table 6.2). The stromal cells are the source of endocrine and other specialized elements. At the time of testicular differentiation there are no indications of sex differences in the appearance of the major cell types[18, 19]. Only in terms of their topographical distribution is there evidence of divergence of male and female germ cells.

Testis

Testicular differentiation involves two main processes: maturation of germ cells leading to the production of spermatozoa, and differentiation of endocrine cells required for the secretion of male hormones. The early development of the testis is remarkable in that endocrine secreting cells appear in the stroma shortly after sexual differentiation takes place. These cells play an important rôle in the development of the reproductive tracts[20]. Spermatogenesis does not begin until sexual maturity is reached, in contrast to the situation in the ovary where oogenesis takes place during early development, but several changes do nevertheless occur in the sex cords and developing tubules of the fetal and postnatal testis involving both germ cells and supporting cells. These changes have an important influence in establishing the pattern of germ cell differentiation.

173

Table 6.2 *Gonadal cells present at different stages of development*

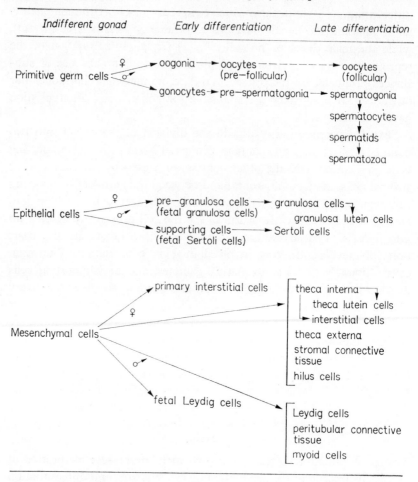

Indifferent gonad	Early differentiation	Late differentiation

Germ cells

Germ cell differentiation in the testis consists of two distinct stages:
(1) *pre-spermatogenesis* (pre-pubertal stage), beginning in the embryo
and directed toward the formation of a pool of spermatogonia at the
end of the prepubertal period; and (2) *spermatogenesis* (stage of sexual
maturation), consisting of repeated cycles of maturation of spermatogonia to spermatozoa, continuing throughout the adult period until
the time of senescence. In spite of conflicting views of early investigators,
it is now clear that there is continuity of development from primitive
germ cells present in the embryo to spermatogonia entering sperma-

togenesis in the adult[21, 22]. The cell types involved in the progressive differentiation of testicular germ cells are indicated in Table 6.2.

Pre-spermatogenesis

The changes involved in pre-spermatogenic germ cell maturation occur in association with certain structural modifications in the developing testis. As the primitive germ cells complete their migration to the presumptive testis, cords of cells extend from the surface epithelium into the underlying mesenchyme. The cords include both epithelial cells and germinal elements referred to as *gonocytes*[23]. The gonocytes are large round to oval cells, singly arranged (Plate 6.5). They are similar in appearance to primitive germ cells, indicating that the gonocytes are primitive germ cells that have moved into the sex cords. As the cords become separated from the surrounding mesenchyme by the formation of a thin basal lamina, the tunica albuginea intervenes between the cords and the surface epithelium. The germ cells now move to the periphery of the cords, the latter beginning to take on the appearance of primitive seminiferous tubules. At the same time, the tunica albuginea separating the surface epithelium from the primitive tubules becomes increasingly thickened. The germ cells which have moved to the tubular periphery are called *pre-spermatogonia*[24] (also fetal spermatogonia[25], immature spermatogonia[26], type II gonocytes[27, 28]). As prepubertal development proceeds, all of the gonocytes eventually move to the tubular periphery to become pre-spermatogonia (Plate 6.6).

At the time of initial sex cord formation gonocytes are randomly distributed. Some are central, while others are nearer the periphery. The cells have large spherical nuclei with varying sized nucleoli. Chromatin granules are sparse and evenly distributed. Cytoplasm contains glycogen and a few scattered spherical mitochondria, and is otherwise poorly developed. Initially, gonocytes are far outnumbered by supporting elements. With continuing mitotic activity, a progressive increase in the number of gonocytes takes place, eventually producing an approximately equal number of germinal and supporting elements. During this time, elongated projections of the gonocyte cytoplasm can be seen extending toward the basal lamina[29]. Contained within the projections are bundles of microfibrils[30] apparently aiding in the movement which establishes the germ cells in a peripheral location.

The cells located at the tubular periphery, the pre-spermatogonia, resemble spermatogonia of the adult, but lack certain features associated with complete maturation. Pre-spermatogonia differ from gonocytes in that they lack glycogen[30] and are generally arranged in pairs of similar appearing cells[31]. Pairing is a result of incomplete mitotic division resulting in the maintenance of bridges between daughter cells[32]. The process is similar to that occurring in the ovary, where repeated incomplete division leads to interconnection of groups of cells. In the testis, mitotic activity apparently ceases after a single such division, resulting in a paired arrangement. The cells fail to divide further, remaining in a prolonged interphase until the initiation of spermatogenesis[33]. During this period, increase in cytoplasmic volume occurs, due to an accumulation of mitochondria, Golgi elements and ribosomes, in preparation for further maturation. Cellular enlargement is associated with an increase in synthesis of DNA and nuclear protein[34]. Also during this period, there is extensive germ cell degeneration[35], producing a marked reduction in the number of germ cells. Mitosis does not resume until spermatogenesis begins. Shortly before the resumption of mitotic activity, pre-spermatogonia undergo several changes leading to their differentiation into *spermatogonia*. These include a decrease in cytoplasmic volume, nuclear condensation and enlargement of the nucleolus[36]. Due to extensive proliferation of the nucleolonema, the nucleolus comes to occupy a considerable portion of the nuclear mass (Plate 6.7). Nucleolar enlargement appears to be related to an increasing rate of RNA synthesis associated with spermatogonial maturation[37]. *In vitro* studies in the rat suggest that maturation of prespermatogonia to spermatogonia requires testosterone[38].

Spermatogenesis

With the appearance of spermatogonia, the pattern of spermatogenesis is established. There appear to be several factors responsible for the initiation of spermatogenesis, including body weight, testis weight and endocrine factors[39]. In some animals, the initial spermatogenic cycle is slightly shorter than later cycles[40], but the basic pattern of cellular growth and differentiation is established from the start[33]. The duration of spermatogenic cycles is constant for a particular species, and in most mammals is of the order of several weeks[41]. Co-ordinated cycles of maturation with typical cell associations are present throughout the seminiferous tubules, producing *waves of spermatogenesis*. As a result, production of spermatozoa from spermatogonia is maintained without interruption once spermatogenesis is initiated. The mechanisms involved in spermatogonial renewal, whereby continued maturation of spermatogonia takes place without exhausting the supply of stem cells, are not at all clear in spite of considerable speculation[42, 43].

Spermatogenesis can be divided into three main periods: (1) a period of multiplication of spermatogonia leading to the formation of *spermatocytes*; (2) a period in which spermatocytes undergo two meiotic divisions leading to the formation of *spermatids*; (3) a period of *spermiogenesis* in which spermatids are transformed into *spermatozoa*.

Spermatogonia are classified as type A and type B. Type A spermatogonia are the direct descendants of the fetal germ cells. They are smaller in size than either gonocytes or pre-spermatogonia. Spermatogonia are arranged at the tubular periphery as rows of cells with a slightly flattened oval shape, dense homogeneous nuclear material and a prominent nucleolus attached to the inner nuclear membrane. The nucleolus is larger than that of the gonocyte, but considerably smaller than that of the preceding pre-spermatogonial maturation stage. Spermatogonia are further distinguished from pre-spermatogonia by the presence of paired mitochondria with dense intermitochondrial granular material. Interconnection of adjacent spermatogonia results from formation of cytoplasmic bridges (Plate 6.8). Rows of spermatogonia are connected by multiple intercellular bridges resulting in the formation of extensive syncytial aggregations[44]. The multiple interconnections result from repeated incomplete mitotic divisions. Spermatogonia undergo several mitotic divisions, the number depending on the species[45].

The initial stages of spermatogenic development are associated with a high rate of RNA and DNA synthesis[31, 46]. RNA synthesis is higher in the type A than in the type B spermatogonia. An increase in heterochromatin accompanies the differentiation of type A to type B spermatogonia, as does an increase in the duration of DNA synthesis[24]. The mitotic chromosomes of the type B cells are larger and shorter than those of the A-type. Type B spermatogonia are small spherical cells with several crust-like aggregates of chromatin attached to the inner nuclear membrane (Plate 6.9). They are located either at the tubular periphery or just inside the type A cells. In an attempt to define the sequence of events in spermatogonial maturation more precisely, some authors have further sub-classified type A and B cells and an intermediate form (In) on the basis of nuclear characteristics, staining reaction and cell shape[47–49]. These features appear to be species dependent.

Spermatocytes are of two types, primary and secondary. *Primary spermatocytes*, produced following the final spermatogonial division, undergo the first meiotic division, which is the longer of the two divisions. The early (preleptotene) spermatocytes resemble type B spermatogonia[50]. Later stages of leptotene, zygotene, pachytene and diplotene are clearly distinguished by changes in cell size and chromatin arrangement. During the course of meiotic prophase, the spermatocytes undergo enlargement to several times the size of spermatogonia. By electron microscopy, tripartite chromatid complexes, referred to as synaptinemal complexes[51], can be seen in spermatocyte nuclei (Plate 6.10). The cells are spherical, with conspicuous aggregates of nuclear chromatin. During the prolonged first meiotic division, DNA synthesis has a long S phase[52]. Intense RNA synthesis is observed during meiotic prophase, but then drops off sharply[53].

Cytoplasm includes spherical mitochondria with a characteristic central clear zone, abundant endoplasmic reticulum and an increasingly complex Golgi apparatus. Phosphatase and oxidative activity are present in the cytoplasm[54]. The cells also have a positive PAS reaction due to the presence of glycoproteins[55].

The second meiotic division occurs in *secondary spermatocytes*. The duration of this division is very short, lasting several hours. RNA synthesis is either barely detectable[46] or absent[53]. The morphologic features of primary and secondary spermatocytes are similar.

The rôle of pituitary gonadotrophins in early stages of spermatogenesis has not been definitely established[56], although it has been noted that modification of gonadotrophin production affects synthesis of DNA by spermatogonia and primary spermatocytes[57] and completion of meiosis in primary spermatocytes[58]. This may be mediated through the production of androgens by the Leydig cells[59].

Spermatids are the haploid cells resulting from the second meiotic division (Plates 6.11 and 6.12). They undergo marked cellular changes, beginning as small spherical cells and ending up as greatly elongated spermatozoa. This process, known as spermiogenesis, requires the presence of testosterone and FSH[38]. Nuclear transformation from the spherical shape of the early spermatid to the elongated shape of the spermatozoon is associated with progressive aggregation of chromatin granules, resulting in an extremely dense homogeneous appearance. In spite of the increased condensation of nuclear chromatin, the amount of DNA per nucleus remains relatively constant during spermiogenesis[60]. Cytoplasmic changes include formation of the *acrosome*, elaboration of the *flagellum* and elimination of a portion of the cytoplasm.

The acrosome, a glycoprotein-rich structure surrounding the sperm nucleus, is considered to play a key rôle in the initial stages of fertilization. It is initially formed by contributions from the Golgi apparatus and endoplasmic reticulum[61, 62]. Acrosome development occurs in several phases referred to as the Golgi phase, the cap phase, the acrosome phase and the maturation phase. In the first phase, pro-acrosomal granules produced in the Golgi region aggregate into a single spherical dense acrosomic granule located adjacent to the nucleus of the young spermatid. Next, dense material radiates along the surface of the nucleus, forming an acrosomal cap. Finally, with progressive elongation, the nucleus is surrounded by several layers of membranes – the nuclear membrane, the inner acrosomic membrane, the acrosome, the outer acrosomic membrane and the cytoplasmic membrane. The acrosome contains several enzymes, including phosphatases[63], hyaluronidase[64], neuraminidase[65] and aryl sulphatase[66]. These enzymes are generally not detected until the spermatocyte or spermatid stage of development[67].

Distal to the nucleus, the spermatid cytoplasm includes a proximal centriole from which develops the locomotor apparatus consisting of an axial filament surrounding a microtubular central canal, a caudal sheath or manchette, and a helical arrangement of mitochondria around the axial filament. The spermatid cytoplasm incorporates tritiated amino acids during all stages of spermiogenesis[37]. The rate of incorporation is high during the first half of spermiogenesis, and then gradually falls to zero.

With the completion of maturation, the cells are released into the tubular lumen as spermatozoa. Portions of the cytoplasm, the so-called 'residual bodies', are eliminated and phagocytized by Sertoli cells, a process which appears to be important in the regulation of spermatogenesis[68]. The residual bodies are rich in lipids and other substances that may be utilized by the Sertoli cells in the production of hormones or other regulatory materials[69]. Final stages of spermatozoal differentiation occur outside the testis in the epididymis and genital ducts. The structural characteristics of spermatozoa, which vary considerably in different species, have been reviewed in detail elsewhere[70].

Supporting cells

The supporting cells of the developing testis are those cells that differentiate into *Sertoli cells* (sustentacular cells). Their principal function throughout testicular development is the support of germ cell nutrition and maturation. A close association of supporting cells and germ cells is established early in development within the primitive sex cords.

Pre-spermatogenic stage

In the pre-spermatogenic stage, supporting cells are arranged in columnar fashion perpendicular to the basal lamina. They are the predominant cell type in the sex cords, outnumbering the germ cells. Although it had once been considered that Sertoli cells were derived from the stromal mesenchyme, morphologic studies have clearly demonstrated that the cells which differentiate into Sertoli cells are the supporting cells derived from the surface epithelium[71]. The general arrangement of the supporting cells within the sex cords and their ultrastructural appearance indicate that the pattern of Sertoli cell maturation (Table 6.2) is established early in development[12]. Nuclei of the supporting cells are elongated, with deep irregular indentations and prominent nucleoli. Cytoplasm includes many polyribosomes, elongated mitochondria with closely spaced transverse cristae, scattered lipid droplets, prominent Golgi areas and a moderate amount of endoplasmic reticulum (Plate 6.13).

Although endoplasmic reticulum and nucleoli are not as well developed as in Sertoli cells of the spermatogenic stage, the cellular organization of the pre-spermatogenic supporting cells indicates that they are capable of synthetic and secretory activity. With histochemical techniques, the cells show high NAD diaphorase, lactate dehydrogenase and glucose-6-phosphate dehydrogenase activity[72]. Structures associated with steroid hormone production are present in the fetal and post-natal supporting cells, although to a limited degree[73].

Pituitary regulation of early testicular growth appears to be mediated through the supporting, or Sertoli cells[74]. Follicle stimulating hormone (FSH) and luteinizing hormone (LH) have a synergistic effect, with FSH playing a leading rôle in the pre-pubertal testis[75]. In the period prior to the initiation of spermatogenesis, there is a sharp increase in DNA synthesis in the Sertoli cell nucleus[76], possibly due to the action of gonadotrophins[57].

Spermatogenic stage

Mitosis stops in Sertoli cells shortly after spermatogenesis begins[76]. With the onset of spermatogenesis, Sertoli cells undergo further nucleolar maturation and development of endoplasmic reticulum[77]. The cells contain a variety of enzymes and there appears little doubt that this is related to activity in support of germ cell maturation during spermato-

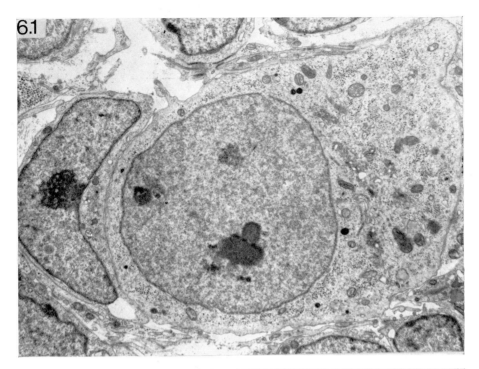

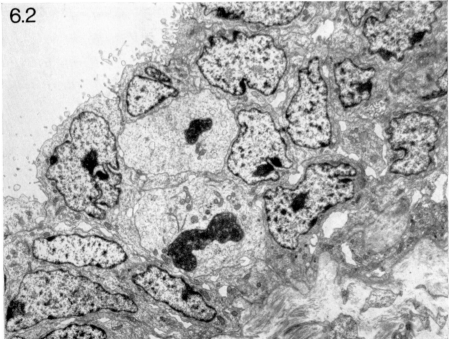

Plate 6.1 Primitive germ cell with closely attached epithelial cell in gonadal mesenchyme, 7 week human embryo. Uranyl acetate and lead citrate. × 6750.

Plate 6.2 Coelomic epithelium on gonadal surface consists of elongated cells with irregular nuclei and includes cells in division. Portion of tunica albuginea is seen at lower right. Uranyl acetate and lead citrate. × 3600.

Plate 6.3 Section of human fetal testis at 10 weeks shows well developed tunica albuginea separating single layered surface epithelium and underlying sex cords. Toluidine blue. × 480.

Plate 6.4 Section of human fetal ovary at 10 weeks shows continuity between surface epithelium and cortical cell groups. Note numerous germ cells with spherical nuclei and prominent nucleoli in the gonadal cortex. Tunica albuginea and mesenchyme are poorly developed. Toluidine blue. × 480.

Plate 6.5 Section of fetal lamb testis, showing large oval gonocyte surrounded by supporting cells in primitive seminiferous tubule. Uranyl acetate and lead citrate. × 3200.

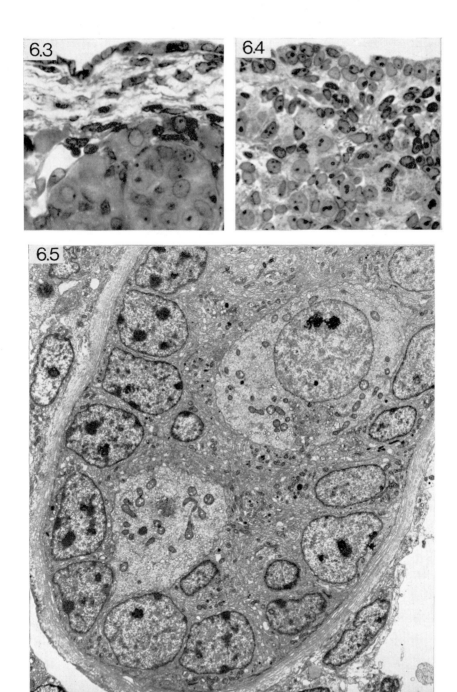

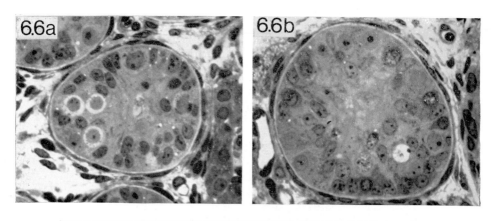

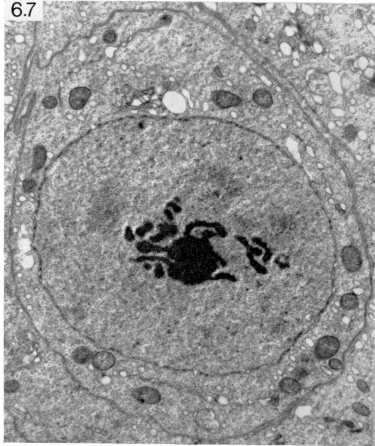

Plate 6.6 Light micrographs of pre-pubertal rabbit testis. a. At 7 weeks, large pre-spermatogonia are present at tubular periphery. b. At 8 weeks, pre-spermatogonia undergo marked nucleolar enlargement and differentiate into spermatogonia. Toluidine blue. × 540.

Plate 6.7 Electron micrograph of rabbit pre-spermatogonium with prominent central nucleolus. Uranyl acetate and lead citrate. × 11 250.

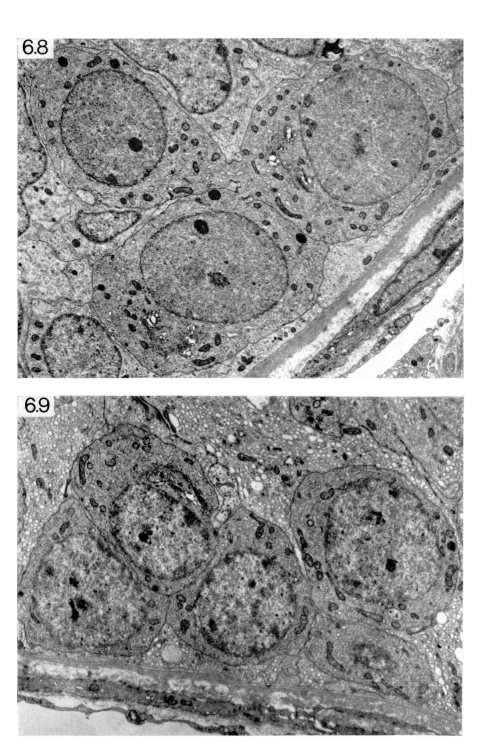

Plate 6.8 Section of rabbit testis shows three type A spermatogonia linked by intercellular bridges. Uranyl acetate and lead citrate. × 5850.

Plate 6.9 Type B spermatogonia, rabbit testis. Uranyl acetate and lead citrate. × 6750.

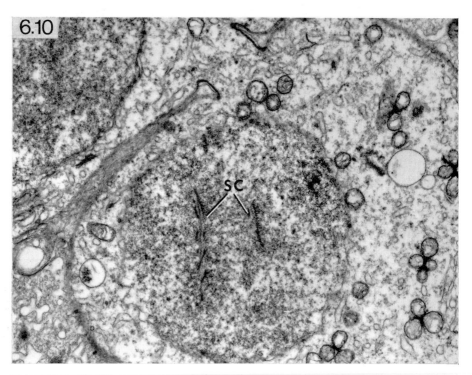

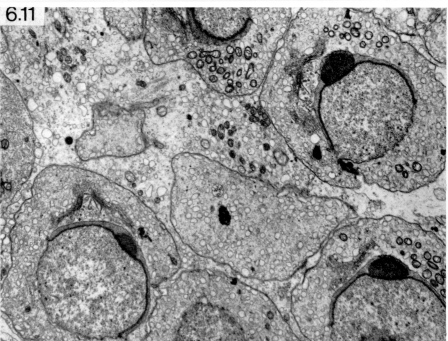

Plate 6.10 Spermatocyte, with tripartite synaptinemal complexes (sc), monkey testis. Uranyl acetate and lead citrate. × 12 150.

Plate 6.11 Spermatids of rabbit testis in the Cap phase. Uranyl acetate and lead citrate. × 5400.

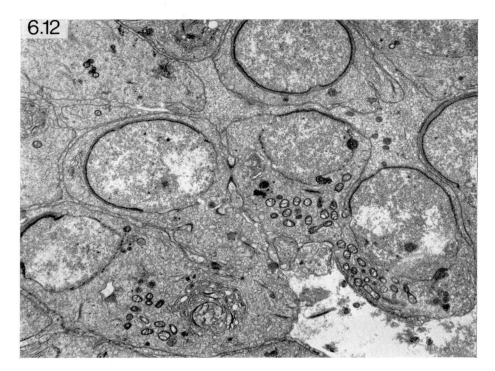

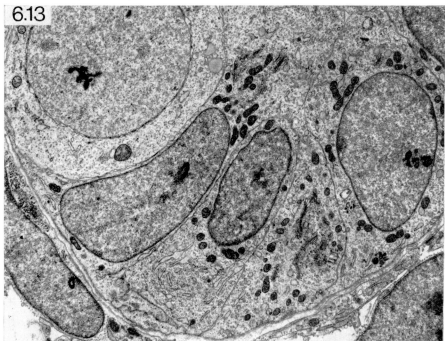

Plate 6.12 Spermatids of rabbit testis in the Acrosome phase. Uranyl acetate and lead citrate. × 5400.

Plate 6.13 Supporting cells (fetal Sertoli cells), with elongated nuclei and rod-shaped mitochondria, contrasting with spherical nucleus and large globular mitochondria of germ cell shown at upper left, 20 day rabbit fetus. Uranyl acetate and lead citrate. × 5400.

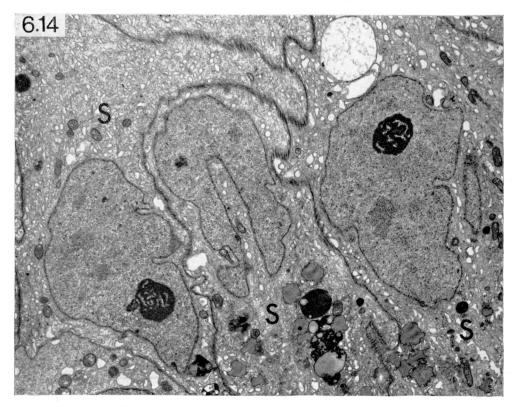

Plate 6.14 Mature Sertoli cells (S), showing complex intercellular junctions, monkey testis. Uranyl acetate and lead citrate. × 5000.

Plate 6.15 Fetal and adult Leydig cells, showing similar ultrastructural features, ▶ including eccentric nucleus, abundant smooth endoplasmic reticulum, tubula, mitochondria and dense lipofuscin granules. a. Fetal Leydig cell, × 6750. b. Adult Leydig cells, × 2880. Uranyl acetate and lead citrate.

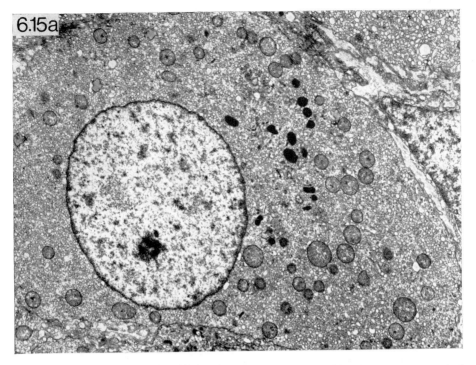

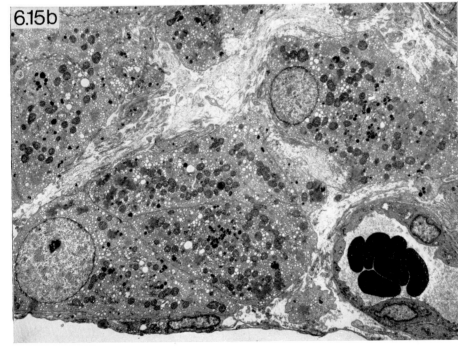

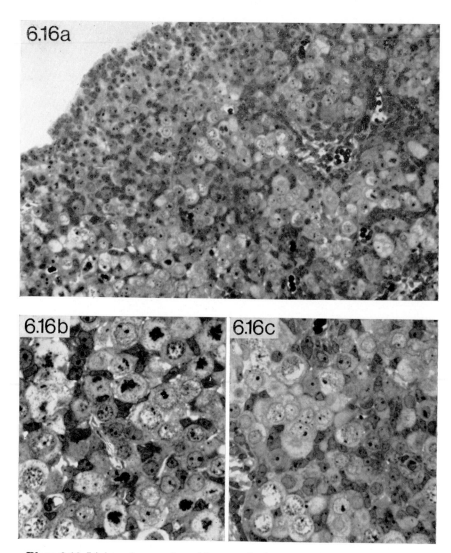

Plate 6.16 Light micrographs of human fetal ovary, 12 weeks gestation. a. Surface epithelium and cortical cell groups consisting of germ cells and supporting cells (pre-granulosa cells), × 135. b. Oogonia in motisis, × 540. c. Oocytes in meiotic prophase, × 540.

Plate 6.17 Oogonia in mitosis, rabbit fetal ovary. Inset, light micrograph, ▶ × 960, toluidine blue. Electron micrograph, × 6400, uranyl acetate and lead citrate.

Plate 6.18 Oocyte in meiosis, human fetal ovary. Uranyl acetate and lead ▶ citrate. × 5200.

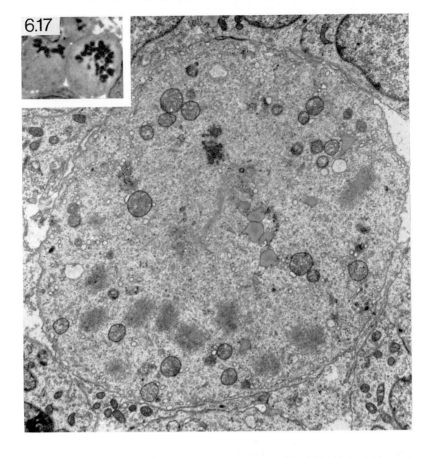

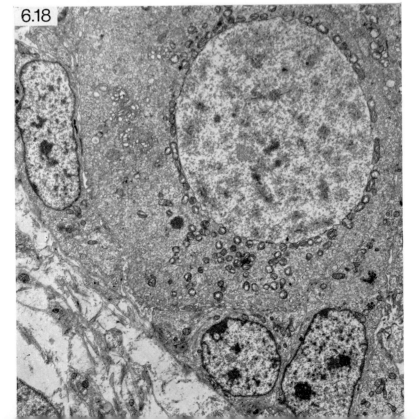

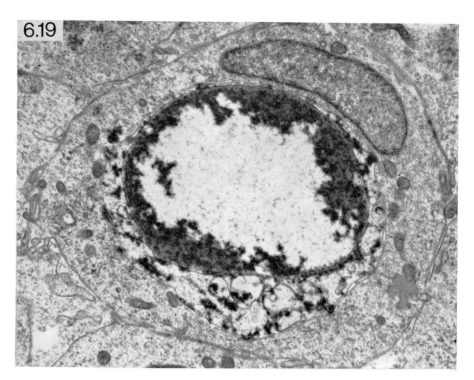

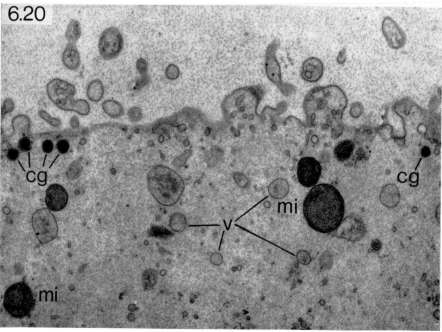

Plate 6.19 Phagocytosis of degenerating germ cell by pre-granulosa cell, human fetal ovary. Uranyl acetate and lead citrate. × 13 500.

Plate 6.20 Section of rabbit oocyte recovered from Fallopian tube, showing cortical granules (cg), spherical mitochondria (mi) and smooth vesicles (v). Uranyl acetate and lead citrate. × 17 550.

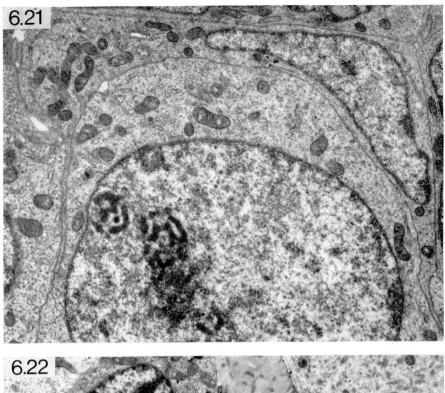

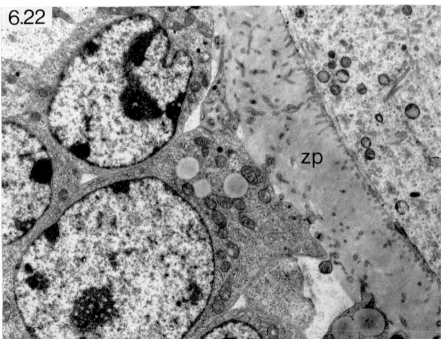

Plate 6.21 Elongated pre-granulosa cell extending along surface of adjacent oocyte, early stage in follicle formation, 11 day old rabbit. Uranyl acetate and lead citrate. × 10 800.

Plate 6.22 Section of follicle of adult mouse showing abundant lipid, mitochondria, ribosomes and rough endoplasmic reticulum in granulosa cell cytoplasm. Slender granulosa cell projections extend across zona pellucida (zp) to make contact with microvilli on surface of oocyte shown at upper right. Uranyl acetate and lead citrate. × 5400.

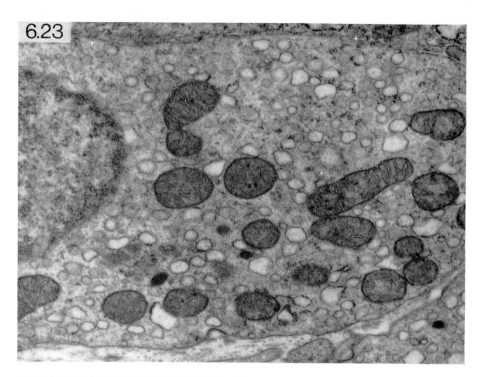

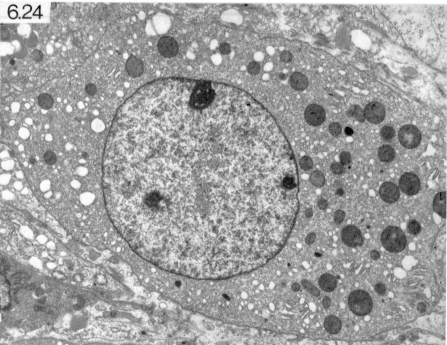

Plate 6.23 Luteinized granulosa cell, rabbit ovary, with prominent smooth endoplasmic reticulum and tubular mitochondria. Uranyl acetate and lead citrate. × 22 500.

Plate 6.24 Interstitial cell, human fetal ovary, with extensive development of smooth endoplasmic reticulum. Uranyl acetate and lead citrate. × 5400.

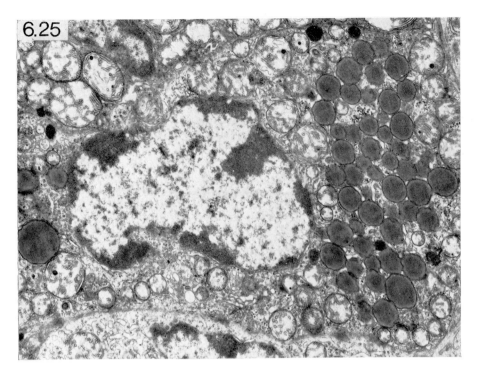

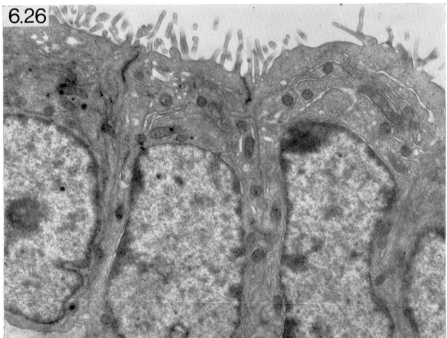

Plate 6.25 Interstitial cell, adult mouse ovary, with numerous lipid droplets and tubular mitochondria filling the cytoplasm. Uranyl acetate and lead citrate. × 11 250.

Plate 6.26 Surface epithelium, adult mouse ovary, consists of single layer of flattened columnar cells. Uranyl acetate and lead citrate. × 11 250.

genesis[78]. Tracer studies indicate that labelled FSH passes from the base of Sertoli cells into cytoplasmic ramifications surrounding germ cells, suggesting a possible rôle of the Sertoli cell as an intermediate site of gonadotrophin storage for action in spermatogenesis[79].

The Sertoli cells also play a rôle in maintaining the overall architecture within the seminiferous tubules. Each Sertoli cell is associated with groups of germinal elements, produced by intertwining of the Sertoli cell cytoplasm around germ cells, particularly spermatids. This is an important element in the co-ordination of spermatogenic maturation. A peculiar feature of mature Sertoli cells is the presence of complex intercellular junctions (Plate 6.14) found only between Sertoli cells[80]. These junctions play a rôle in restricting intercellular flow of material within the seminiferous tubules[81]. Thus, the Sertoli cells act as a permeability barrier.

Sertoli cells show a continued and gradual increase in lipids with age, starting just prior to puberty[82]. The lipid is depleted during spermatogonial division, suggesting that stored lipid is utilized in the maintenance of spermatogenesis[83].

Whether or not Sertoli cells are capable of steroid hormone secretion remains to be determined. Although cytoplasmic structures and enzymes associated with steroid biosynthesis are present, hormonal activity by Sertoli cells has not been demonstrated. Suggestive evidence for such activity comes from the observation that Sertoli cell tumours in certain animals are associated with oestrogenic effects[84]. In spite of the current lack of definite proof of steroid production by Sertoli cells, it seems likely that further investigation will provide such evidence[85, 86].

Stromal cells

The stromal cells derived from the gonadal mesenchyme can be divided into two main compartments: (1) *peritubular tissue*, including an inner layer of smooth muscle and an outer layer of fibroblasts; and (2) *intertubular tissue*, consisting of mesenchymal cells capable of differentiating into hormone secreting Leydig cells (interstitial cells).

Peritubular tissue

The peritubular tissue, or *boundary tissue*, surrounds the seminiferous tubules. It has a dual function: contraction of the tubules, thus regulating the flow of spermatozoa along the tubular lumen; and control of the access of materials from the blood stream into the seminiferous tubules. In the adult rat, the boundary tissue consists of two cellular layers and two non-cellular layers arranged as follows: an inner non-cellular layer consisting of a network of collagen fibrils between two basement membranes, an inner cellular layer of smooth muscle cells, an outer non-cellular layer consisting of scattered collagen fibrils associated with a single basement membrane, and an outer cellular layer of fibroblasts[87, 88]. At the time of sex differentiation, only a single basal lamina surrounds the cords. Later, a layer of mesenchymal cells appears around the basal lamina. With the formation of distinct tubules, additional layers of non-cellular and cellular elements appear, but without clear demarcation or evidence of specialized differentiation. By the time the initial spermatogenic cycle is established, four distinct layers are evident and intracytoplasmic myofilaments have made their appearance within the inner cellular layer[87]. The development of the myoid cells is associated with the initiation of spermatogenesis and is probably under gonadotrophin control[89]; contractions can be induced in immature rat testicular tissue in tissue culture by human chorionic gonadotrophin[90].

Intertubular tissue

The intertubular, or *interstitial tissue* undergoes a biphasic pattern of development. At the time of testicular differentiation, the interstitial tissue separating the sex cords consists entirely of undifferentiated mesenchymal cells. Shortly thereafter, many of the cells begin to differentiate into *Leydig cells*. This process continues for a specified time within the fetal testis, and in some animals in the neonatal testis. This is followed by a period of regression in which the cells revert to an appearance intermediate between that of fibroblasts and Leydig cells. About the time that spermatogenesis is initiated, there is a re-differentiation of Leydig cells.

The differentiation of fetal Leydig cells from stromal mesenchyme has been described electron microscopically[73] and histochemically[91]. Initially, localized aggregates of smooth endoplasmic reticulum appear in association with elements of rough endoplasmic reticulum in the cytoplasm of the mesenchymal cells. Smooth endoplasmic reticulum is apparently derived from the rough portion in the course of Leydig cell differentiation[73]. As this occurs, the cells are transformed from elongated mesenchymal elements into large oval epithelioid cells (Fig. 6.1). The nucleus becomes spherical and moves to an eccentric position, while the cytoplasm is filled with extensive formations of smooth endoplasmic reticulum, large tubular mitochondria, prominent Golgi areas and scattered dense bodies (Plate 6.15). All of these features are charac-

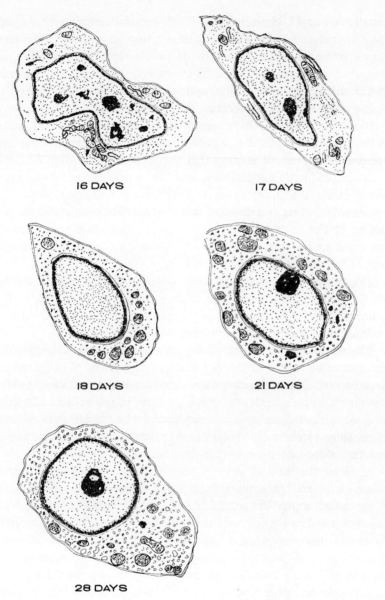

16 DAYS

17 DAYS

18 DAYS

21 DAYS

28 DAYS

Fig. 6.1 Differentiation of mesenchymal cells into Leydig cells, rabbit fetal testis.

teristic of steroid-secreting cells. Histochemical techniques for localizing 3-β-hydroxysteroid dehydrogenase have demonstrated this important enzyme of steroid biosynthesis in the interstitial tissue of fetal testes[92, 93]. The fetal Leydig cells are also rich in lactic dehydrogenase, NAD diaphorase and alcoholic dehydrogenase and their cytoplasm stains with Sudan dyes indicating the presence of lipids[72].

Biochemical studies have shown that fetal testes, utilizing similar pathways to those of the adult testis, are capable of synthesizing testosterone[94, 95]. Thus, it is clear that the fetal differentiation of Leydig cells has functional significance. Since the appearance of Leydig cells is followed by the differentiation of the male reproductive tracts, it appears likely that secretions of fetal Leydig cells control the differentiation of the genital tract[96]. A possibility also exists that hormones produced by the fetal testis may be responsible for the sexual maturation of the central nervous system[97, 98].

The factors stimulating the initial differentiation of the Leydig cells are not known, but maternal chorionic gonadotrophins are felt to play a primary rôle[91]. Fetal pituitary gonadotrophins may also be involved in maintenance of the cells in a differentiated state[99].

The reason for regression of the Leydig cells is not clear, but it is thought to be related to changes in circulating levels of gonadotrophic hormones[100]. One of the early signs of regression is cytoplasmic vacuolization[101]. Subsequently, there is a decrease in cell volume and in the amount of smooth endoplasmic reticulum. The regression is not complete, since biochemical observations indicate presence of many of the enzymes associated with steroidogenic activity[102].

At about the time of puberty, the RNA content of the interstitial tissue increases[103], and the second phase of Leydig cell differentiation is completed under the regulation of gonadotrophic hormones[104]. At puberty, the Leydig cell acquires the capacity to convert androstenedione into testosterone, presumably by an increase in the amount of the enzyme 17-β-hydroxysteroid dehydrogenase[105]. Subsequently, clusters of Leydig cells appear, particularly around capillaries. Adult Leydig cells resemble fetal Leydig cells, except that certain crystalline structures, such as Reinke crystals in the human testis, are found only in adult cells.

Tunica albuginea

The tunica albuginea is that portion of the testicular stroma lying beneath the surface epithelium. This layer is much thicker in the testis than in the ovary. In early stages of differentiation, it is quite cellular, consisting of fibroblasts and capillaries with prominent endothelial lining cells. Leydig cells are present in the inner portions. The prominence and cellularity of the tunica albuginea at the time of testicular differentiation help to distinguish the testis from the ovary. In the course of development, cellularity is reduced and the layer consists primarily of dense extracellular collagen, but smooth muscle cells capable of contractile activity are also present[106]. The hilar portion of the tunica albuginea, known as the mediastinum testis, is a specialized area in which the rete tubules extend in from the mesonephros at the time of testicular differentiation. The tubules join with the terminal portions of the seminiferous tubules, the straight tubules (tubuli recti), to form the proximal portion of the male genital tract.

Surface epithelium

The surface epithelium (germinal epithelium, coelomic epithelium) is reduced to a single layer shortly after sexual differentiation occurs and remains so throughout the remainder of development. Once the sex cords form, the tunica albuginea intervenes to form a complete separation between the cords and the surface epithelium. The flattened columnar epithelium later becomes the inner layer of the tunica vaginalis.

Ovary

Ovarian differentiation involves two main processes: formation and maturation of oocytes, and differentiation of endocrine cells required for the secretion of female sex hormones. In contrast to the testis, stromal endocrine cells of the ovary do not undergo extensive differentiation during early development. On the other hand, oogenesis takes place much earlier than spermatogenesis, oocyte formation occurring in the fetal or neonatal period*. This results in establishment of the oocyte population well before the time of sexual maturation.

Germ cells

Germ cell differentiation in the ovary consists of three stages: (1) a *Pre-follicular stage*, occurring entirely in the early developmental period and resulting in formation of a fixed stock of follicular oocytes;

* In a few species, such as certain prosimians[107], oogenesis continues into adulthood.

(2) a *follicular stage*, lasting months to years, during which time the oocytes increase in size but undergo no further maturation; and (3) an *ovulatory stage*, of short duration, in which the final stages of oocyte maturation take place. Although it had once been thought that all of the early germ cells underwent degeneration, it is now clear that there is continuity of development from the surviving primitive germ cell to the mature oocyte[108]. Evidence for neoformation of germ cells from the surface epithelium has been clearly refuted[109, 110].

Pre-follicular stage

The changes involved in pre-follicular germ cell maturation occur in association with modifications of the overall structure of the developing ovary. As the primitive germ cells complete their migration to the presumptive ovary, groups of cells move from the surface epithelium into the underlying mesenchyme. The cell groups are arranged in dense sheets consisting of germinal elements and supporting cells (Plate 6.16a). Groups of germ cells undergoing mitosis in synchrony are called *oogonia* (Plate 6.16b). After several such divisions, the cells begin meiosis and are then referred to as *oocytes* (Plate 6.16c). Generally, this process begins in the deepest portions of the cortex. Eventually, all cells enter meiosis, after which the cell groups become broken up into individual follicles also starting in the innermost portions. The newly formed follicles, or *primary follicles*, consist of a single oocyte and a unilaminar wall of supporting cells. A progressively increasing amount of stromal connective tissue separates the follicles from one another. Also at this time, the upper portions of the cords become completely separated from the overlying surface epithelium as the tunica albuginea develops between them. Pre-follicular development thus involves three stages of germ cell maturation, primitive germ cells, oogonia and oocytes (Table 6.2).

Primitive germ cells are large oval cells with a slightly eccentric nucleus and cytoplasmic protrusions suggesting an amoeboid appearance. Cytoplasm contains abundant glycogen and scattered spherical mitochondria with eccentric cristae. Initially, germ cells are far outnumbered by supporting elements. With continued mitotic activity, the relative number of germ cells progressively increases. During this time germ cells come to lie in close association. The cytoplasm shows loss of glycogen and an increased number of mitochondria. The cells lose their amoeboid appearance and nuclei are more centrally placed within the generally symmetrical cellular structure.

186

The cells are now referred to as oogonia. The principal differences from primitive germ cells are lack of glycogen, grouped arrangement, increased number of mitochondria and increased rate of mitosis (Plate 6.17). Intercellular bridges appear between adjacent cells, a result of incomplete cytokinesis in the course of mitotic division[111]. Repeated incomplete division produces an increasing number of cellular interconnections with each subsequent mitosis. The result is a syncytial grouping of germ cells, which develop in synchrony.

The maturation of oogonia and their transformation into oocytes is referred to as oogenesis*. It is during the period of oogenesis that genomic information is transcribed in the germ cells[112]. DNA synthesis takes place throughout the oogonial stage[113] and in the pre-meiotic interphase[114]. The interval from the end of oogonial DNA synthesis to the beginning of premeiotic DNA synthesis is of the order of several hours[115].

Oocytes undergo two meiotic divisions, but only a portion of the first occurs in the pre-follicular stage of development. The initial stages of meiotic prophase, leptotene, zygotene, pachytene and diplotene, are completed before follicle development takes place. The concentration of radioactive DNA incorporated in oocytes remains relatively constant during this time[110]. Persistence of cellular interconnection enables continued synchronous maturation, as cytoplasmic organelles are exchanged across the bridges[116, 117]. It is also possible that gene products wander through the bridges[118].

Meiotic oocytes can be recognized under the electron microscope by the presence of synaptinemal complexes similar to those seen in spermatocytes (Plate 6.18). The cells retain a spherical shape and continue to show a relative paucity of cytoplasmic organelles. Mitochondria characteristically exhibit parallel eccentric cristae curved toward the periphery.

Many of the oocytes undergo degeneration during this period[119]. Cellular degeneration begins with peripheral condensation of chromatin and accumulation of lysosome-like structures in the cytoplasm[18, 120]. Degenerating cells are eliminated by extrusion into the coelomic cavity in some species[121] and by phagocytosis by supporting cells in others (Plate 6.19)[18, 120, 122]. Degeneration occurs in three distinct waves, affecting oogonia in mitosis, oocytes in pachytene and oocytes in diplotene[119]. Surviving oocytes are separated away from their neighbours

* The term, oogenesis, has also been used to include the entire period of oocyte development.

by elongated extensions of surrounding supporting cells, resulting in the formation of *primary follicles*[123]. At this time the oocytes are at the dictyate stage of meiosis.

Follicular stage

The stage of follicular development is prolonged, extending from early development to the time of ovulation in the adult. Throughout this period the oocyte nucleus remains in the dictyate stage, apparently under the influence of the surrounding granulosa cells[124]. During this time the oocyte grows considerably in size, primarily due to an increase in cytoplasmic volume resulting from accumulation of mitochondria and other organelles and proliferation of the Golgi system. Some of the organelles are produced *de novo*, while others are produced from the nuclear envelope by a process of blebbing[125]. The general structure of the cytoplasm remains relatively simple, with most of the organelles clustered in a limited region around the nucleus (Balbiani body).

The oocyte produces large quantities of RNA and protein during the growth phase[126]. Part of this material is utilized for growth, while the remainder is stored. Cytochemical and autoradiographic studies indicate that messenger RNA is stored in granular bodies in the oocyte cytoplasm[127].

Many oocytes undergo degeneration during this period by a process called *follicular atresia*[128]. The atretic changes affect the entire follicle, beginning in the oocyte and spreading to involve the surrounding granulosa cells. The factors responsible for atresia are poorly understood. The cytologic changes in atretic oocytes are similar to those seen in degenerating pre-follicular oocytes and, as in the earlier stage, involve a major segment of the oocyte population. Thus, of the large pool of oocytes formed in the fetal and neonatal period, only a relatively few survive to reach maturity.

Ovulatory stage

Final stages in oocyte maturation begin just prior to the time of ovulation, dependent on the influence of luteinizing hormone[129]. One of the first changes is a rapid development of microvilli on the surface of the oocyte[130]. The microvilli establish contact with cytoplasmic extensions of the granulosa cells immediately adjacent to the oocyte. Mitochondria increase further in number and endoplasmic reticulum undergoes considerable development. Organelles are no longer concentrated in the perinuclear region. Dense bodies, referred to as cortical granules

(Plate 6.20), are synthesized in the Golgi region and become aligned along the periphery of the oocyte cytoplasm[131]. The cortical granules are considered to play a rôle in the process of fertilization, since they disappear after sperm penetration, but their precise function is unknown[132].

Resumption of meiosis begins with condensation of the nucleolus, followed by its gradual disappearance and fragmentation of nuclear membranes[133]. Completion of the first meiotic division results in separation of the first polar body. Subsequent completion of the second meiotic division occurs, as a consequence of sperm penetration. Thus, final stages of oocyte development are completed only at the time of fertilization.

Supporting cells

The supporting cells of the developing ovary are those cells that differentiate into *granulosa cells* (follicle cells). Their principal function throughout ovarian development is the support of germ cell nutrition and maturation. A close association of the two cell types is established early in development.

Pre-follicular stage

In the pre-follicular stage, the supporting cells are referred to as *pre-granulosa cells*, or fetal granulosa cells. They are closely associated with adjacent germ cells (Fig. 6.2), in a manner resembling the relationship between fetal Sertoli cells and germ cells in the testis. The cytoplasmic organization of pre-granulosa cells is similar to that of mature granulosa cells, consisting of many ribosomes, elongated mitochondria with closely stacked transverse cristae, prominent Golgi areas, a large amount of rough endoplasmic reticulum and scattered lipid material[134]. Nuclei are irregular and elongated, with prominent nucleoli.

Pre-granulosa cells have some of the features of steroid hormone producing cells, but lack the abundant smooth endoplasmic reticulum generally found in hormonally active cells. Enzymes involved in steroid biosynthesis have been found in the fetal ovary[135, 136]. Although early ovarian development has generally been thought to be independent of pituitary regulation[137], pre-granulosa cells do appear to be responsive to exogenous gonadotrophin stimulation to a limited degree[138].

The origin of granulosa cells is not entirely clear. While it is possible that some pre-granulosa cells differentiate from mesenchymal tissue[139], it appears likely that the majority, if not all of the granulosa cell population is derived from the surface epithelium rather than from the mesenchyme[7, 8, 108, 130]. However, the evidence for this is not secure, since it consists primarily of morphologic observations indicating similarities between epithelial cells and pregranulosa cells.

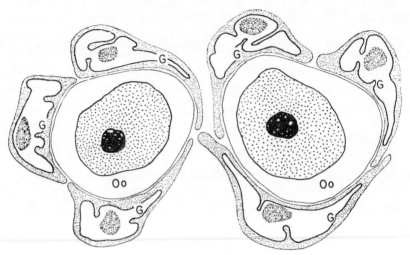

Fig. 6.2 Arrangement of germ cells (Oo) and pregranulosa cells (G) in prefollicular ovary.

During oogenesis, tight apposition is maintained between adjacent pre-granulosa cells and germ cells[140]. Cytoplasmic projections resembling microvilli are present on the surface of the pre-granulosa cells and some of these extend for a considerable distance toward the nuclei of oocytes (Fig. 6.3), without, however, establishing direct cytoplasmic continuity*. Pinocytosis appears to be the principal manner by which materials are transferred to the germ cells during the pre-follicular stage of development[142]. This is also a possible method by which pre-granulosa cells may influence early germ cell maturation[143].

During oogenesis, pre-granulosa cells become progressively more elongated, extending around the border of adjacent oocytes, ultimately separating them off from their neighbours to form primary follicles (Plate 6.21). The energy required for the encirclement of oocytes is derived from accumulating lipid stores found in the cytoplasm of pre-granulosa cells. Accumulation of lipid is a result at least in part of

* In invertebrates, cytoplasmic bridges connect germ cells and surrounding follicle or nurse cells[141].

phagocytosis of degenerating germ cells[120]. Ingested cellular material is broken down within the pre-granulosa cell cytoplasm, contributing to the lipid stores.

Follicular stage
Once follicle formation begins, the cells surrounding the oocytes are referred to as granulosa cells (Table 6.2). Initially, each oocyte is surrounded by a single layer of granulosa cells, forming a primary follicle. In comparison to pre-granulosa cells, granulosa cells are cuboidal rather than elongated and the nucleus is generally more regular and symmetrical in outline, with fewer and less shallow indentations.

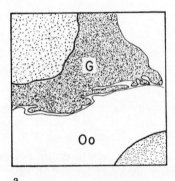 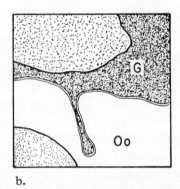

a.   b.

Fig. 6.3 a. and b. Cytoplasmic projections of pre-granulosa cells (G) extend along cell borders of germ cells (Oo) and produce deep indentations in germ cell cytoplasm.

The granulosa cells contribute to the formation of the zona pellucida[144], the acellular homogeneous material separating the oocyte from the granulosa cells. The granulosa cells maintain their close relationship with the oocyte by sending long cytoplasmic processes across the zona pellucida to interdigitate with microvilli on the surface of the oocyte[145]. The cytoplasm of the granulosa cells is rich in lipid material and rough endoplasmic reticulum (Plate 6.22). Mitochondria are packed with parallel transverse cristae, resembling the arrangement in pre-granulosa cells.

Proliferation of granulosa cells results in formation of progressively enlarging multilayered follicles. Granulosa cell proliferation in small follicles is regulated by follicle stimulating hormone (FSH) produced by the pituitary gland[146]. The effect of FSH involves utilization of already formed RNA for protein synthesis[147] and initiation of some process leading to DNA replication[148]. FSH also produces accumulation of 3-β-hydroxysteroid dehydrogenase in the granulosa cells[149]. In larger follicles, granulosa cells are under the influence of another pituitary hormone, luteinizing hormone (LH)[146]. LH stimulation produces an accumulation of a variety of oxidative and steroidogenic enzymes[149]. Studies on isolated granulosa cells reveal that LH induces a marked increase in their rate of respiration[150]. Granulosa cells are capable of converting lactate to pyruvate[151], and of supplying pyruvate to the oocyte. This relationship is important in promoting meiotic maturation at the time of ovulation, and is apparently regulated by LH[152].

Luteinization

At the time of ovulation, granulosa cells undergo a great increase in size, associated with a proliferation of cytoplasmic organelles involved in steroid hormone production[153, 154]. This process, referred to as *luteinization*, involves extensive development of smooth endoplasmic reticulum, formation of tubular mitochondria, proliferation of the Golgi apparatus and further accumulation of lipid material (Plate 6.23). Luteinized granulosa cells (*granulosa lutein cells*) are involved in the production of progestational hormones required for implantation of the fertilized ovum[155]. *In vitro* studies suggest that LH has a key role in initiating luteinization[156]. Luteinization is a direct result of rupture of the follicle (i.e., ovulation), which may release a luteostatic substance present in the oocyte responsible for maintenance of granulosa cells in a non-luteinized state[157].

Stromal cells

The stromal cells derived from the gonadal mesenchyme can be divided into two main compartments: (1) *perifollicular tissue*, including an inner layer of hormone secreting cells (theca interna) and an outer layer of fibroblasts (theca externa); and (2) *interfollicular tissue*, consisting of mesenchymal elements capable of differentiating into hormone secreting cells (interstitial cells). In many species, the interfollicular stromal cells remain undifferentiated for the most part, and are referred to simply as ovarian stroma or cortical stroma.

Perifollicular tissue

Theca interna cells first appear at the time of follicle formation as elongated cells applied to the outer surface of the follicle. At this time, their appearance is that of undifferentiated mesenchyme. Gradually, there is an accumulation of polyribosomes, endoplasmic reticulum and lipid droplets[158], which is believed to be under the control of gonadotrophic hormones[159]. Theca cells remain outside the follicular basement membrane, the *membrane granulosa* (membrane propria), and are in direct contact with blood vessels (granulosa cells are not, since vessels do not penetrate the membrana granulosa). In the pre-ovulatory follicle, theca interna cells appear to be the principal source of estrogenic hormones, although granulosa cells are also involved[160].

When luteinization occurs, granulosa and theca interna cells are transformed in a similar manner. The luteinized theca cells (*theca lutein cells*) have an ultrastructural appearance characteristic of steroid hormone producing cells and are indistinguishable in this respect from luteinized granulosa cells. At the time of luteinization, blood vessels extend into the granulosal layer, and thus luteinized granulosa and theca cells both have direct access to the circulation.

Theca externa cells are those perifollicular elements that retain the appearance of fibroblasts throughout the course of differentiation. Ordinarily, they function only in providing architectural support. Interspersed among the theca externa cells are smooth muscle elements, located particularly in the vicinity of blood vessels[161]. Little is known about the formation and function of these cells, but it is thought they may play a rôle in promoting ovulation[162].

Interfollicular tissue
The pattern of growth and differentiation of the extrafollicular ovarian stroma shows considerable species variability. In some rodents and other mammals, this tissue gives rise to *interstitial cells*, which have a major endocrine function[163]. In other species, scattered interstitial cells may be present, but the bulk of stromal tissue consists of dense fibrous connective tissue. This tissue has the capacity to undergo luteinization under certain conditions[164], but generally functions primarily as a supportive framework for the ovary.

In those animals with well developed interstitial tissue, there is a biphasic pattern of development reminiscent of that occurring in the testis. An initial phase, referred to as the primary interstitial tissue, appears in the late fetal and early neonatal period (Plate 6.24), and then disappears. The cells differentiate from mesenchymal elements in the vicinity of blood vessels[165]. Differentiation involves accumulation of smooth endoplasmic reticulum, tubular mitochondria, Golgi elements and lipid droplets[166]. During this time, 3-β-hydroxysteroid dehydrogenase activity develops within the interstitial cells[167, 168]. The secondary interstitial tissue differentiating in the post-natal period appears to be derived from theca interna cells of degenerating follicles[169], under the influence of LH[170]. The secondary interstitial cells, capable of producing steroid hormones[171], contain abundant lipid material (Plate 6.25).

An additional cell type, *hilus cells*, showing an exact correspondence with Leydig cells of the testis including androgenic activity, has been noted to occur in the region of the rete ovarii at the hilus of the ovary, particularly around sympathetic nerves[172].

Tunica albuginea

The tunica albuginea is that portion of the ovarian connective tissue lying immediately beneath the surface epithelium. In the adult ovary, this layer consists primarily of extracellular collagen, with scattered fibroblasts and capillaries. It thus forms a protective coat around the ovarian cortex. The development of the tunica albuginea from the gonadal mesenchyme begins at the time of sex differentiation. Initially, it is a thin poorly defined layer, partially separating the surface epithelium from the developing sex cords. Once the movement of germ cells and epithelial cells into the sex cords is completed, the tunica albuginea thickens, producing complete separation of the surface epithelium and the underlying follicle-filled cortex.

Surface epithelium

The surface epithelium (germinal epithelium) is a derivative of the coelomic epithelium. During sex cord formation and oogenesis, the epithelium is several layers in thickness, including columnar epithelial cells and germ cells. The epithelial cells are of tall columnar type, with irregular nuclei and complex cytoplasmic organization, including many ribosomes, elongated mitochondria with closely spaced transverse cristae, prominent Golgi areas, scattered rough endoplasmic reticulum and occasional dense bodies[7]. Along the free surface, microvilli and terminal bars are present, as are found in other mesothelial cells. These structures are related to restriction of fluid passage. Except for the presence of these specialized structures and the absence of lipid, the epithelial cells are identical in appearance to pre-granulosa cells. Mitochondrial features of the two are the same, contrasting with the spherical mitochondria of the germ cells. Thus, from a morphologic standpoint, there is substantial evidence that germ cells do not originate from the surface epithelium, while it is likely that granulosa cells do. Once oogenesis is completed and all of the germ cells have left the surface epithelium, the epithelium is reduced to a single layer of columnar cells (Plate 6.26). Although the surface epithelium is not truly 'germinal', it does function in receiving and transporting germ cells prior to and during oogenesis.

Summary

From the foregoing it can be seen that gonadal cell differentiation and growth include a variety of developmental changes, many of which are common to both ovary and testis. Among these are the following:

Germ cells arise extragonadally in the yolk sac region and migrate to the genital ridges in the early embryo.

Primitive germ cells present in the embryonic gonads are the sole precursors of adult germ cells.

Germ cells in male and female gonads undergo the same basic sequence of events: mitosis, meiosis and maturation.

Mitosis involves synchronous division of localized groups of cells.

Germ cells are connected by intercellular bridges during the stages of mitosis and meiosis.

Meiosis consists of two nuclear divisions leading to the formation of haploid gametes.

Cellular degeneration eliminates large numbers of germ cells at various stages of development.

A close relationship between supporting cells and germ cells is present at all stages of development.

Supporting cells have a direct influence on the maturation of germ cells.

The supporting cells are under the influence of pituitary gonadotrophic hormones.

Stromal cells derived from the gonadal mesenchyme undergo special modifications adapted for the synthesis and secretion of steroid sex hormones.

Similarities in cellular growth and differentiation in the ovary and testis indicate clear cellular homologies. These are outlined in Table 6.3. In addition to these homologies, there are also important differences in the development of the two types of gonads, including the following:

Table 6.3 *Cellular homologies in the ovary and testis*

Ovary	Testis
Germ cells	
Primitive germ cells	Primitive germ cells
Oogonia	Spermatogonia
Oocytes	Spermatocytes
Supporting cells	
Pre-granulosa cells	Supporting cells (fetal Sertoli cells)
Granulosa cells	Sertoli cells
Stromal cells	
Theca interna*	Internal cellular layer
Theca externa	External cellular layer
Interstitial cells	Leydig cells*
Hilus cells	Intertubular connective tissue
Stromal connective tissue	

* From a functional point of view, theca interna and Leydig cells might be considered homologous.[8]

Oogenesis takes place early in development, while spermatogenesis begins after sexual maturation.

Meiosis is arrested for a prolonged period of time in the ovary, but not in the testis.

Oocyte maturation occurs only at the time of ovulation, while in the testis germ cell maturation is a continuous process in the adult.

In the ovary, formation of oocytes does not occur after the neonatal period, while in the testis new stem cells are constantly produced in the adult.

Oocytes become isolated from one another in follicles, while testicular germ cells remain in groups in the seminiferous tubules.

Supporting cells in the adult ovary have a major endocrine function, while endocrine activity of supporting cells in the testis remains to be demonstrated.

Stromal cells undergo extensive development and have a major endocrine function in the fetal and neonatal testis, but there is no evidence of corresponding changes in the early ovary, except in certain species.

A prominent tunica albuginea is present in the early testis, but is less well developed and appears later in the ovary.

The surface epithelium is reduced to a single layer of cells at the time of sexual differentiation in the testis, while remaining several layers in thickness in the female gonad.

The special features associated with ovarian and testicular differen-tiation are among the least well understood aspects of gonadal development. Mechanisms responsible for gonadal sexual differentiation and the induction of sex-specific hormonal regulatory functions are not at all clear. These are important areas of ongoing investigation.

Acknowledgements

This work was supported in part by research grant HD 05727 from the National Institutes of Health. The author wishes to thank Mrs Lucy A. Conner for assisting in preparation of the electron micrographs, Mrs Carmen Sanchez for preparing the light micrographs, Mrs Irene V. Rohan for typing the manuscript and Mr Lawrence Parsons for providing the illustrations.

References

1. WITSCHI, E. (1948), 'Migration of the germ cells of human embryos from the yolk sac to the primitive gonadal folds', *Carnegie Contributions to Embryology*, **32**, 67–80.
2. MCKAY, D. G., HERTIG, A. T., ADAMS, E. C. and DANZIGER, S. (1953), 'Histochemical observations on the germ cells of human embryos', *Anatomical Record*, **117**, 201–219.
3. CHIQUOINE, A. D. (1954), 'The identification, origin and migration of the primordial germ cells in the mouse embryo', *Anatomical Record*, **118**, 135–146.
4. MINTZ, B. and RUSSELL, E. S. (1957), 'Gene-induced embryological modifications of primordial germ cells in the mouse', *Journal of Experimental Zoology*, **134**, 207–237.
5. GONDOS, B., BHIRALEUS, P. and HOBEL, C. J. (1971), 'Ultrastructural observations on germ cells in human fetal ovaries', *American Journal of Obstetrics and Gynecology*, **110**, 644–652.
6. FALIN, L. I. (1969), 'The development of genital glands and the origin of germ cells in human embryogenesis', *Acta anatomica*, **72**, 195–232.
7. GONDOS, B. (1969), 'Ultrastructure of the germinal epithelium during oogenesis in the rabbit', *Journal of Experimental Zoology*, **172**, 465–480.
8. GILLMAN, J. (1948), 'The development of the gonads in man, with a consideration of the role of fetal endocrines and the histogenesis of ovarian tumors', *Carnegie Contributions to Embryology*, **32**, 81–131.
9. GALLIEN, L. G. (1965), 'Genetic control of sexual differentiation in verte-brates'. In *Organogenesis*. Eds. DeHaan, R. L. and Ursprung, H., Ch. 23, New York: Holt, Rinehart and Winston.
10. DONAHUE, R. P. (1972), 'Control of meiosis and germ cell selection'. In *Perspectives in Cytogenetics: The Next Decade*. Eds. Wright, S. W., Crand-all, B. F. and Boyer, L., Ch. I, Charles C. Thomas: Springfield, Ill.

11. BARR, M. L. (1959), 'Sex chromatin and phenotype in man', *Science*, **130**, 679–685.

12. GONDOS, B. and CONNER, L. A. (1973), 'Ultrastructure of developing germ cells in the fetal rabbit testis', *American Journal of Anatomy*, **136**, 23–42.

13. JOST, A. (1958), 'Embryonic sexual differentiation (morphology, physiology, abnormalities)'. In *Hermaphroditism, Genital Anomalies and Related Endocrine Disorders*, Ch. 2, Williams and Wilkins: Baltimore.

14. WITSCHI, E. (1951), 'Embryogenesis of the adrenal and the reproductive glands', *Recent Progress in Hormone Research*, **6**, 1–27.

15. JOST, A. (1970), 'Hormonal factors in the sex differentiation of the mammalian foetus', *Philosophical Transactions of the Royal Society of London B*, **259**, 119–130.

16. LILLIE, F. (1916), 'The theory of the free-martin', *Science*, **43**, 611–613.

17. PRICE, D. and ORTIZ, E. (1965), 'The role of fetal androgen in sex differentiation in mammals'. On *Organogenesis*. Eds. DeHaan, R. L. and Ursprung, H., Ch. 25, Holt, Rinehart and Winston: New York.

18. FRANCHI, L. L. and MANDL, A. M. (1962), 'The ultrastructure of oogonia and oocytes in the foetal and neonatal rat', *Proceedings of the Royal Society of London B*, **157**, 99–114.

19. FRANCHI, L. L. and MANDL, A. M. (1964), 'The ultrastructure of germ cells in foetal and neonatal male rats', *Journal of Embryology and experimental Morphology*, **12**, 289–308.

20. JOST, A. (1953), 'Problems of fetal endocrinology: the gonadal and hypophyseal hormones', *Recent Progress in Hormone Research*, **8**, 379–418.

21. MANCINI, R. E., NARBAITZ, R. and LAVIERI, J. C. (1960), 'Origin and development of the germinative epithelium and Sertoli cells in the human testis: Cytological, cytochemical and quantitative study', *Anatomical Record*, **136**, 477–489.

22. HUCKINS, C. and CLERMONT, Y. (1968), 'Evolution of gonocytes in the rat testis during late embryonic and early postnatal life', *Archives d'Anatomie, d'Histologie et d'Embryologie*, **51**, 343–354.

23. CLERMONT, Y. and PEREY, B. (1957), 'Quantitative study of the cell population of the seminiferous tubules in immature rats', *American Journal of Anatomy*, **100**, 241–267.

24. COUROT, M., HOCHEREAU-DE REVIERS, M.-T. and ORTAVANT, R. (1970), 'Spermatogenesis'. In *The Testis*. Eds. Johnson, A. D., Gomes, W. R. and Vandemark, N. L., Vol. I, Ch. 6, Academic Press: New York.

25. VAN WAGENEN, G. and SIMPSON, M. E. (1965), *Embryology of the Ovary and Testis. Homo Sapiens and Macaca Mulatta*, Yale University Press: New Haven.

26. SAPSFORD, C. S. (1962), 'Changes in the cells of the sex cords and the seminiferous tubules during the development of the testis of the rat and mouse', *Australian Journal of Zoology*, **10**, 178–192.

27. HILSCHER, W. and MAKOSKI, H.-B. (1968), 'Histologische und autoradiographische Untersuchungen zur "Präspermatogenese" und "Sperm-

atogenese" der Ratte', *Zeitschrift für Zellforschung und mikroskopische Anatomie*, **86**, 327–350.

28. ERICKSON, B. H., REYNOLDS, R. A. and BROOKS, F. T. (1972), 'Differentiation and radioresponse (dose and dose rate) of the primitive germ cell of the bovine testis', *Radiation Research*, **50**, 388–400.

29. GONDOS, B. and HOBEL, C. J. (1971), 'Ultrastructure of germ cell development in the human fetal testis', *Zeitschrift für Zellforschung und mikroskopische Anatomie*, **119**, 1–20.

30. WARTENBERG, H., HOLSTEIN, A. F. and VOSSMEYER, J. (1971), 'Zur Cytologie der pränatalen Gonaden-Entwicklung beim Menschen. II. Elektronenmikroskopische Untersuchungen uber die Cytogenese von Gonocyten und fetalen Spermatogonien im Hoden', *Zeitschrift für Anatomie und Entwicklungsgeschichte*, **134**, 165–185.

31. HILSCHER, B., HILSCHER, W., DELBRÜCK, G. and LEROUGE-BÉNARD, B. (1972), 'Autoradiographische Bestimmung der S-Phasen-Dauer der Gonocyten bei der Wistarratte durch Einfach- und Doppelmarkierung', *Zeitschrift für Zellforschung und mikroskopische Anatomie*, **125**, 229–251.

32. GONDOS, B. and ZEMJANIS, R. (1970), 'Fine structure of spermatogonia and intercellular bridges in *Macaca nemestrina*', *Journal of Morphology*, **131**, 431–446.

33. COUROT, M. (1962), 'Développement du testicule chez l'agneau. Etablissement de la spermatogenese', *Annales de Biologie animale, Biochimie et Biophysique*, **2**, 25–41.

34. SAPSFORD, C. S. (1965), 'The synthesis of DNA and nuclear protein by gonocytes in the testes of normal and x-irradiated rats', *Australian Journal of Biological Sciences* **18**, 653–664.

35. BEAUMONT, H. and MANDL, A. M. (1963), 'A quantitative study of primordial germ cells in the male rat', *Journal of Embryology and experimental Morphology*, **11**, 715–740.

36. VILAR, O. (1970), 'Histology of the human testis from neonatal period to adolescence'. In *The Human Testis*. Eds. Rosemberg, E. and Paulsen, C. A., Ch. 11, Plenum Press: New York.

37. MONESI, V. (1965), 'Synthetic activities during spermatogenesis in the mouse', *Experimental Cell Research*, **39**, 197–224.

38. STEINBERGER, E. and STEINBERGER, A. (1969), 'The spermatogenic function of the testes'. In *The Gonads*. Ed. McKerns, K. W., Ch. 22, Appleton-Century-Crofts: New York.

39. COUROT, M. (1971), 'Établissement de la spermatogénèse chez l'agneau (Ovis aries). Étude expérimentale de son contrôl gonadotrope; importance des cellules de la lignée sertolienne. *Doctoral Thesis, University of Paris*.

40. HUCKINS, C. (1965), 'Duration of spermatogenesis in pre- and postpubertal Wistar rats', *Anatomical Record*, **151**, 364 (Abstract).

41. ROOSEN-RUNGE, E. C. (1962), 'The process of spermatogenesis in mammals', *Biological Reviews*, **37**, 343–377.

42. OAKBERG, E. F. (1971), 'Spermatogonial stem-cell renewal in the mouse', *Anatomical Record*, **169**, 515–532.

43. CLERMONT, Y. (1972), 'Kinetics of spermatogenesis in mammals: Seminiferous epithelium cycle and spermatogonial renewal', *Physiological Reviews*, **52**, 198–236.

44. DYM, M. and FAWCETT, D. W. (1971), 'Further observations on the numbers of spermatogonia, spermatocytes, and spermatids connected by intercellular bridges in the mammalian testis', *Biology of Reproduction*, **4**, 195–215.

45. ROOSEN-RUNGE, E. C. (1969), 'Comparative aspects of spermatogenesis', *Biology of Reproduction*, Suppl. 1, 24–39.

46. MONESI, V. (1964), 'Ribonucleic acid synthesis during mitosis and meiosis in the mouse testis', *Journal of Cell Biology*, **22**, 521–532.

47. LEBLOND, C. P. and CLERMONT, Y. (1952), 'Definition of the stages of the cycle of the seminiferous epithelium of the rat', *Annals of the New York Academy of Sciences*, **55**, 548–573.

48. CLERMONT, Y. (1969), 'Two classes of spermatogonial stem cells in the monkey (*Cercopithecus aethiops*)', *American Journal of Anatomy*, **126**, 57–72.

49. ROWLEY, M. J., BERLIN, J. D. and HELLER, C. G. (1971), 'The ultrastructure of four types of human spermatogonia', *Zeitschrift für Zellforschung und mikroskopische Anatomie*, **112**, 139–157.

50. BURGOS, M. H., VITALE-CALPE, R. and AOKI, A. (1970), 'Fine structure of the testis and its functional significance'. In *The Testis*. Eds. Johnson, A. D., Gomes, W. R. and Vandemark, N. L., Vol. I, Ch. 9, Academic Press: New York.

51. MOSES, M. (1956), 'Chromosomal structures in crayfish spermatocytes', *Journal of Biophysical and Biochemical Cytology*, **2**, 215–217.

52. MONESI, V. (1962), 'Autoradiographic study of DNA synthesis and the cell cycle in spermatogonia and spermatocytes of mouse testis using tritiated thymidine', *Journal of Cell Biology*, **14**, 1–18.

53. UTAKOJI, T. (1966), 'Chronology of nucleic acid synthesis in meiosis of the male Chinese hamster', *Experimental Cell Research*, **42**, 585–596.

54. AMBADKAR, P. M. and GEORGE, J. C. (1964), 'Histochemical localization of certain oxidative enzymes in the rat testis', *Journal of Histochemistry and Cytochemistry*, **12**, 587–590.

55. DALCQ, A. M. (1967), 'Sur la cytochimie chex les rongeurs', *Comptes Rendus*, **264**, 2386–2391.

56. STEINBERGER, A., FICHER, M. and STEINBERGER, E. (1970), 'Studies of spermatogenesis and steroid metabolism in cultures of human testicular tissue'. In *The Human Testis*. Eds. Rosemberg, E. and Paulsen, C. A., Ch. IV, Plenum Press: New York.

57. ORTAVANT, R., COUROT, M. and HOCHEREAU-DE REVIERS, M. T. (1972), 'Gonadotrophic control of tritiated thymidine incorporation in the germinal cells of the rat testis', *Journal of Reproduction and Fertility*, **31**, 451–453.

58. LOSTROH, A. J. (1969), 'Regulation by FSH and ICSH (LH) of reproductive function in the immature male rat', *Endocrinology*, **85**, 438–445.

59. CLERMONT, Y. and HARVEY, S. C. (1967), 'Effects of hormones on spermatogenesis in the rat'. In *Endocrinology of the Testis*. Eds. Wolsten-

holme, G. E. W. and O'Connor, M., Ch. 9, Little, Brown and Company: Boston.

60. GLEDHILL, B. L., GLEDHILL, M. P., RIGLER, R. and RINGERTZ, N. R. (1966), 'Changes in deoxyribonucleoprotein', *Experimental Cell Research*, **41**, 652–665.

61. SUSI, F. R., LEBLOND, C. P. and CLERMONT, Y. (1971), 'Changes in the Golgi apparatus during spermiogenesis in the rat', *American Journal of Anatomy*, **130**, 251–268.

62. FRANKLIN, L. E. (1971), 'An association of endoplasmic reticulum with the Golgi apparatus in golden hamster spermatids', *Journal of Reproduction and Fertility*, **27**, 67–71.

63. ALLISON, A. C. and HARTREE, E. F. (1970), 'Lysosomal enzymes in the acrosome and their possible role in fertilization', *Journal of Reproduction and Fertility*, **21**, 501–515.

64. SRIVASTAVA, P. N., ADAMS, C. E. and HARTREE, E. F. (1965), 'Enzymic action of acrosomal preparations on the rabbit ovum in vitro', *Journal of Reproduction and Fertility*, **10**, 61–67.

65. SRIVASTAVA, P. N., ZANEVELD, L. J. D. and WILLIAMS, W. L. (1970), 'Mammalian sperm acrosomal neuraminidases', *Biochemical and biophysical Research Communications*, **29**, 575–582.

66. SEIGUER, A. C. and CASTRO, A. E. (1972), 'Electron microscopic demonstration of arylsulfatase activity during acrosome formation in the rat', *Biology of Reproduction*, **7**, 31–42.

67. BISHOP, D. W. (1968), 'Testicular enzymes as fingerprints in the study of spermatogenesis'. In *Reproduction and Sexual Behavior*. Ed. Diamond, M., Ch. 18, Indiana University Press: Indianapolis.

68. LACY, D. (1960), 'Light and electron microscopy and its use in the study of factors influencing spermatogenesis in the rat', *Journal of the Royal Microscopy Society*, **79**, 209–225.

69. LACY, D. (1967), 'The seminiferous tubule in mammals', *Endeavour*, **26**, 101–108.

70. FAWCETT, D. W. (1958), 'The fine structure of the mammalian spermatozoon', *International Review of Cytology*, **7**, 195–234.

71. GIER, H. T. and MARION, G. B. (1970), 'Development of the mammalian testis'. In *The Testis*. Eds. Johnson, A. D., Gomes, W. R. and Vandemark, N. L., Vol. I, Ch. 1, Academic Press: New York.

72. JIRASEK, J. E. (1967), 'The relationship between the structure of the testis and differentiation of the external genitalia and phenotype in man'. In *Endocrinology of the Testis*. Eds. Wolstenholme, G. E. W. and O'Connor, M., Ch. 1, Little, Brown and Company: Boston.

73. BLACK, V. E. and CHRISTENSEN, A. K. (1969), 'Differentiation of interstitial cells and Sertoli cells in fetal guinea pig testes', *American Journal of Anatomy*, **124**, 211–238.

74. COUROT, M. (1967), 'Endocrine control of the supporting and germ cells of the impuberal testis', *Journal of Reproduction and Fertility*, Supplement 2, 89–101.

75. COUROT, M. (1970), 'Effect of gonadotropins on the seminiferous tubules

of the immature testis'. In *The Human Testis*. Eds. Rosemberg, E. and Paulsen, C. A., Ch. IV, Plenum Press: New York.

76. OKADA, K. (1970), 'Experimental study on the proliferation and the maturation of germ-cell, Sertoli cell and Leydig cell of the rat testis', *Japanese Journal of Urology*, **61**, 1125–1146.

77. FLICKINGER, C. J. (1967), 'The postnatal development of the Sertoli cells of the mouse', *Zeitschrift für Zellforschung und mikroskopische Anatomie*, **78**, 92–113.

78. BRÖKELMANN, J. (1963), 'Fine structure of germ cells and Sertoli cells during the cycle of the seminiferous epithelium in the rat', *Zeitschrift für Zellforschung und mikroskopische Anatomie*, **59**, 820–850.

79. CASTRO, A. E., SEIGUER, A. C. and MANCINI, R. E. (1970), 'Electron microscopic study of the localization of labeled gonadotropins in the Sertoli and Leydig cells of the rat testis', *Proceedings of the Society for Experimental Biology and Medicine*, **133**, 582–586.

80. NICANDER, L. (1967), 'An electron microscopical study of cell contacts in the seminiferous tubules of some mammals', *Zeitschrift für Zellforschung und mikroskopische Anatomie*, **83**, 375–397.

81. DYM, M. and FAWCETT, D. W. (1970), 'The blood-testis barrier in the rat and the physiological compartmentation of the seminiferous epithelium', *Biology of Reproduction*, **3**, 308–326.

82. LYNCH, K. M. and SCOTT, W. W. (1950), 'The lipid content of the Leydig cell and Sertoli cell in the human testis as related to age, benign prostatic hyperplasia, and prostatic cancer', *Journal of Urology*, **64**, 767–776.

83. LOFTS, B. (1961), The effect of follicle-stimulating hormone and luteinizing hormone on the testis of hypophysectomized frogs (*Rana temporaria*). *General and Comparative Endocrinology*, **1**, 179–189.

84. HUGGINS, C. and MOULDER, P. V. (1945), 'Estrogen production by Sertoli cell tumour of the testis', *Cancer Research*, **5**, 501–514.

85. CHRISTENSEN, A. K. and MASON, N. R. (1965), 'Comparative ability of seminiferous tubules and interstitial tissue of rat testes to synthesize androgens from progesteron-4-^{14}C *in vitro*', *Endocrinology*, **76**, 646–659.

86. PARVINEN, M. (1970), 'The fate of exogenous tritiated cholestered, pregnenolone, progesterone and testosterone in the interstitial and tubular tissue of the rat testis'. In *Morphological Aspects of Andrology*. Eds. Holstein, A. F. and Horstmann, E., Section A, Springer Verlag: Berlin.

87. LEESON, C. R. and LEESON, T. S. (1963), 'The postnatal development and differentiation of the boundary tissue of the seminiferous tubule of the rat', *Anatomical Record*, **147**, 243–260.

88. ROSS, M. H. (1970), 'The Sertoli cell and the blood-testicular barrier: an electronmicroscopic study'. In *Morphological Aspects of Andrology*. Eds. Holstein, A. F. and Horstmann, E., Section A, Springer Verlag: Berlin.

89. KORMANO, M. (1970), 'The development and function of the peritubular tissue in the rat testis'. In *Morphological Aspects of Andrology*. Eds. Holstein, A. F. and Horstmann, E., Section A, Springer Verlag: Berlin.

90. SUVANTO, O. (1970), 'Behaviour of rat seminiferous tubules in organ culture, with special reference to their contractility'. In *Morphological*

Aspects of Andrology. Eds. Holstein, A. E. and Horstmann, E., Section A, Springer Verlag: Berlin.

91. NIEMI, M., IKONEN, M. and HERVONEN, A. (1967), 'Histochemistry and fine structure of the interstitial tissue in the human foetal testis'. In *Endocrinology of the Testis*. Eds. Wolstenholme, G. E. W. and O'Connor, M., Ch. 2, Little, Brown and Company: Boston.

92. HITZEMAN, J. W. (1962), 'Development of enzyme activity in the Leydig cells of the mouse testis', *Anatomical Record*, **143**, 351–362.

93. SCHLEGEL, R. J., FARIAS, E., RUSSO, N. C., MOORE, J. R. and GARDNER, L. I. (1967), 'Structural changes in fetal gonads and gonaducts during maturation of an enzyme, steroid 3β-ol dehydrogenase, in the gonads, adrenal cortex and placenta of fetal rats', *Endocrinology*, **81**, 565–572.

94. BLOCH, E. (1967), 'The conversion of 7-^{3}H pregnenolone and 4-^{14}C-progesterone to testosterone and androstenedione by mammalian fetal testes *in vitro*', *Steroids*, **9**, 415–430.

95. LIPSETT, M. B. and TULLNER, W. W. (1965), 'Testosterone synthesis by the fetal rabbit gonad', *Endocrinology*, **77**, 273–277.

96. ZAAIJER, J. J. P. and PRICE, D. (1971), 'Early secretion of the androgenic hormones by human fetal gonads and adrenal glands in organ culture and possible implications for sex differentiation'. In *Hormones in Development*. Eds. Hamburgh, M. and Barrington, E. J., Ch. 40, Appleton-Century-Crofts: New York.

97. HARRIS, G. W. and LEVINE, S. (1965), 'Sexual differentiation of the brain and its experimental control', *Journal of Physiology*, **181**, 379–400.

98. BARRACLOUGH, C. A. (1966), 'Modification in the CNS regulation of reproduction after exposure of prepubertal rats to steroid hormone', *Recent Progress in Hormone Research*, **22**, 503–528.

99. JOST, A. (1947), 'Expériences de décapitation de l'embryon de lapin', *Comptes Rendus*, **225**, 322–324.

100. HOOKER, C. W. (1970), 'The intertubular tissue of the testis'. In *The Testis*. Eds. Johnson, A. D., Gomes, W. R. and Vandemark, N. L., Vol. I, Ch. 8, Academic Press: New York.

101. PELLINIEMI, L. J. and NIEMI, M. (1969), 'Fine structure of the human foetal testis. I. The interstitial tissue', *Zeitschrift für Zellforschung und mikroskopische Anatomie*, **99**, 507–522.

102. INANO, H. and TAMAOKI, B. (1966), 'Bioconversion of steroids in immature rat testes *in vitro*', *Endocrinology*, **79**, 579–590.

103. JARLSTEDT, J. and STEWARD, V. W. (1968), 'Content of ribonucleic acid in rat interstitial cells at different ages', *Endocrinology*, **82**, 1063–1065.

104. DeKRETSER, D. M. (1967), 'Changes in the fine structure of the human testicular interstitial cells after treatment with human gonadotropins', *Zeitschrift für Zellforschung und mikroskopische Anatomie*, **83**, 344–358.

105. TANNER, J. M. (1969), 'Growth and endocrinology of the adolescent'. In *Endocrine and Genetic Diseases of Childhood*. Ed. Gardner, L. I., Sect. 1., W. B. Saunders: Philadelphia and London.

106. DAVIS, J. R., LANGFORD, G. A. and KIRBY, P. J. (1970), 'The testicular

capsule'. In *The Testis*. Eds. Johnson, A. D., Gomes, W. R. and Vandemark, N. L., Vol. I, Ch. 5, Academic Press: New York.

107. BUTLER, H. (1969), 'Post pubertal oogenesis in Prosimiae', *Recent Advances in Primatology*, **2**, 15–21.

108. FRANCHI, L. L., MANDL, A. M. and ZUCKERMAN, S. (1962), 'The development of the ovary and the process of oogenesis'. In *The Ovary*. Eds. Zuckerman, S., Mandl, A. M. and Eckstein, P., Vol. I, Ch. 1, Academic Press: New York.

109. ZUCKERMAN, S. (1951), 'The number of oocytes in the mature ovary', *Recent Progress in Hormone Research*, **6**, 63–109.

110. KENNELLY, J. J. and FOOTE, R. H. (1966), 'Oocytogenesis in rabbits. The role of neogenesis in the formation of the definitive ova and the stability of oocyte DNA measured with tritiated thymidine', *American Journal of Anatomy*, **118**, 573–590.

111. GONDOS, B. and ZAMBONI, L. (1969), 'Ovarian development: The functional importance of germ cell interconnections', *Fertility and Sterility* **20**, 176–189.

112. DAVIDSON, E. H. and HOUGH, B. R. (1972), 'Utilization of genetic information during oogenesis'. In *Oogenesis*. Eds. Biggers, J. D. and Schuetz, A. W., Ch. 7, University Park Press: Baltimore.

113. PETERS, H., LEVY, E. and CRONE, M. (1965), 'Oogenesis in rabbits', *Journal of Experimental Zoology*, **158**, 169–180.

114. PETERS, H. (1970), 'Migration of gonocytes into the mammalian gonad and their differentiation', *Philosophical Transactions of the Royal Society of London B*, **259**, 91–101.

115. KENNELLY, J. J., FOOTE, R. H. and JONES, R. C. (1970), 'Duration of premeiotic deoxyribonucleic acid synthesis and the stages of prophase 1 in rabbit oocytes', *Journal of Cell Biology*, **47**, 577–584.

116. ZAMBONI, L. and GONDOS, B. (1968), 'Intercellular bridges and synchronization of germ cell differentiation during oogenesis in the rabbit', *Journal of Cell Biology*, **36**, 276–282.

117. RUBY, J. R., DYER, R. F. and SKALKO, R. G. (1969), 'The occurrence of intercellular bridges during oogenesis in the mouse', *Journal of Morphology*, **127**, 307–340.

118. BEATTY, R. A. (1970), 'The genetics of the mammalian gamete', *Biological Reviews*, **45**, 73–119.

119. BAKER, T. G. (1963), 'A quantitative and cytological study of germ cells in human ovaries', *Proceedings of the Royal Society of London B*, **158**, 417–433.

120. GONDOS, B. (1972), 'Germ cell degeneration in the developing rabbit ovary'. In *Cell Differentiation*. Eds. Harris, R., Allin, P. and Viza, D., Section IV, Munksgaard: Copenhagen.

121. JONES, E. and KROHN, P. (1961), 'The relationships between age, numbers of oocytes and fertility in virgin and multiparous mice', *Journal of Endocrinology*, **21**, 469–495.

122. GONDOS, B., BHIRALEUS, P. and HOBEL, C. J. (1971), 'Ultrastructural observations on germ cells in human fetal ovaries', *American Journal of Obstetrics and Gynecology*, **110**, 644–652.

123. ODOR, D. L. and BLANDAU, R. J. (1969), 'Ultrastructural studies on fetal and early postnatal mouse ovaries. I. Histogenesis and organogenesis', *American Journal of Anatomy*, **124**, 163–186.

124. THIBAULT, C. G. (1972), 'Final stages of mammalian oocyte maturation'. In *Oogenesis*. Eds. Biggers, J. D. and Schuetz, A. W., Ch. 21, University Park Press: Baltimore.

125. BAKER, T. G. (1970), 'Electron microscopy of the primary and secondary oocyte', *Advances in the Biosciences*, **6**, 7–23.

126. SMITH, L. D. (1972), 'Protein synthesis during oocyte maturation'. In *Oogenesis*. Eds. Biggers, J. D. and Schuetz, A. W., Ch. 12, University Park Press: Baltimore.

127. WEAKLEY, B. S. (1971), 'Basic protein and ribonucleic acid in the cytoplasm of the ovarian oocyte in the golden hamster', *Zeitschrift für Zellforschung und mikroskopische Anatomie*, **112**, 69–84.

128. INGRAM, D. L. (1962), 'Atresia'. In *The Ovary*. Eds. Zuckerman, S., Mandl, A. M. and Eckstine, P., Vol. I, Ch. 4, Academic Press: London.

129. BAKER, T. G. and NEAL, P. (1972), 'Gonadotrophin-induced maturation of mouse Graafian follicles in organ culture'. In *Oogenesis*. Eds. Biggers, J. D. and Schuetz, A. W., Ch. 20, University Park Press: Baltimore.

130. ZAMBONI, L. (1972), 'Comparative studies on the ultrastructure of mammalian oocytes'. In *Oogenesis*. Eds. Biggers, J. D. and Schuetz, A. W., Ch. 1, University Park Press: Baltimore.

131. SZOLLOSI, D. (1967), 'Development of cortical granules and the cortical reaction in rat and hamster eggs', *Anatomical Record*, **159**, 431–446.

132. AUSTIN, C. R. (1968), *Ultrastructure of Fertilization*. Holt, Rinehart and Winston, New York.

133. ZAMBONI, L. and MASTROIANNI, L. (1966), 'Electron microscopic studies on rabbit ova. II. The penetrated tubal ovum', *Journal of Ultrastructural Research*, **14**, 118–132.

134. BJÖRKMAN, N. (1962), 'A study of the ultrastructure of the granulosa cells of the rat ovary', *Acta anatomica*, **51**, 125–147.

135. GOLDMAN, A. S., YAKOVAC, W. C. and BONGIOVANNI, A. M. (1966), 'Development of activity of 3β-hydroxysteroid dehydrogenase in human fetal tissues and in two anencephalic newborns', *Journal of Clinical Endocrinology and Metabolism*, **26**, 14–22.

136. WENIGER, J.-P., CHOURAQUI, J. and ZEIS, H. (1972), 'Sur l'activité sécrétoire de l'ovaire embryonnaire de mammifére', *Annales d'Endocrinologie*, **33**, 243–250.

137. HERTZ, R. (1963), 'Pituitary independence of the prepubertal development of the ovary of the rat and rabbit and its pertinence to hypo-ovarianism in women'. In *The Ovary*. Eds. Grady, H. G. and Smith, D. E., Ch. 7, Williams and Wilkins: Baltimore.

138. GONDOS, B. (1970), 'Morphologic evidence for steroidogenic activity in the prefollicular rabbit ovary', *Excerpta Medica International Congress Series*, **210**, 247.

139. PETERS, H. (1969), 'The development of the mouse ovary from birth to maturity', *Acta endocrinologica*, **62**, 98–116.

140. GONDOS, B. (1970), 'Granulosa cell-germ cell relationship in the developing rabbit ovary', *Journal of Embryology and experimental Morphology*, **23**, 419–426.

141. BROWN, E. H. and KING, R. C. (1964), 'Studies on the events resulting in the formation of an egg chamber in *Drosophila melanogaster*', *Growth*, **28**, 41–81.

142. NORREVANG, A. (1968), 'Electron microscopic morphology of oogenesis', *International Review of Cytology*, **23**, 113–186.

143. RAVEN, CHR. P. (1961), *Oogenesis: The storage of developmental information*. Pergamon Press, Oxford.

144. MERKER, H. J. (1961), 'Elektronenmikroskopische Untersuchungen über die Bildung der Zona Pellucida in den Follikeln des Kaninchenovars', *Zeitschrift für Zellforschung und mikroskopische Anatomie*, **54**, 677–688.

145. ANDERSON, E. and BEAMS, H. W. (1960), 'Cytologic observations on the fine structure of the guinea pig ovary with special reference to the oogonium, primary oocyte and associated follicle cells', *Journal of Ultrastructural Research*, **3**, 432–446.

146. RYLE, M. (1969), 'Morphological responses to pituitary gonadotrophins by mouse ovaries *in vitro*', *Journal of Reproduction and Fertility*, **20**, 307–312.

147. BEN-OR, S. and BROZA, R. (1970), 'Effect of gonadotrophins on protein synthesis in the ovary of the mouse'. In *Gonadotrophins and Ovarian Development*. Eds. Butt, W. R., Crooke, A. C. and Ryle, M., Part II, E. & S. Livingstone: Edinburgh.

148. ESHKOL, A., HARDY, B., and PARIENTE-CORIAT, C. (1970), 'Changes in the rate of nucleic acid synthesis in juvenile mouse ovaries after FSH stimulation'. In *Gonadotrophins and Ovarian Development*. Eds. Butt, W. R., Crooke, A. C. and Ryle, M., Part II, E. & S. Livingstone: Edinburgh.

149. BRANDAU, H. (1970), 'Histochemical localization of enzyme activities in normal and gonadotrophin stimulated mouse ovaries'. In *Gonadotrophins and Ovarian Development*. Eds. Butt, W. R., Crooke, A. C. and Ryle, M., Part II, E. & S. Livingstone: Edinburgh.

150. AHRÉN, K., HAMBERGER, L. and RUBINSTEIN, L. (1969), 'Acute *in vivo* and *in vitro* effects of gonadotrophins on the metabolism of the rat ovary'. In *The Gonads*. Ed. McKerns, K. W., Ch. 12, Appleton-Century-Crofts: New York.

151. DONAHUE, R. P. and STERN, S. (1968), 'Follicular cell support of oocyte oocyte maturation: production of pyruvate *in vitro*', *Journal of Reproduction and Fertility*, **17**, 395–398.

152. BIGGERS, J. D. (1970), 'Metabolism of the oocyte'. In *Oogenesis*. Eds. Biggers, J. D. and Schuetz, A. W., Ch. 13, University Park Press: Baltimore.

153. ENDERS, A. C. (1962), 'Observations on the fine structure of lutein cells', *Journal of Cell Biology*, **12**, 101–113.

154. BLANCHETTE, E. J. (1966), 'Ovarian steroid cells. I. Differentiation of the

lutein cell from the granulosa follicle cell during the preovulatory state and under the influence of exogenous gonadotrophins', *Journal of Cell Biology*, **31**, 501–516.

155. RYAN, K. J. and SMITH, O. W. (1965), 'Biogenesis of steroid hormones in the human ovary', *Recent Progress in Hormone Research*, **21**, 367–402.

156. HAMBERGER, L. (1968), 'Influence of gonadotrophins on the respiration of isolated cells from the prepubertal rat ovary', *Acta physiologica Scandinavica*, **74**, 410–425.

157. EL-FOULY, M., COOK, B., NEKOLA, M. and NALBANDOV, A. V. (1970), 'Role of the ovum in follicular luteinization', *Endocrinology*, **87**, 288–293.

158. CHRISTENSEN, A. K. and GILLIM, S. W. (1969), 'The correlation of fine structure and function in steroid-secreting cells, with emphasis on those of the gonads'. In *The Gonads*. Ed. MCKERNS, K. W., Ch. 16, Appleton-Century-Crofts: New York.

159. MERKER, H. J. and DIAZ-ENCINAS, J. (1969), 'Das elektronenmikroskopische Bild des Ovars juveniler Ratten und Kaninchen nach Stimulierrung mit PMS und HCG. I. Theka und Stroma (Interstitielle Drüse)', *Zeitschrift für Zellforschung und mikroskopische Anatomie*, **94**, 605–623.

160. RYAN, K. J. (1962), 'Synthesis of hormones in the ovary'. In *The Ovary*. Eds. Grady, H. G. and Smith, D. E., Ch. 5, Williams & Wilkins: Baltimore.

161. OSVALDO-DECIMA, L. (1970), 'Smooth muscle in the ovary of the rat and monkey', *Journal of Ultrastructure Research*, **29**, 218–237.

162. LIPNER, H. J. and MAXWELL, B. A. (1960), 'Hypothesis concerning the role of follicular contractions in ovulation', *Science*, **131**, 1737–1738.

163. HARRISON, R. J. (1962), 'The structure of the Ovary. Mammals'. In *The Ovary*. Eds. Zuckerman, S., Mandl, A. M. and Eckstine, P., Ch. 2C, Academic Press: London.

164. MORRIS, J. M. and SCULLY, R. E. (1958), *Endocrine Pathology of the Ovary*. C. V. Mosby: St Louis.

165. DEANESLY, R. (1970), 'Oogenesis and the development of the ovarian interstitial tissue in the ferret', *Journal of Anatomy*, **107**, 165–178.

166. STEGNER, H.-E. (1970), 'Electron microscopic studies of the interstitial tissue in the immature mouse ovary'. In *Gonadotrophins and Ovarian Development*. Eds. Butt, W. R., Crooke, A. C. and Ryle, M., Section II, E. & S. Livingstone: Edinburgh.

167. PRESL, J., JIRASEK, J., HORSKY, J. and HENZL, M. (1965), 'Observations on steroid-3β-ol dehydrogenase activity in the ovary during the early postnatal development of the rat', *Journal of Endocrinology*, **31**, 283–284.

168. MORI, H. and MATSUMOTO, K. (1970), 'On the histogenesis of the ovarian interstitial gland in rabbits. I. Primary interstitial gland', *American Journal of Anatomy*, **129**, 289–306.

169. FLERKÓ, B., HAJÓS, F. and SÉTÁLÓ, GY. (1967), 'Electron microscopic observations on rat ovaries in different stages of development and steroidogenesis', *Acta Morphologica Academiae Scientiarum Hungaricae*, **15**, 163–183.

170. DAVIES, J. and BROADUS, C. D. (1968), 'Studies on the fine structure of ovarian steroid-secreting cells in the rabbit. I. The normal interstitial cells', *American Journal of Anatomy*, **123**, 441–474.
171. HILLIARD, J., ARCHIBALD, D. and SAWYER, C. H. (1963), 'Gonadotropic activation of preovulatory synthesis of progestin in the rabbit', *Endocrinology*, **72**, 59–66.
172. STEINBERG, W. H. (1949), 'The morphology, androgenic function, hyperplasia and tumors of the human ovarian hilus cells', *American Journal of Pathology*, **25**, 493–522.

Bibliography

BIGGERS, J. D. and SCHUETZ, A. W. (1972), *Oogenesis*, University Park Press: Baltimore.

BUTT, W. R., CROOKE, A. C. and RYLE, M. (1970), *Gonadotrophins and Ovarian Development*, E. & S. Livingstone: Edinburgh.

GRADY, H. G. and SMITH, D. E. (1963), *The Ovary*, Williams & Wilkins: Baltimore.

HOLSTEIN, A. F. and HORSTMANN, E. (1970), *Morphological Aspects of Andrology*, Grosse Verlag: Berlin.

JOHNSON, A. D., GOMES, W. R. and VANDEMARK, N. L. (1970), *The Testis*, Academic Press: New York and London.

McKERNS, K. W. (1969), *The Gonads*, Appleton-Century-Crofts: New York.

WOLSTENHOLME, G. E. W. and O'CONNOR, M. (1967), *Endocrinology of the Testis*, Little, Brown and Company: Boston.

ZUCKERMAN, S., MANDL, A. M. and ECKSTEIN, P. (1962), *The Ovary*, Academic Press: New York and London.

7 The Development of the Cells of the Blood

P. F. HARRIS

Pre-natal Development

Introduction

Before birth haemopoiesis occurs in four centres. The yolk sac is the *primary haemopoietic centre*. Experimental evidence suggests that its main function is to export stem cells via the embryonic circulation to the *secondary centres*, liver, spleen and bone marrow. The liver itself also exports stem cells to the spleen and bone marrow. Haemopoiesis spreads through the various centres in three phases, increasing in one as it recedes in another. It occurs first in the yolk sac, then the liver, and finally in spleen and bone marrow. This change in location reveals a fundamental property of the primitive stem cell, the ability to migrate.

Development of blood islands – the earliest haemopoietic centres

The early embryo is essentially bilaminar, consisting of a dorsal (epiblast) layer and a ventral endodermal layer. The epiblast further differentiates into ectoderm and mesoderm. Thus, the embryo becomes trilaminar with dorsal (ectoderm), middle (mesoderm) and ventral (endoderm) layers. It is the mesoderm which gives rise to all haemopoietic cells (Fig. 7.1).

The earliest centres of blood formation are the blood islands in the wall of the yolk sac. The mesodermal cells make intimate contact with endodermal cells lining the yolk sac and this contact is vital for the further differentiation of cells in the blood islands into primitive haemic stem cells. The endoderm seems to create a micro-environment which is essential for differentiation. A similar dependence upon micro-environment occurs post-natally in the differentiation of stem cells in bone marrow.

Functional significance and location of blood islands

In birds and mammals blood islands are the primordial haemopoietic centres. They have two functions: (1) production of primitive blood

209

cells (mostly erythroid) and (2) the export of multipotent stem cells to establish haemopoiesis in the secondary centres.

Blood islands occur in two sites. Some, the *intra-embryonic* islands, remain in close contact with the yolk sac wall within the embryo and

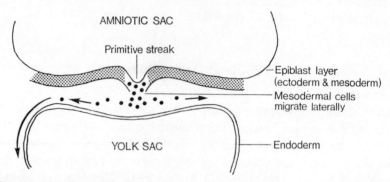

Fig. 7.1 Mesodermal origin of primitive blood islands.

these are probably the most important. Others migrate laterally around the yolk sac to develop in *extra-embryonic* sites, some remaining close to the yolk sac but others migrating more distantly to surround the allantois in the body stalk or even into the mesoderm of the chorion of the placenta (Fig. 7.2).

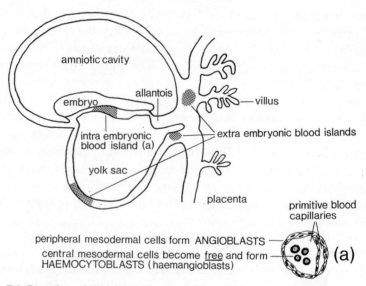

Fig. 7.2 Location of blood islands and their basic components.

Cytology of blood islands – development of primitive stem cells

In blood islands mesodermal cells differentiate into central (haemangioblast) and peripheral (angioblast) groups. Angioblasts develop into flat endothelial cells lining the walls of the primitive capillaries in the blood islands. The central haemangioblasts become detached from the walls, their cell membranes initially remaining in close contact with desmosome-like structures between them. Soon they develop intense basophilia of the cytoplasm and separate into free cells (Plate 7.1), a

Stage	PRE–NATAL		POST–NATAL
Site of erythropoiesis	YOLK SAC	LIVER SPLEEN BONE MARROW	BONE MARROW
Type of haemoglobin	HbE (up to 12 weeks)	HbF (8–36 weeks +) →	HbA (after 12 weeks)
Export of stem cells	YOLK SAC	LIVER SPLEEN BONE MARROW	BONE MARROW → Thymus Spleen Yellow marrow

Fig. 7.3 Types of haemoglobin in developing red cells and stem cell export centres.

change particularly dependent upon the proximity of endodermal cells lining the yolk sac (Plate 7.2).

These central cells are now primitive haemocytoblasts having migratory capacity.

Haematogenous migration of stem cells from the yolk sac and their potentialities

The yolk sac blood islands are particularly rich in stem cells. The scale of their export is considerable. Their blood levels are ten times greater before, than after birth. They seed the secondary haemopoietic centres, including lymphoid tissues, establishing within them their own stem cell populations (Fig. 7.3). This stem cell migration has been elegantly demonstrated by Moore and Owen[1] using a sex-chromosome marker technique in developing chick embryos whose circulations were joined together. By their ability to re-populate haemopoietic tissues of lethally irradiated mice and to give rise to mixed colonies when assayed by splenic colonization[2] these cells are shown to be truly *multipotent*. Some are more restricted in potential, forming only granulocytes and macrophages[3].

Thus a fundamental event occurs at the inception of haemopoiesis. *Fixed* mesenchymal cells transform into *free* multipotent stem cells, enabling a cell migration stream to become established. Derangement, excision[39] or failure in development of yolk sac blood islands, profoundly affects subsequent haemopoietic development.

Local differentiation of stem cells in yolk sac blood islands

Differentiation of stem cells remaining in the yolk sac is relatively restricted. In vertebrates all cell types except lymphocytes can be produced, but in the mammal the yolk sac is predominantly erythropoietic. This may be explained by environment. Local oxygen tension may be low. Also, intensive cellular proliferation in the embryo at this stage requires urgent development of an efficient oxygen transport system. Further, in the sterile conditions of uterine development there is no stimulus for granulocytopoiesis. All stages of erythroblast occur. The red cells are nucleated and larger than post-natal RBC. On entering the circulation the nucleus is retained but ribosomes and most mitochondria disappear. Once haemopoiesis is established in the liver, it recedes in the yolk sac, possibly as a result of feedback or numerical exhaustion following export of stem cells.

Haemopoiesis in the liver

With recession of the yolk sac, the liver assumes a major haemopoietic rôle.

Origin of haemopoietic stem cells in the liver

Three sources for stem cells are conventionally proposed, each an intrinsic component of the liver anlage: (a) endoderm giving rise to hepatocytes, (b) endothelium of developing sinusoids, and (c) mesenchyme surrounding vitelline veins in the septum transversum. Despite recent electron microscopic studies of developing endodermal cells, evidence points to an extrinsic origin for the stem cells which migrate into the liver from the yolk sac. They develop extravascularly close to the endoderm forming hepatocytes. Presumably, as in the yolk sac, the endodermal cells create a suitable microenvironment.

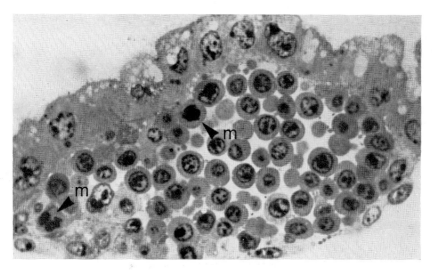

Plate 7.1 Cells in primitive blood islands. The central cells are free and enclosed by peripheral haemangioblasts. × 752. *m* = mitotic figure. (Reproduced from *Regulation of Hematopoiesis*, ed. Gordon, 1970.)

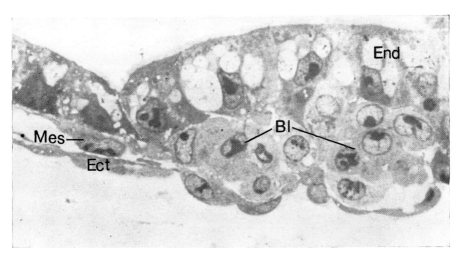

Plate 7.2 Intimate relation of mesodermal cells in developing blood islands to endodermal cells. × 752.

Bl = cells in primitive blood island		*Ect* = ectoderm
Mes = mesoderm		*End* = endoderm

(Reproduced from *Regulation of Hematopoiesis*, ed. Gordon, 1970.)

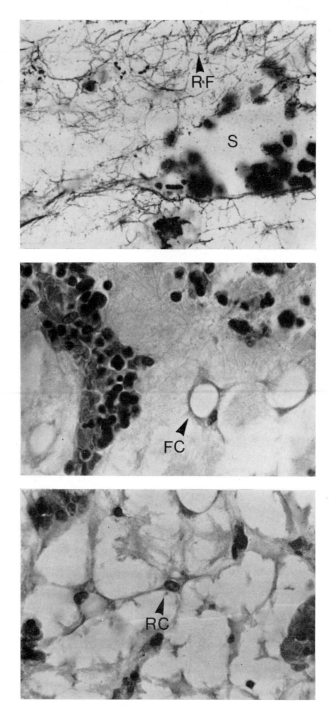

Plate 7.3 The micro-skeleton of bone marrow. × 480.

RF = reticulin fibres   *FC* = fat cell

S   = sinusoid   *RC* = reticulum cell

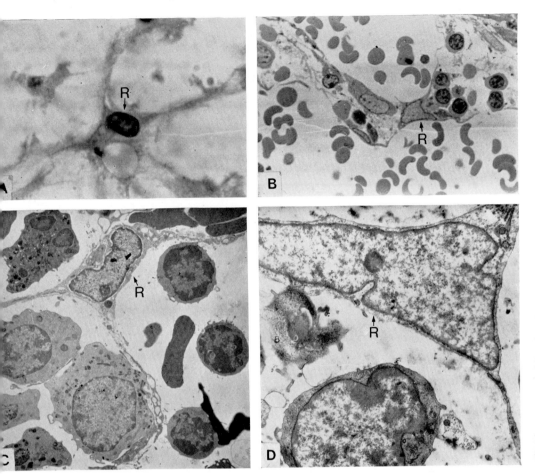

Plate 7.4 Reticulum cells in bone marrow. A, × 980; B, × 840; C, × 2450; D, × 7000. *R* = reticulum cell.

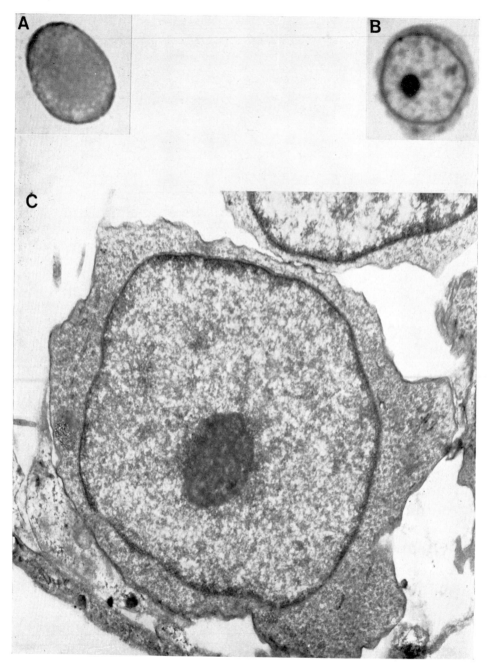

Plate 7.5 Bone marrow – multipotent stem cell. A, bone marrow smear – Leishman stain. × 1350. B, bone marrow – thin section (Araldite) – toluidine blue. × 1800. C, bone marrow – electron micrograph. × 9000. Particularly noticeable are (i) the high nuclear/cytoplasmic ratio; (ii) leptachromatic chromatin; (iii) conspicuous nucleolus and (iv) abundant ribosomes with absence of other organelles.

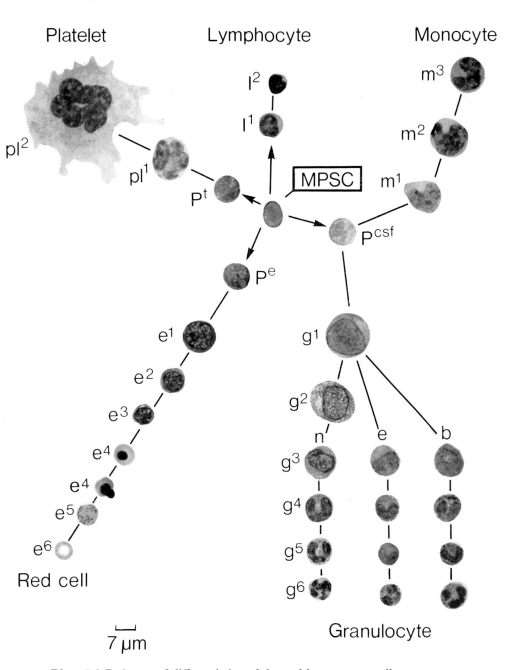

Plate 7.6 Pathways of differentiation of the multipotent stem cell.

MPSC = multipotent stem cell
Pe = progenitor cell sensitive to erythropoietin
Pcsf = progenitor cell sensitive to colony stimulating factor
Pt = progenitor cell sensitive to thrombopoietin
e¹ = proerythroblast
e² = basophilic (early) erythroblast

e³ = polychromatic (intermediate) erythroblast
e⁴ = orthochromatic (late) erythroblast
e⁵ = reticulocyte
e⁶ = red cell
g¹ = myeloblast
g² = promyelocyte
g³ = myelocyte
g⁴ = metamyelocyte
g⁵ = stab (band) cell

g⁶ = polymorphonuclear cell
n = neutrophil
e = eosinophil
b = basophil
m¹ = monoblast
m² = promonocyte
m³ = monocyte
pl¹ = megakaryoblast
pl² = megakaryocyte

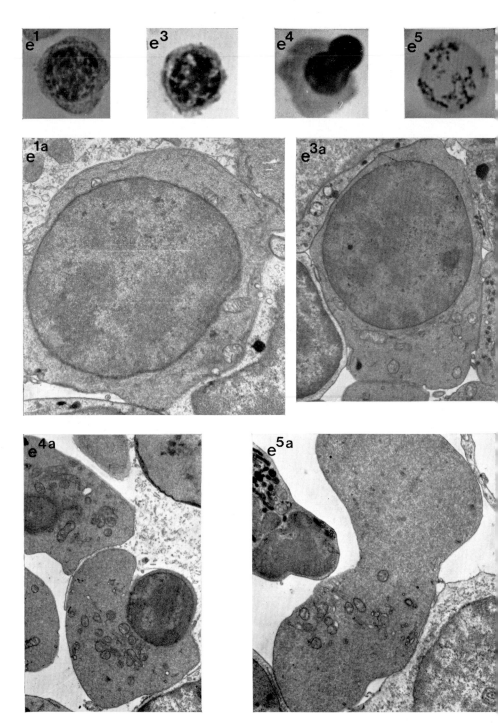

Plate 7.7 Stages in development of the red cell.

e^1 = proerythroblast. × 1050
e^{1a} = proerythroblast. × 5250
e^3 = polychromatic erythroblast. × 1050
e^{3a} = polychromatic erythroblast. × 5250
e^4 = orthochromatic erythroblast with extruding nucleus. × 1050

e^{4a} = orthochromatic erythroblast with extruding nucleus. × 5250
e^5 = reticulocyte. × 1050
e^{5a} = reticulocyte. × 5250

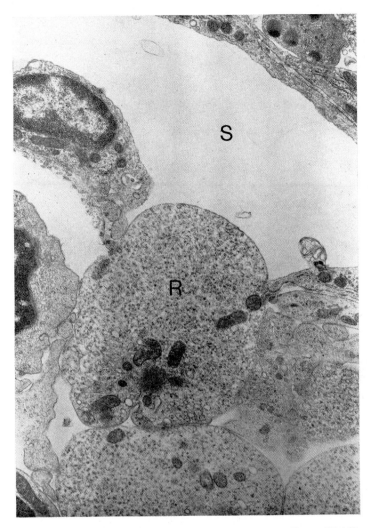

Plate 7.8 Reticulocyte passing through sinusoidal wall. × 17 000.
S = sinusoid; *R* = reticulocyte. (Reproduced from *Regulation of Hematopoiesis*, ed. Gordon, 1970.)

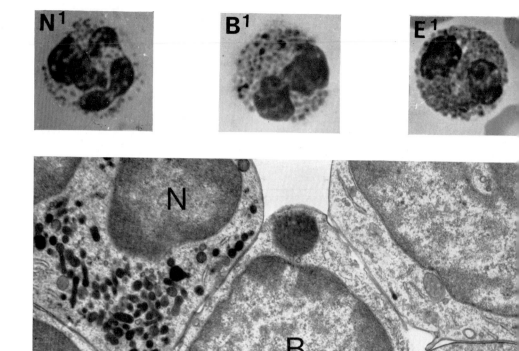

Plate 7.9 Three types of granular leucocytes. N = neutrophil; B = basophil; E = eosinophil. The three cells N^1, B^1 and E^1 are stained with Leishman stain. $\times$ 1350. N, B and E are electron micrographs. $\times$ 6750.

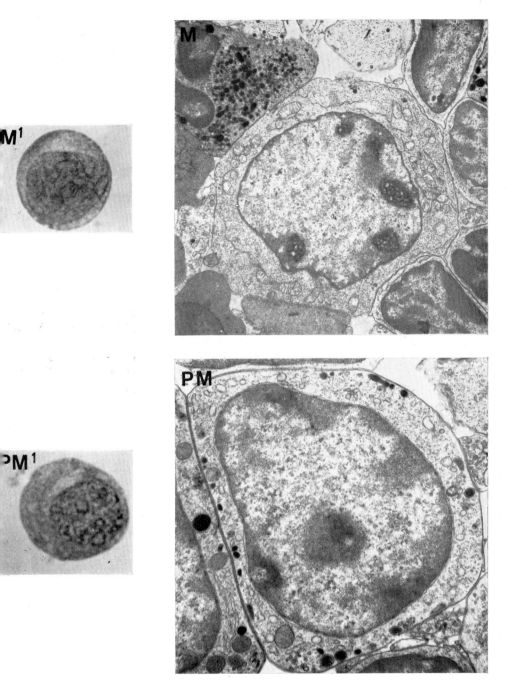

Plate 7.10 Myeloblast; promyelocyte. M = myeloblast; PM = promyelocyte
M^1 and PM^1 stained with Leishman stain. × 1500. M and PM are electron
micrographs. × 7500.

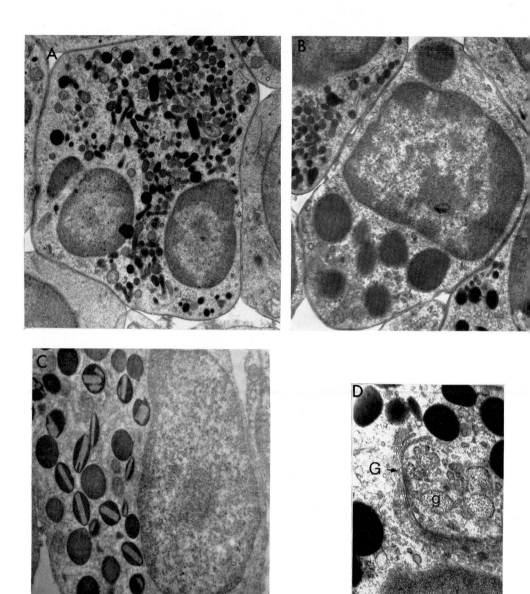

Plate 7.11 Ultrastructure of neutrophil, basophil and eosinophil granules.

A, neutrophil. × 6000. Note the varying densities of the primary, secondary and tertiary granules.

B, basophil. × 6000. The granules are large and electron dense.

C, eosinophil. × 6000. Two types of granule are visible. Type 1 are homogeneous and very electron dense. Type 2 are more angular and have a crystalloid internum. (*By courtesy of* Dr George Hudson.)

D, formation of eosinophil granules in the region of the Golgi zone. × 7600.

G = Golgi zone; g = developing granule.

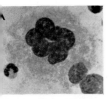

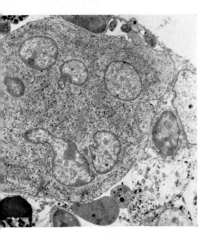

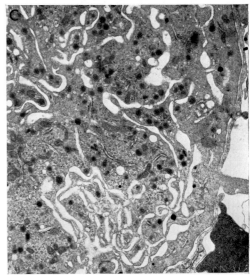

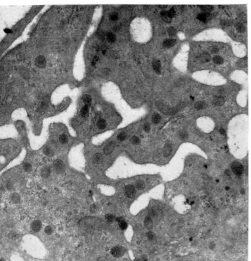

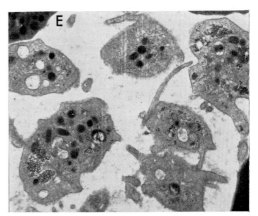

Plate 7.12 Platelet formation from megakaryocyte.

A, megakaryocyte in bone marrow smear stained with Leishman stain. × 840.

B, megakaryocyte, an electron micrograph. × 2100. Note the multi-lobed nucleus and electron-dense granules.

C, megakaryocyte, an electron micrograph. × 8400. Note the electron-dense granules, glycogen and canaliculi.

D, megakaryocyte, an electron micrograph. × 17 500. The canaliculi are well-developed and the cytoplasm is about to fragment and be shed as platelets.

E, platelets, an electron micrograph. × 14 000. Note the electron-dense granules and glycogen.

Types of blood cells produced in the liver
Although liver stem cells are multipotent as shown by transfusion into lethally irradiated recipients, in the micro-environment of the liver priority is given to erythropoiesis and this persists for most of the hepatic phase. Some of the red cells produced are nucleated with diameters larger than adult RBC and their haemoglobin is of fetal type. Granulocyte precursors also appear in the liver, becoming particularly conspicuous at the end of the phase. Lymphocytes are not formed.

Recession of hepatic haemopoiesis
Hepatic haemopoiesis, already waning at birth, ceases rapidly afterwards. What causes the decline? The micro-environment may deteriorate, causing stem cells to migrate into bone marrow or quantitative exhaustion may follow persistent export to spleen and marrow.

Splenic haemopoiesis and origin of its stem cells
Splenic haemopoiesis is relatively late. Typical stem cells appear in and around vascular spaces which form in condensed mesenchyme of the dorsal mesogastrium. There is evidence that these cells migrate into the spleen from the yolk sac (in birds) or liver, and that they do not arise from local mesenchyme or endothelial cells lining the vascular spaces.

Types of blood cells produced in the spleen
Erythropoiesis appears first, followed by granulocytopoiesis. Erythropoiesis predominates for most of the pre-natal phase. At birth red cells and granulocytes recede and a lymphoid character develops, lymphocytes accumulating around central arterioles in the white pulp. Marker chromosome techniques indicate that early splenic lymphocytes arise *extrinsically* from the thymus.

Bone marrow haemopoiesis
Haemopoiesis occurs predominantly in bones formed by endochondral ossification. As the cartilage in the centre of the developing bone calcifies, its cells die. The resulting cavity fills with mesenchymal-type cells and capillaries derived from the surrounding periosteum. The capillaries dilate into extremely thin-walled sinusoids and stem cells now appear within them. They were thought to arise locally from mesenchyme or sinusoidal endothelium but marker chromosome techniques

show that multipotent stem cells, together with slightly more differentiated cells migrate into the sinusoids from yolk sac and spleen in birds, or the liver in mammals.

Types of blood cells produced in fetal marrow
Species variation occurs. In the chick, although occurring simultaneously, erythropoiesis for reasons unknown occurs within sinusoids whilst granulocyte production is extra-sinusoidal. In mammalian marrow, granulopoiesis and erythropoiesis are extra-sinusoidal, granulopoiesis predominating before birth.

Haemoglobin variations during pre-natal erythropoiesis
Haemoglobin consists of four polypeptide chains, each with Haem attached. There are five types of haemoglobin in man; two *embryonic* (HbE) with paired α and ϵ chains or 4 ϵ chains; *fetal* (HbF) with paired α and γ chains; two *adult* (HbA), one having α and β chains and the other, HbA_2, with paired α and δ chains. HbE and HbF have a higher O_2 affinity than HbA. This is important because the round nucleated fetal RBC are less efficient in carrying O_2 and also fetal tissues have a high O_2 requirement. Haemoglobin type changes with alteration in location of erythropoiesis. Thus, HbE is associated with yolk sac erythropoiesis and HbF with liver, spleen and early marrow erythropoiesis. At birth HbF predominates but is rapidly replaced by HbA (Fig. 7.3). Why the haemoglobin type changes is unknown. Translation of information by mRNAs from different genetic loci is involved. Haemoglobin A and HbF can be synthesized simultaneously in the same erythroblast. Thus a new clone of cells is not necessary in the changeover from HbF to HbA.

Post-natal Growth and Differentiation of Blood Cells

After birth, blood cell production becomes restricted to bone marrow. However, like the yolk sac in pre-natal haemopoiesis, it has the vital ability to export stem cells.

Sites of haemopoiesis and the effects of age
There are three stages to consider, the early post-natal period, then the immature animal and finally the mature animal. Shortly after birth stem cells are confined to marrow cavities of developing bones. Here their numbers increase substantially and haemopoiesis advances rapidly. Haemopoiesis ceases in the liver but lymphocytopoiesis increases in the

spleen. An exception which is important because of its use in the laboratory is the mouse, the spleen of which continues to produce red cells, granulocytes and platelets into adulthood. In the immature animal, haemopoiesis occurs in most of the skeleton including throughout the shafts of long bones, pelvic and pectoral girdles, vertebral column, ribs, sternum and flat bones of skull. The active marrow appears red. As adulthood approaches recession of haemopoiesis occurs in medullary cavities. There is a marked increase in fat cells, visible macroscopically as a change from red to yellow colour. In this recession active marrow becomes concentrated centripetally in the body, disappearing from the limb bones until in the young adult only the most proximal parts of the femur and humerus contain red marrow. Haemopoiesis becomes confined to the skull, ribs, sternum, vertebral column and limb girdles. In man, the iliac crest and sternum are accessible for taking marrow samples.

The reasons for the recession are unknown. Age changes in the microcirculation of the marrow may contribute to the conversion of red into yellow marrow. The recession does not necessarily mean that the absolute volume of red marrow decreases since enlargement of medullary cavities accompanies skeletal growth. Although a change in the cellularity of active marrow has not been found with advancing age[4] little is known concerning changes in generation times, life-spans and total body levels of RBC and leucocytes with advancing age. In man, no significant differences have been found in red and white cell counts with increasing age, but in mice and dogs, RBC levels significantly decline as also do leucocyte levels in cattle.

Effects of haemopoietic stress on sites of haemopoiesis

Increased haemopoietic demands evoke *cellular* and/or *locational* responses. The cellular response involves increased cell production and accelerated release (v.i.). In the locational response haemopoiesis reappears in previously active centres. The degree of the stimulus and the animal's age are important. In adults, yellow marrow, particularly in limb bones, reverts to red marrow. In extreme stimulation haemopoiesis occurs in spleen and liver. In young animals most of the skeleton is already actively haemopoietic so that extension soon occurs into spleen, liver and even kidney.

How does this extension occur? Although dormant stem cells *in situ* might become re-activated, haematogenous seeding of the sites by stem cells from active marrow probably occurs.

Micro-environment; the concept of soil and seed

Multipotent stem cells circulate post-natally in the blood and can be concentrated in the buffy (leucocyte) coat. When injected into lethally irradiated recipients they re-populate the marrow and lymphoid tissues. Why do they only colonize haemopoietic tissues? They may be compared with 'seed' requiring a special 'soil' in which to grow, this being provided by the micro-environment (stroma) of haemopoietic tissues.

Components of the micro-environment

The micro-environment of the marrow is like the meshes of a sponge, with free haemopoietic cells lying between them. Contributing to the meshes are two components, a *microcirculation* (sinusoids) and a

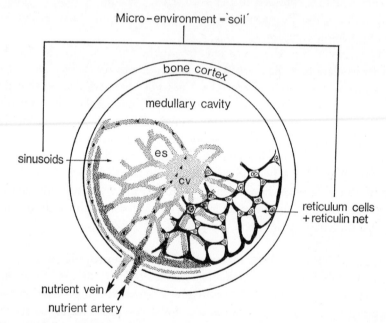

Fig. 7.4 Bone marrow structure – the micro-environment. *es* = extra-sinusoidal space; *cv* = central vein.

micro-skeleton formed by reticulum cells and their derivatives, fat cells and reticulin fibres (Fig. 7.4).

The micro-circulation. The micro-circulation consists of a vast thin-walled network of sinusoids drained by veins and fed by arteries derived from the main nutrient artery of the medullary cavity or from those supplying the cortex. Separate areas of marrow are supplied by specific arteries[5]. In the marrow they branch abruptly into thin-walled arterioles which lead directly into sinusoids. Sinusoids have a zonal distribution but all eventually converge upon a thin-walled central vein (Fig. 7.4).

Sinusoids are extremely thin-walled and, being contractile, vary in calibre. They dilate rhythmically with pulse pressure. What is the structure of the sinusoidal wall? Functional considerations suggest that it should not be a complete barrier. Under the light microscope only a simple layer of flat endothelial cells is seen. The electron microscope reveals minimal contact between the attenuated plasma membranes with occasional brief overlap. Significantly, small gaps (<60 nm) occur between the plasma membranes. Adventitial cells, a discontinuous basement membrane[6] and junctional complexes[5] all reinforce the endothelial cells.

The micro-skeleton. The micro-skeleton has two components: (a) reticulin fibres produced by reticulum cells or fibroblasts, and (b) reticulum cells and fat cells (Plate 7.3) The fibres form an extensive network particularly concentrated around blood vessels, sinusoids and fat cells. They are argyrophilic and have the structure of immature collagen. Fat cells are formed from reticulum cells. Reticulum cells are not easily seen in red marrow, being dispersed amongst many free haemopoietic cells. They are evident in yellow marrow or following arrest of haemopoiesis in red marrow by X-irradiation, since they are extremely radio-resistant. Reticulum cells are stellate fixed cells intimately related to the reticulin fibres along which their tenuous processes extend. They are best seen in marrow sections rather than smears (Plate 7.4). The cell is 15–25 μm in diameter and has a large pale round nucleus. One or two nucleoli may be visible. Cytoplasm is abundant with little reaction to stains and the cell margins are often difficult to define. The ultrastructure of reticulum cells reflects their functions, including phagocytosis. The nuclear chromatin resembles that in undifferentiated cells, being finely dispersed with minimal condensation on the nuclear membrane. A nucleolus is usually seen. Long cytoplasmic processes extend between adjacent free cells and along sinusoidal walls. A Golgi zone is usually conspicuous. Mitochondria are few and small. Extensive micropinocytosis may be evident and also numerous phagosomes. There are many inclusions especially ferritin, haemosiderin and disintegrating RBC, also erythroblast nuclei, paracrystalline bodies and myelin figures.

Functions of the micro-environment

Physical factors such as blood flow, adhesiveness of cells due to zeta potentials, and chemotaxis resulting from haemopoietic substances being concentrated in reticulum cells may create an environment attracting stem cells and promoting differentiation.

Sinusoids. Integrity of sinusoids is vital. Damage to them may cause aplastic anaemia. By regulating blood flow they control access of haemopoietic substances, including humoral factors controlling differentiation, also the access of stem cells to bone marrow and the export of newly-formed cells. Blood cells are very plastic and negotiate the sinusoidal wall in various ways. Granulocytes seem to favour diapedesis whilst megakaryocytes shed their platelets by lying astride the small gaps normally present in the wall. Rapid enlargement of cell clones next to sinusoidal walls may cause them to yield, so releasing the cells. The presence of extrasinusoidal RBC suggests that blood may flow like a tide out of the sinusoids, taking newly-formed cells back with it when it returns. Marrow cells transfused into lethally irradiated recipients easily penetrate sinusoidal walls, the walls being disrupted by X-irradiation. Sinusoidal endothelial cells form part of the reticulo-endothelial system. They show marked evidence of pinocytosis and avidly ingest bacteria, red cells and carbon. The ingested material later appears within macrophages in the parenchyma. It may be transferred to them or the endothelial cells themselves may be mobilized. Sinusoidal cells might well contribute to haemopoiesis by transferring haemopoietic substances.

The micro-skeleton. Although reticulin fibres may support marrow cells, they are probably particularly important in supporting the micro-circulation. In the disease 'myelofibrosis' reticulin overproduction crowds out haemopoietic cells, thus leading to marrow failure. Reticulum cells have various functions including the production of reticulin fibres, transformation into fat cells, phagocytosis, 'nurse cell' functions and the physical attraction of cells. Red marrow always contains some fat cells and yellow marrow has large numbers. Their function is not simply space-filling. They can be rapidly mobilized as an energy source and they contain haemopoietic agents such as butyl alcohol. Fat cell formation is conspicuous in marrow recovering from X-irradiation. Large numbers accumulate at what appears to be a vital period, just before massive regeneration of erythrocytes and granulocytes[7]. Reticulum cells phagocytose cell fragments and nuclei extruded from erythroblasts. Often they are surrounded by groups of other marrow cells, especially erythroid cells, also plasma cells, lymphocytes and 'transitional cells'. In this particular location they have been called 'nurse cells'[8]. What are they doing? They may store and process various factors, including iron, which they transfer to surrounding haemopoietic cells. Reticulum cells may produce haemopoietic substances. Those lining the marrow cavity appear to synthesize a factor which particularly stimulates granulocyte and monocyte formation. The initial attraction of early erythroblasts around 'nurse cells' is followed by repulsion as they mature. This probably results from alterations in adhesive forces between cell membranes. Reticulum cells appear to form a gelatinous mucopolysaccharide which accumulates in marrow after X-irradiation. This may physically trap transfused stem cells, helping them to colonize the marrow. Finally, recent research does not support the long-established view that reticulum cells are multipotent stem cells. They show little, if any, proliferative activity even following considerable haemopoietic stimulation.

Summary of the micro-environment
Marrow sinusoids and reticulum cells are vital for blood cell development. Without intact sinusoids, haemopoietic factors and circulating stem cells could not enter marrow, nor newly-formed cells emerge. By transferring haemopoietic materials to surrounding cells and producing haemopoietic factors and mucopolysaccharides which influence stem cell proliferation, reticulum cells act as 'nurse cells'.

Theories of haemopoiesis
Until recently, theories of haemopoiesis were based solely upon the observations of pioneer histologists whose work, although meticulous, was basically morphological. Research aimed at protecting animals from the effects of whole-body X-irradiation provided the impetus for development of the modern theories of haemopoiesis. The fundamental discovery was made that bone marrow or blood leucocyte transfusions will protect lethally irradiated animals by re-populating their haemopoietic tissues. This led to the current concept of a free multipotent stem cell which circulates in the peripheral blood. Sophisticated modern techniques have amplified our knowledge of the stem cell. These include chromosome markers to study cell migration and seeding of tissues, radioisotopes to label cells and study their turnover, electron-microscopy for ultrastructure and techniques for assaying stem cells.

Haemopoiesis raises three fundamental questions. Are blood cells formed intra- or extra-sinusoidally, do they arise from free or fixed stem cells and is there one multipotent precursor or one for each cell line?

Intra-sinusoidal or extra-sinusoidal origin?
In mammals all blood cells develop extra-sinusoidally. They lie free between the reticulin fibres. Only when mature do they normally pass through the sinusoidal wall. During accelerated cell production, for example anaemia, and in certain leukaemias, numerous immature cells enter the circulation. Factors preventing immature cells from being released are largely unknown, but may include physico-chemical agents controlling cellular attraction and adhesion. The structure and coatings of cell membranes also appear to be important. In avians it is still generally held that red cells develop intra-sinusoidally from endothelial cells[9] but leucocytes are thought to originate extra-sinusoidally from reticulum cells.

Free or fixed location of primitive stem cells?
Early haematologists advanced two opposing theories concerning the mobility of primitive stem cells. The first proposed a fixed population of stem cells, sinusoidal cells giving rise to red cells and reticulum cells to leucocytes. With marrow depletion reticulum cells proliferated and formed free cells which divided and differentiated into granulocytes, monocytes and lymphocytes. These free cells had a high nuclear/cytoplasmic ratio and intense cytoplasmic basophilia and this accords with modern descriptions of the multipotent stem cell. Although the reticulum cell is no longer accepted as the stem cell, in the embryo *fixed* mesenchymal cells are the anlage of *free* haemocytoblasts in the blood islands. In the second theory, the small lymphocyte or a cell closely resembling it was suggested as the stem cell[10, 11]. It followed that it must be a free cell circulating widely in the blood. It was thought to have all the potentialities of primitive mesenchyme.

There is now firm evidence that the primitive haemopoietic stem cell is, in fact, a free cell. This evidence will now be reviewed. If bone marrow in one limb is shielded whilst the rest of the body is lethally irradiated, death is usually prevented. Stem cells from the shielded marrow migrate via the bloodstream to enter and repopulate the irradiated haemopoietic organs. Marrow cell transfusions from histocompatible donors will similarly protect. Chromosome marker techniques prove that the cells repopulate the host's haemopoietic tissues. The technique has been used clinically in attempts to rectify marrow failure in man. Normal blood contains cells which actively synthesize DNA[12]. They therefore have proliferative potential. Uncommonly, actual mitoses may be seen in them. They originate from the marrow, their numbers rising during demands for increased haemopoiesis[13]. Transfusions of the leucocytic fraction from normal blood protect compatible recipients from the effects of lethal irradiation. Their marrow becomes colonized by stem cells which generate red cells, platelets and leucocytes[14]. This proves that there are multipotent stem cells circulating freely in the peripheral blood. Parabiosis is a more physiological way of demonstrating stem cell mobility. The circulations of two histocompatible animals are cross-linked (parabiosis) and their haemopoiesis is suppressed by X-irradiation, except for lead-shielding of marrow in the hind limb of one animal. Isotopic labelling shows that stem cells migrate from the shielded marrow to re-populate marrow of the non-shielded parabiont[15]. These shielding, transfusion and parabiotic experiments all demonstrate that in post-natal haemopoiesis there is a free, circulating stem cell which normally resides in bone marrow.

One multipotent stem cell or separate stem cells

There are two theories of haemopoiesis. The *monophyletic* theory proposes that all blood cells arise from a multipotent stem cell, and there is now unequivocal evidence for such a cell. According to its microenvironment, it can generate all the different types of blood cells found in bone marrow and can also seed primary lymphoid organs such as the thymus. Synonyms for the multipotent stem cell include – haemocytoblast, transitional cell, lymphoidocyte and CFU (colony-forming unit, v.i.). In the *polyphyletic* theory each type of blood cell has its own primitive stem cell. It is now untenable in mammals. Cunningham, Sabin and Doan[9] modified the theory and suggested that red cells came from sinusoidal endothelium and granulocytes, monocytes and lymphocytes from reticulum cells.

Evidence for the monophyletic theory. In rodents, transfusions of buffy-coat leucocytes into lethally irradiated recipients restores both leucopoiesis *and* erythropoiesis, donor-type red cells being identified by Hb markers. If the stem cell is multipotent then demands for increased production of one particular group might be expected to curtail differentiation along other lines. This competition for stem cells becomes evident during very marked erythropoietic or antigenic stimulation. Thus, granulocyte and monocyte production fall in severely anaemic guinea-pigs and mice. In rabbits, decreased erythropoiesis, as revealed by Hb synthesis, accompanies increased lymphocyte production during a primary immune response. The detection of distinctive 'marker' chromosomes in karyotypes prepared from dividing cells in haemopoietic and lymphoid tissues, and cell cultures, unequivocally indicates a common ancestor for red cells, granulocytes and lymphocytes. The markers may be natural, for example the differences between the sex chromosomes, or pathological, for example those induced by X-irradiation and the Ph chromosome in chronic granulocytic leukaemia. Examples of the use of chromosome markers to demonstrate the existence of a multipotent stem cell include the following (1) Sex chromosome analyses show that stem cells will migrate from yolk sacs in female chick embryos to populate liver, spleen and marrow in male parabionts. Similarly, murine yolk sac cells from females when transfused into X-irradiated males will re-populate their marrow, spleen and lymphoid tissues. (2) In human heterozygotes positive correlations occur between the mosaic composition of erythroid, granulocytic and lymphoid tissues in G-6-P dehydrogenase activity. (3) In human chronic granulocytic leukaemia the atypical Philadelphia (Ph) chromosome is present not only in granulocytic progenitors but also in erythroblasts. (4) Stem cells carrying irradiation induced chromosome abnormalities when transfused into suitable recipients form splenic colonies of granulocytes, red cells and megakaryocytes, the early forms of which all contain the abnormal chromosome. (5) The T6 chromosome has been extensively employed as a marker (see Micklem and Loutit[16]) to investigate seeding potentialities of stem cells in marrow and lymphoid tissue suspensions. It was discovered fortuitously in an X-irradiated mouse and originated from an unequal translocation between two autosomes. Transfusion of T6 marrow cells or fetal liver cells into lethally irradiated recipients results in the marker appearing in the regenerated erythroid, granulocytic and megakaryocyte populations, and to a lesser extent in lymphoid tissues, especially the thymus. However, lymph-nodes appear to also have a small self-maintaining population. The genetic anaemia in W-series mice[17] can also be rectified using T6 fetal liver cells.

Functional characteristics of haemopoietic stem cells

A broad classification of stem cells based on recent research, but which may require future modification is shown in Fig. 7.5. In order of least differentiation there are *three grades*, (1) multipotent cell, (2) progenitor cell, and (3) blast cell.

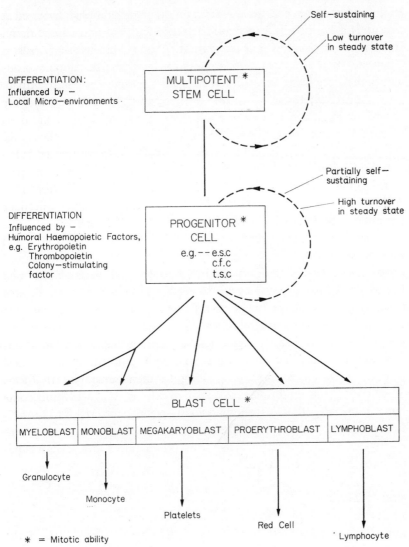

Fig. 7.5 Stem cell classification.

Multipotent stem cell

Although the multipotent stem cell is self-maintaining, this capacity is not unlimited. Thus serial transplantation through successive recipients eventually results in 'burn out' of re-populating ability. The capacity also varies with maturity, being maximal for yolk sac stem cells but less for neonatal liver cells. It is even less in post-natal marrow cells. However, advancing age in an animal does not further curtail the capacity.

223

In the yolk sac stem cells proliferate rather than differentiate, the need at this early stage being to increase the stem cell pool and export them. In steady-state haemopoiesis many multipotent stem cells are quiescent. They are triggered into active cycle with increased haemopoietic demand, the state being reversible if demand slackens. Only stem cells in active cycle will differentiate into progenitor cells, and a potentially hazardous situation has been discovered for this stage. Thus, there is no fixed ratio between stem cells remaining behind to maintain their pool and those differentiating into progenitor cells. Local micro-environmental differences such as ability to contact reticulum cells, variations in mucopolysaccharides and availability of differentiating factors may decide the issue. However, differentiation of the earliest stem cells does not appear to require humoral factors such as erythropoietin.

Progenitor cell

Progenitor cells differ from the multipotent cell because the vast majority are in active cell cycle. Nevertheless, owing to their further differentiation, their numbers have to be partially sustained from the multipotent pool. Although only in the infancy of differentiation, the progenitor cell is already committed to develop along one particular pathway. Humoral haemopoietic factors determine the pathways and progenitor cells are named according to their sensitivity to them. They are (1) the erythropoietin sensitive cell (E.S.C.) which differentiates into red cells (2) thrombopoietin sensitive cell (T.S.C.) which forms megakaryocytes, and (3) colony forming cell (C.F.C.) sensitive to colony stimulating factor which forms granulocytes and macrophages (monocytes) in agar-culture.

Blast cell

Blast cells are more differentiated than progenitor cells. They are the first clearly-recognizable precursors of specific cell lines. Thus, (1) the myeloblast produces granulocytes (2) the pro-erythroblast forms red cells (3) the monoblast produces monocytes (4) the lymphoblast forms lymphocytes and (5) the megakaryoblast ultimately forms platelets. Most blast cells are in active cell cycle. Because of their continuing differentiation the population is sustained by differentiation of antecedent progenitor cells. Differentiation appears to be influenced by factors similar to those controlling progenitor cells.

Identity and structure of early stem cells – the multipotent stem cell and progenitor cell

The identities and structure of early stem cells have only recently been elucidated. For many years, the small lymphocyte has been proposed as the multipotent stem cell. However, it could not be the typical small lymphocyte of lymphoid tissues or thoracic duct lymph. Thus, transfusions of these cells mostly give either equivocal or no protection against lethal irradiation. However, *bone marrow* does contain a cell which superficially resembles a lymphocyte in its size and high nuclear/cytoplasmic (N:C) ratio but which also has blast features. This *lymphocyte-like 'transitional' cell*[18] is a very early stem cell. Transfusions of bone marrow rich in these cells effectively restore haemopoiesis[19] and large numbers accumulate at a vital stage in the recovery of bone marrow from sub-lethal γ-irradiation. Serial marrow transplants[20] and transfusion of specific cell fractions[21, 22] also implicate a lymphocyte-like or transitional cell as the multipotent stem cell.

Morphology

In smears the transitional cell is round or oval-shaped and 9–18 μm in diameter. Smaller forms are probably daughter cells resulting from mitosis, and may subsequently enlarge. Two distinctive features are a very high N:C ratio and leptochromatic nuclear chromatin (Plate 7.5). The markedly basophilic cytoplasm forms a very thin, and sometimes incomplete rim around the nucleus, or it may be concentrated in a nuclear hof. As might be expected, the ultrastructure of the transitional cell is remarkably simple. The high N:C ratio is confirmed. Nuclear chromatin is extremely translucent with minimal condensation on the nuclear membrane. One or two nucleoli are present. The cytoplasm contains numerous ribosomes and some polyribosomes. Apart from a few mitochondria, organelles are conspicuous by their absence.

Progenitor cells. Morphological differences between transitional cells and progenitor cells are minimal. Some transitionals have slightly more cytoplasm than others, and variations occur in the depth of cytoplasmic basophilia. The nucleus remains leptochromatic. These features may indicate transformation into progenitor forms. Recent studies of human fetal marrow reveal that although some cells have the morphology of transitional cells, they are peroxidase positive and sudanophilic, indicating differentiation along granulocytic and monocytic pathways.

Some physical differences in early stem cells and their relation to function
Differential and velocity centrifugation show that in density and volume
post-natal stem cells are a mixed population. In early fetal life most stem
cells are of low density but later a high-density population appears. The
fetal multipotent stem cell is larger than its post-natal counterpart
probably because of different proliferation rates, fetal cells cycling
rapidly whilst adult cells are resting or only in slow cycle. Progenitor
cells are larger than multipotent cells and of greater density. Because
many are in active cycle they are extremely radio-sensitive.

Methods for studying and assaying stem cells
Stem cell assay
Stem cells from various haemopoietic tissues can be assayed. The *in
vivo* technique of splenic colonization[23] is used for studying multipotent stem cells. When cell suspensions are injected into previously
irradiated mice, they seed in their spleens forming visible colonies after
7–10 days. These can be counted, measured and analysed histologically.
Metcalf and Moore[3] perfected an *in vitro* agar-culture technique to
study progenitor cells. Cells are grown in petridishes in a mixture of
cell culture medium, agar and humoral colony-stimulating factor (CSF).
After 7–10 days the colonies are counted and measured. At present the
method is only suitable for assaying the type of progenitor which is the
common precursor of granulocytes and macrophages. Macrophages
tend to predominate in the older colonies and this may indicate deterioration. Progenitor cells are cortisone-sensitive and also asparagine-
dependant.

Erythropoietin sensitivity
Erythropoiesis is suppressed in mice by hyper-transfusion with packed
RBC. Erythropoietin-sensitive progenitor cells are then activated by
injecting erythropoietin, the stimulation being assessed quantitatively
by blood reticulocyte counts or ^{59}Fe uptake.

Radio-active isotopes
Isotopes, together with autoradiography, are widely employed to study
stem cell kinetics. Tritiated thymidine is incorporated into DNA and is
used to identify proliferating cells and to study life-spans and cell cycle.
Either single dose ('pulse' or 'flash') labelling or repeated infusion can
be employed. Stem cell turnover can be studied by giving 'suicide'
doses of high specific activity. Following its uptake, stem cells are killed

by intracellular irradiation. Those with rapid turnover are selectively destroyed, the extent being assayed by splenic colonization or agar-culture. Red cell kinetics are studied using ^{59}Fe and ^{55}Fe, the latter also being used in autoradiography. Kinetics of platelet production are studied using the amino acid ^{75}Se selenomethionine.

Cell markers

Cell markers are essential for studying the fate of transfused stem cells and their progeny. Radio-isotopic labelling of DNA has been used but the technique is limited by the labelled cells being dispersed amongst the host's cells and also halving of the label with each mitosis. Unique chromosomes like the T6 mouse chromosome and also sex chromosomes have been extensively employed as markers. Dividing cells are arrested in metaphase by colcemid and then swollen by suspension in hypotonic citrate solution. After fixation with ethyl alcohol and acetic acid they are spread on a slide and the dispersed chromosomes stained with orcein.

Isolation of specific cells

Stem cells can be concentrated and isolated from cell suspensions by differential centrifugation or selective filtration. Centrifugation in a column of graded concentrations of bovine serum albumin or sucrose causes haemopoietic cells to segregate into layers according to their specific gravities. When filtered through columns of glass wool or beads, granulocytes and monocytes adhere but stem cells pass through.

Gamma and X-irradiation and radiomimetic drugs

Animals whose haemopoietic tissues have been suppressed by lethal whole-body irradiation can be transfused with stem cells and protection measured by survival rates, recovery of blood cell counts and splenic colonization. The fate of the transfused cells can be traced using marker chromosomes. Stem cell migration and autorepopulation of haemo-poietic tissues can be studied by partial shielding of marrow or spleen. Radiomimetic drugs can be used to arrest haemopoiesis before trans-fusing stem cells. Colcemid, vincristine and vinblastine only destroy cells in mitosis but actinomycin-D and 5-fluorouracil kill at any stage in the cell cycle.

Parabiosis

If the circulations of two animals are united either naturally, as in uniovular twins, or surgically, stem cell migration can be studied especially when chromosome markers are used to identify the migrating cells.

Stem cell migration

There is evidence that stem cell migration which is so vital in the pre-natal development of haemopoietic organs, persists into adult life, the bone marrow becoming the exporting centre. Mammalian peripheral blood contains small numbers of DNA-synthesizing cells (5–10 mm^3). They are mostly transitional, monocytic and blast-like cells whose turnover rates resemble those of multipotent cells and progenitor cells. Undoubtedly they are the stem cells present in the buffy coat of blood and originate from the marrow[13]. Their numbers in the peripheral blood increase whenever there is a stimulus for increased cell production, including irradiation, antigenic stimulation, anaemia and the immediate post-natal period. Why are these DNA-synthesizing cells present in the blood? Are they accidentally released with mature cells, their numbers particularly increasing during the boosted release which accompanies increased haemopoiesis? They may, however, have considerable functional significance. During haemopoietic stress, boosting of circulating stem cells would facilitate seeding of tissues which are becoming depleted or in need of an influx of stem cells, for example lymphoid tissues reacting to stress or antigens and the extension of erythropoiesis into spleen and yellow marrow during anaemia. Even under normal conditions the spleen and lymphoid organs appear to be continually renewed by haematogenous stem cells originating from bone marrow. These may either be actual lymphoid progenitors or multipotent cells which differentiate appropriately on contacting the local micro-environment.

Factors concerned in blood cell differentiation

Differentiation is the transformation of uncommitted multipotent cells into cells with specialized functions. Once started there is no return and the fundamental property of mitosis disappears. During differentiation specific organelles develop, such as lysosomes in neutrophils, also specific proteins like red cell haemoglobin. Both intrinsic (genetic) and extrinsic (environmental) factors affect differentiation.

Intrinsic factors affecting differentiation

In mice there are two genetic anaemias associated with mutations at either the W locus or the Sl locus. Each anaemia has a different basis.

Cells from W-anaemic mice will not grow in normal mice and this suggests an intrinsic defect of actual regulators of differentiation present in the W locus. Cells from S1-anaemic mice do grow in normal mice and this suggests an environmental defect. A product originating from the S1-gene may stimulate resting multipotent cells into active cycle.

Extrinsic factors – rôle of micro-environmental and humoral (hormonal) factors

Both micro-environmental and humoral haemopoietic factors influence cellular mechanisms concerned with differentiation. Differentiation involves gene activation and suppression. Prior to differentiation, genes concerned with initiating the chemical events of differentiation are suppressed. Differentiation commences when the genome of stem cells becomes receptive to specific humoral (hormonal) factors, cyclic 3, 5-AMP acting as mediator. These factors activate previously repressed genes controlling specific mRNA production from DNA templates and synthesis of specialized proteins, such as enzymes and Hb, results.

Micro-environment. The influence of local micro-environment on differentiation is well shown in the pattern of colonization of the spleen by multipotent stem cells transfused into irradiated mice. Erythroid colonies predominate over other types. Moreover they appear throughout the red pulp but granulocytic ones are predominantly subcapsular. Each colony arises from one stem cell. In bone marrow, erythroid islands and megakaryocytes often lie close to sinusoidal walls whereas granulocytes, although clonal, are more ubiquitous.

Variations in marrow vascularity and blood flow may influence differentiation[24]. Thus very vascular marrow with high velocity blood flow is associated with increased erythropoiesis whereas decreased flow promotes granulocyte formation and inhibits erythropoiesis.

In the early embryo we have already seen how specificity of environment is vital for further differentiation of early stem cells, for example the proximity of endodermal cells in the yolk sac wall or of those which form hepatocytes in the liver. Endodermal cells appear to have an equivalent rôle to stromal reticulum cells in post-natal haemopoietic tissues. The reticulum cells form part of the microskeleton of bone marrow, spleen and lymphoid tissues. The micro-environment involves intimate contact and interaction between individual stem cells and reticulum cells. Stem cell receptivity to local environment probably varies with particular stages of their cell cycle.

Micro-environment cells may have various rôles. By influencing the physico-chemical environment of developing cells they may cause them to remain in their clones until mature. Reticulum cells appear to form mucopolysaccharides which coat the surfaces of developing haemopoietic cells and these may influence cell adhesion. McCuskey, Meineke and Townsend[24] have correlated changes in marrow mucopolysaccharides with different types of haemopoiesis. Thus sulphated mucopolysaccharides accumulate during early stem cell proliferation and in granulocytopoiesis, but neutral mucopolysaccharides accumulate during erythropoiesis. Cell degradation products resulting from phagocytosis may be transferred by reticulum cells to haemopoietic cells for re-utilization. Humoral haemopoietic factors may not only be localized and transmitted but actually produced, e.g. colony stimulating factor. In summary we have seen how factors such as state of blood flow and type of mucopolysaccharide are associated with particular functional environments, E.M.E. (erythropoietic micro-environment) or G.M.E. (granulopoietic micro-environment).

Humoral regulators and feedback mechanisms. Changes in blood cell levels provide the 'feedback' which regulates compensatory responses in haemopoiesis. Thus, subnormal levels stimulate cell production and hasten cell release, whilst elevated levels have the opposite effect. Response to feedback is mediated through humoral (hormonal) factors, the levels of which vary with functional needs, e.g. tissue O_2 tension. Humoral factors have now been positively identified which regulate RBC, platelet, granulocyte and monocyte production. They may stimulate or inhibit stem cells and steady-state haemopoiesis reflects a balance between these two actions. In leukaemia, imbalance occurs between stimulating and inhibitory factors. Currently more is known about stimulatory factors than those which inhibit. Humoral stimulatory factors have various basic actions. (1) Dormant (G_0) multipotent stem cells enter into active cycle and change into progenitor cells. (2) Progenitor cells are caused to differentiate along one particular pathway. The stage of their cell cycle may determine whether progenitors respond, receptivity being maximal during G 1 and early S-phase. (3) By shortening the cell cycle progenitor turnover is increased. (4) More mitoses occur in blast cells and other incompletely differentiated forms. (5) Maturation is hastened as also is the discharge of newly-formed cells into the sinusoids. (6) Marrow blood flow is increased by local vascular response. At the cellular level humoral factors act on the genome, causing production of mRNAs, tRNAs, and rRNAs which control differentiation. However, differentiation can occur without humoral agents. For example, in human polycythaemia vera, in early embryonic yolk sac erythro-poiesis and in virus-induced murine anaemia, autonomous genome activation or a substitute genome of viral origin appear to initiate erythropoiesis.

Erythropoietin. This is a glyco-protein (M.W. 60 000–70 000) present in the plasma and urine of many vertebrates, including birds, fishes and mammals. Increased levels occur in anaemia and hypoxia. It is formed thus:

ENZYME   SUBSTRATE

Erythrogenin $\longrightarrow$ *Plasma globulin* $\longrightarrow$ *Erythropoietin*
(Renal erythropoietic factor,   (Erythropoietic
REF)   stimulating factor, ESF)

Produced in:
Kidney (principally, but not   *Liver*
exclusively)
? granule cells of Juxta-
Glomerular apparatus, *or*
visceral cells of Bowman's
capsule.
Kidney monitors blood O_2 tension.

The actions of erythropoietin are those already given for humoral stimu-latory factors. The stimulation or depression of haemopoiesis caused by certain hormones (e.g. androgens, cortico-steroids, oestrogen and thyroxine) probably results from their actions on erythrogenin or plasma globulin.

Thrombopoietin. Thrombopoietin is a glyco-protein present in the plasma β-globulin fraction. Its origin is uncertain. Blood thrombo-poietin and platelet levels are reciprocal, and thrombopoietin is phago-cytosed by platelets. Perhaps this is how feedback operates. Thrombo-poietin has the general actions of humoral stimulatory factors. It in-creases the size (amount of cytoplasm) and number of megakaryocytes, and accelerates platelet release into the circulation. By causing poly-ploidy it increases the amount of cytoplasm and therefore the number of platelets which can be shed.

Factors controlling blood leucocyte levels. Blood leucocyte levels are enhanced by two groups of substances. One group stimulates division and maturation of leucocyte precursors and the other hastens leucocyte discharge from marrow reserves.

Colony stimulating factor (CSF). This factor was detected by Bradley and Metcalf[25] in bone marrow colonies growing on agar. It resembles erythropoietin and probably has an equivalent rôle in leucocytopoiesis. It is a thermo-labile, non-dialysible glyco-protein (M.W. 45 000) in the α-globulin fraction of plasma. Its half-life is only 3–4 h. High levels occur in the embryo but normal serum and urine have only small amounts. Levels increase in infections and leukaemia. CSF occurs in many tissues and organs, including liver, salivary glands, thymus and spleen. Its extensive distribution suggests a widespread origin, for example loose connective tissue and its associated macrophages. Although neutrophils contain CSF this is probably the result of absorption. Bone marrow stromal cells produce small amounts which may control a basal production of monocytes and granulocytes. Antigenic material considerably boosts CSF production and the presence of endogenous or foreign protein, especially bacterial endotoxin, causes its local release. CSF stimulates granulocyte and monocyte production by enhancing the transformation of multipotent stem cells into specific progenitors and by increasing mitosis in immature granulocytes.

Other factors may control granulocyte production. Leukopoietin G (granulopoietin) is present in murine and human plasma and appears to originate in bone marrow. It stimulates granulocytopoiesis and resembles, or is identical with CSF. Granulocytic chalone (M.W. 40 000) and antichalone (M.W. 30 000) are tissue specific substances formed by mature granulocytes. They control granulocyte proliferation by a balanced effect, chalone turning on DNA synthesis in immature precursors whilst antichalone switches it off. This 'switch' is probably operated through short-lived RNAs. Mitotic rates in lymphoid tissues appear to be controlled by a lymphopoietin formed by the thymus. Blood lymphocyte levels control its production.

Factors hastening leucocyte release. Certain plasma factors increase blood leucocyte levels by mobilizing cells from marrow reserves. These include a leucocytosis inducing factor (LIF) which is a thermo-labile α-globulin that increases marrow blood flow, a neutrophil inducing factor (NIF) and a neutrophil releasing factor (NRF). A lymphocytosis stimulating factor (LSF) occurs in chronic lymphocytic leukaemia.

Humoral factors inhibiting haemopoiesis. Normal serum contains factors which suppress the production and maturation of granulocytes and red cells. They include a lipo-protein which inhibits CSF. Their origin remains unknown. Suppressing factors occur in blood diseases such as leukaemia and in patients with thymic tumours the serum contains a factor which causes anaemia by suppressing stem cells.

Other factors controlling growth and differentiation

During differentiation DNA replication and mitosis are influenced by the build-up of specific proteins, which, on reaching a critical concentration, cause DNA synthesis to cease. Thus, in developing megakaryocytes when thrombosthenin reaches a certain level ploidy development is terminated. In erythroblasts Hb synthesis influences the stage at which mitosis ceases, thus controlling the size of RBC. Too rapid synthesis causes premature curtailment of mitosis and macrocytes result. With prolonged synthesis extra mitoses occur and microcytes appear.

Lymphocytes may help to maintain the steady state in proliferating tissues by neutralizing tissue specific factors released from dividing cells[26]. Lymphocytes may also act as trephocytes by donating DNA which is re-utilized by developing cells. Finally, normal haemopoiesis requires an adequate intake of protein and of certain vitamins (B_6, B_{12}, C and K) and minerals (Fe, Mn, Co and Cu).

Structural and functional changes during differentiation

Structurally the multipotent stem cell is very simple, having very few organelles and little cytoplasm. During differentiation specialized functions develop. These are accompanied by changes in structure and it is these which form the basis of the morphological classification of blood and marrow cells. For ordinary light microscopy, smears are preferable to sections, but for electron microscopy ultra-thin sections of marrow or buffy-coat pellets are studied.

Cell size and shape

During differentiation of multipotent stem cells into blast cells their size increases considerably (by as much as 75%) but after the blast stage cells become progressively smaller. The megakaryocyte is a notable exception, being one of the largest cells in the marrow. Profound changes in shape occur during red cell maturation. They are related to function and will be reviewed later. Marked changes in shape are not evident in developing leucocytes but recent studies indicate that subtle changes do occur with development of cell motility.

Nucleus and nucleoli

Blast cells actively synthesize protein and therefore have prominent nucleoli. With specialization, when specific proteins and organelles accumulate the nucleoli decline. The small lymphocyte is an exception. It retains its nucleolus and under antigenic stimulation can transform into a blast-like cell. Mitotic ability disappears as differentiation progresses, but it persists until a comparatively late stage in erythroid cells. The mechanism causing the disappearance may be similar to that which inhibits DNA synthesis, namely the progressive accumulation to a critical level of specific proteins such as haemoglobin.

Although nuclei of early stem cells appear to be regular and are generally round or oval in shape, electron microscopy sometimes reveals narrow nuclear indentations. Lymphocytic and erythroid nuclei retain the rounded shape but granulocytes and megakaryocytes become multi-lobed. Why this radical change occurs is unknown. The degree of lobing is not related to development of ploidy or cytoplasmic volume in the megakaryocyte or to the age of granulocytes, as is still widely taught. During the change from multipotent cell to blast cell the nucleus enlarges, but after the blast stage it becomes progressively smaller. The megakaryocyte is an exception. Its large nuclei are related to polyploidy. Also, when precursor cells divide the nuclei of their daughter cells are initially small but later on may enlarge.

In certain vertebrates, including man, nuclear extrusion occurs towards the end of erythrocyte development. The extrusion is preceded by pyknosis of the nuclear chromatin. The reason for nuclear extrusion is unknown but it might considerably diminish metabolic requirements of RBC, and would facilitate development of the biconcave form, thereby improving red cell efficiency in carrying oxygen and also facilitating gaseous diffusion into RBC.

Cytoplasm

Characteristically, early stem cells contain very little cytoplasm and their N:C ratio is therefore high. As differentiation advances cytoplasm increases. This is very important in megakaryocytes since the volume of their cytoplasm determines the number of platelets produced. Small lymphocytes are an exception, especially those in bone marrow and thymus. Their N:C ratio remains very high. During differentiation cytoplasmic basophilia decreases due to a fall in ribosomal RNA. As basophilia declines, specific granules begin to appear in granulocytes and haemoglobin accumulates in erythroid cells. The plasma cell is an exception. It synthesizes γ-globulin and the cytoplasm contains many ribosomes and extensive rough endoplasmic reticulum.

Differentiation involves an increase in the variety and number of organelles present in blood cells. These include mitochondria, ribosomes, Golgi apparatus, lysosomes and specific granules. Certain organelles may become particularly conspicuous, indicating specialization of cell function. For example a Golgi zone appears in plasma cells, megakaryocytes and phagocytic cells and lysosomes form in cells with phagocytic ability, especially monocytes and neutrophils. Lysosomes comprise some of the granules of leucocytes and this forms the basis of the histochemical classification of leucocytes.

Specific substances increase during differentiation. Their histochemical demonstration enables pathways of differentiation to be traced. Thus, haemoglobin can be detected by Lepehne's peroxidase reaction or by micro-radiography since the globin molecule is radio-opaque. Glycogen accumulates in developing neutrophils and can be demonstrated by the PAS technique. The wide spectrum of enzymes present in lysosomal granules can also be demonstrated histochemically.

Development of the red cell
Stages of differentiation
Erythrocyte development is conventionally described under six stages (Plate 7.6). They are rather arbitrary since there is a continuous morphological spectrum between blast precursors and red cells. These stages include the pro-erythroblast (the specific blast precursor), basophilic (early), polychromatic (intermediate) and orthochromatic (late) erythroblasts and the reticulocyte (young erythrocyte).

During differentiation (1) cell size decreases, (2) the nucleus becomes increasingly pyknotic as the chromatin condenses and is finally extruded from the late erythroblast, (3) there is a decrease in cytoplasmic basophilia with reciprocal increase in Hb, (4) remains of RNA persist in young red cells (reticulocytes) and can be stained supra-vitally. Reticulocyte levels are a sensitive index of erythropoietic activity.

Morphological features of differentiation
The features are shown in Plates 7.6 and 7.7.

Pro-erythroblast. Light microscopy. This is the largest erythroid cell (15–18 μm diameter). The large round central nucleus is completely surrounded by intensely basophilic cytoplasm. Densely stained chromatin largely obscures the one or two nucleoli present. The N:C ratio is high.

Ultrastructure. The cytoplasm is particularly characteristic, being packed with free ribosomes and also polyribosomes in which globin formation occurs. Mitochondria are large and frequent. They contain enzymes and substrate needed for haem synthesis. Ferritin is present,

235

its Fe being available for haem production. The nucleoli are very prominent and synthesize RNA concerned in haemoglobin production. The nuclear chromatin still retains the dispersed pattern typical of multipotent stem cells.

Basophilic erythroblast. Light microscopy. The cell diameter is reduced to 13–14 μm. Nuclear chromatin is more coarsely clumped. The cytoplasm becomes less basophilic and this marks the beginning of Hb accumulation.

Ultrastructure. The nuclear chromatin shows increased condensation. This results from chromosomal inactivation by haemoglobin which precipitates chromosomal DNA. Since RNA synthesis ceases during this stage the nucleolus disappears, and ribosomes and polyribosomes are less evident.

Polychromatic erythroblast. Light microscopy. Cell diameter is reduced to 10–11 μm. The nucleus remains central but is smaller and stains very densely. The cytoplasm stains faintly pink due to increasing amounts of haemoglobin.

Ultrastructure. A conspicuous feature is a marked increase in cytoplasmic density due to haemoglobin accumulation. Further condensation of nuclear chromatin is evident and ribosomes are much less frequent.

Orthochromatic erythroblast. Light microscopy. Cell diameter is 8–9 μm and just exceeds that of a RBC. The dark pyknotic nucleus is usually central. Sometimes it is seen being extruded but the mechanism is obscure. The nucleus appears to enter a small evagination of cytoplasm which then bursts. Macrophages may assist by actively extracting the nucleus. With more haemoglobin present cytoplasm is distinctly pink.

Ultrastructure. There is increased density of the nucleus due to further chromatin condensation, and of the cytoplasm due to Hb accumulation.

Reticulocyte. Light microscopy. In diameter it is marginally larger ($\times 1 \cdot 13$) than a mature red cell. It is not nucleated. Remains of cytoplasmic RNA react with Romanowsky stains to produce small blue dots (punctate basophilia) or a bluish tint (polychromasia). Supra-vital staining with Brilliant Cresyl Blue causes RNA (ribosomal) remnants to precipitate in a reticulated manner (Plate 7.7).

Ultrastructure. The cytoplasm is extremely electron dense due to haemoglobin. Ribosomes and mitochondria are few in number and disappear with full maturation into RBC.

The cytoplasm adjacent to the cell membrane is particularly viscid and this may help to maintain the biconcave form of the red cell. Motility in reticulocytes is suggested by the presence of pseudopodia bulging through sinusoidal walls. Reticulocytes are very plastic and tolerate considerable deformation when passing through the sinusoidal wall. (Plate 7.8).

The red cell. Light microscopy. Mammalian RBC are biconcave, anucleate discs. In lower vertebrates (e.g. birds and amphibia) they have a larger diameter, are biconvex and retain their nucleus. Being very plastic they can tolerate severe deformation as they are squeezed through constricted capillaries or straddle bifurcations in small vessels. In man their mean diameter is 7·2 μm (range 6·7–7·7) and mean thickness is 2·1 μm (range 1·7–2·5 μm). Mean corpuscular volume (mcv) varies with species, being 83–97 μm^3 in man, 67–72 μm^3 in dogs and only 52–58 μm^3 in rats. Red cell mcv increases with advancing years. The mean corpuscular haemoglobin content (mchc) is remarkably constant (27–32 pg). Below a certain haemoglobin content red cells loose their stability. Deformation is not tolerated and lysis occurs. Haemoglobin accounts for 95% of red cell protein but only 5% is adjacent to the plasma membrane.

Ultrastructure. The cytoplasm is very electron dense with an absence of ribosomes and mitochondria. Cell inclusions (Heinz bodies) may be present. They consist of denatured protein and cause loss of plasticity with increased susceptibility to phagocytosis. In biconvex nucleated RBC bundles of microtubules form a peripheral cytoskeleton. They stiffen and maintain cell shape. Biconcave RBC do not have this cytoskeleton.

Red cell membrane. The integrity and biconcave form of the membrane must be maintained for normal function. The membrane is a trilaminar complex of lipo-protein, each layer being 2·5 nm thick. The two outer layers are electron dense but the inner is translucent. The middle layer is a bimolecular lipid layer with polar groups facing outwards covered by protein, and non-polar (hydrophobic) groups facing inwards. Small mosaic areas about 20 nm diameter occur on the surface membrane. They are attached on their inner aspects to a fibrous layer stabilized by lipids. Scanning electron microscopy reveals spherical particles on the inner and outer surfaces of the cell membrane. The outer ones are more frequent and have diameters 6–10 nm. Those on the inner surface have diameters 4–12 nm.

Functional aspects of red cell differentiation

Haem synthesis. Pathways concerned in haem synthesis are shown in Fig. 7.6. A.L.A. synthesis is essential for haem formation and haem levels exert a feedback control on its formation either by preventing formation of the enzyme A.L.A. synthetase or by inhibiting synthetase. Haem levels may also control the enzyme A.L.A. dehydrase which catalyses the formation of P.B.G. from A.L.A.

Globin synthesis. DNA synthesis precedes Hb formation. Messenger RNA transcribes the DNA patterns of the different globins (α, β and γ)

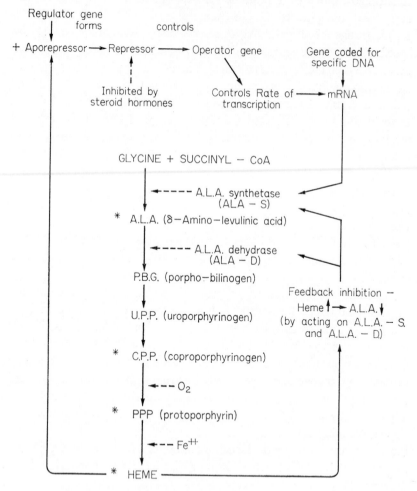

* Isolated with mitochondrial fraction of cell

g. 7.6 Haem synthesis and its regulation.

from separate gene loci onto polyribosomes. When polyribosomes decrease, globin synthesis rapidly declines. The enzyme RNA-ase destroys polyribosomes and breaks down mRNAs used in globin synthesis. The *in vitro* synthesis time for globin is 30–40 s, α and β chain synthesis being fastest. Intracellular globin levels may act as feedback, controlling further globin synthesis. Globin synthesis slightly precedes haem synthesis and this explains the intense cytoplasmic basophilia, abundant polyribosomes and marked radio-opacity of pro-erythroblasts.

Haemoglobin. Oxygen transport is restricted in early erythroblasts but rises rapidly in the later stages when high Hb levels are achieved. HbF may persist or reappear post-natally, especially during severe anaemias caused by bleeding or hereditary haemolysis. Presumably its high O_2 affinity compensates for severe anoxic conditions. The HbF may be generated by erythroid clones which synthesize γ chains or by mRNA for γ chain synthesis or de-repression of the γ-chain locus on genes. Reticulocytes do not synthesize RNA but Hb synthesis does continue in them, suggesting that their mRNA is long-lived.

Erythroblasts. These are extremely active metabolically. They synthesize DNA, also RNA to form haem and globin which are then combined into Hb. Three metabolic pathways are used (Embden-Myerhof, citric-acid cycle and hexose-monophosphate shunt).

Reticulocytes. Reticulocytes are more sticky than RBC. Enzymes in their mitochondria catalyse Fe incorporation into porphyrin. Haemoglobin synthesis continues until ribosomes disappear.

RBC. The biconcave shape and size of the red cell adapt it for optimal gas transport. If the shape becomes biconvex (spherocytosis) or the size decreases (microcytosis) this causes decreased O_2-carrying capacity and increased fragility. Metabolic activity is very high. Glucose is transported across the membrane and by the Embden-Myerhof pathway provides energy which is stored as high-energy phosphate bonds. This energy is used for transporting nutrients and ions across the cell membrane and maintaining its integrity. The anucleate state conserves energy. With increasing age enzyme systems become exhausted. Premature aging, expressed as shortened life-span and increased fragility (haemolytic disease) occurs in certain enzyme deficiencies. Thus, a sex-linked recessive congenital haemolytic anaemia occurs in Negroes and Mediterranean races when exposed to certain drugs. It is associated with G-6P-D deficiency which impairs $NADPH_2$ production.

Kinetics of red cell production – models of erythropoiesis
In steady-state erythropoiesis most multipotent stem cells remain resting (G_0 state), and do not contribute to the progenitor (E.S.C.) pool. The E.S.C. is in active mitotic cycle, one daughter cell remaining to replenish the progenitor pool whilst the other differentiates into a pro-erythroblast. Mitotic ability occurs from the progenitor through to the polychromatic stage and assuming mitosis occurs at each stage each

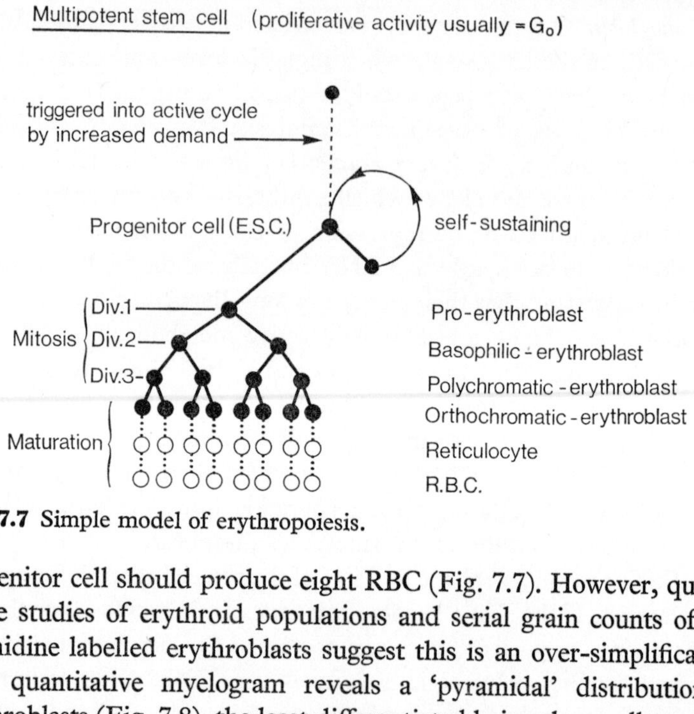

Multipotent stem cell (proliferative activity usually $= G_0$)

triggered into active cycle by increased demand ⟶

Progenitor cell (E.S.C.)   self-sustaining

Mitosis { Div.1 ——   Pro-erythroblast
Div.2 —   Basophilic - erythroblast
Div.3-   Polychromatic - erythroblast
Orthochromatic - erythroblast

Maturation {   Reticulocyte
R.B.C.

Fig. 7.7 Simple model of erythropoiesis.

progenitor cell should produce eight RBC (Fig. 7.7). However, quantitative studies of erythroid populations and serial grain counts of ^{3}H-thymidine labelled erythroblasts suggest this is an over-simplification. The quantitative myelogram reveals a 'pyramidal' distribution of erythroblasts (Fig. 7.8), the least differentiated being the smallest group whilst the more mature erythroblasts are the most numerous. This would confirm the model in Fig. 7.7 if the different stages of erythroblast were in the ratio of 1:2:4:8. However, the quantitative myelogram shows the actual ratios as 1:5:20:10. This could be explained by a slowing-down of maturation as differentiation proceeds; and there is evidence that this is so[27], the duration of the pro-erythroblast stage being 5 h, basophilic stage 35 h, polychromatic stage 80 h, orthochromatic 18 h and reticulocyte 40 h. However, Stohlman[28] suggests that all stages have a basic intermitotic time of 10–15 h.

In fact the number of mitoses during the various stages of differentiation varies between 3 and 7 depending upon erythropoietic stimulus and species. Assuming every cell matures, between eight (from three divisions) and 128 (from seven divisions) RBC should be produced. However, two further factors must be considered (1) the rate of haemoglobin build-up and, (2) ineffective erythropoiesis caused by death of a certain proportion of erythroblasts. When a critical Hb level is reached proliferation ceases. Too rapid synthesis curtails mitosis. Therefore smaller numbers of RBC are produced but they are abnormally large (macrocytes). Prolonged synthesis permits more mitoses. Thus, more RBC are produced but their diameters are smaller than normal (microcytes). In addition, extra mitoses in E.S.C., pro-erythroblasts, and basophilic erythroblasts result from erythropoietin activity which shortens G_1 and allows more mitoses to occur. Taking all these factors into consideration Lajtha estimates that 20 reticulocytes are produced from one E.S.C. (Fig. 7.9).

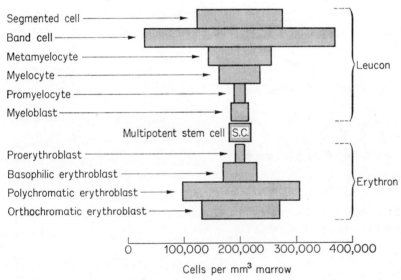

Fig. 7.8 Absolute levels in the main cell groups of normal guinea pig bone marrow.

Differentiation from E.S.C. to reticulocyte takes 5 days (range 3–7 days, depending upon species and haemopoietic demand). As erythroblasts mature the rate of differentiation slows down, the last 24–48 h being spent as reticulocytes. Mitosis (M) takes $\frac{1}{2}$–1 h, whilst DNA synthesis (S) lasts 2·5–7 h. Post-mitotic rest period (G_1) is variable and is curtailed by erythropoietin during anaemia to facilitate increased cell production.

Development of granulocytes

Stages of differentiation

Although six stages are conventionally described in granulocyte development (Plate 7.6), there is a complete morphological spectrum between them all. The stages are (1) myeloblast, the earliest recognizable precursor, (2) pro-myelocyte, traditionally regarded as the precursor of all three forms of myelocyte but its granule content suggests that it is exclusively a neutrophil precursor, (3) myelocyte – the stage when specific neutrophil, eosinophil or basophil granules first appear,

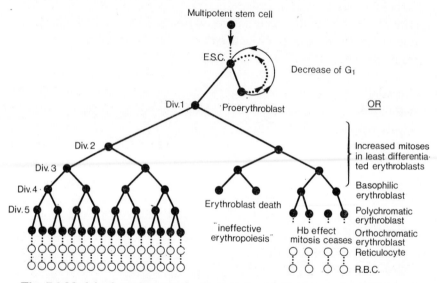

Fig. 7.9 Model of erythropoiesis allowing for ineffective erythropoiesis and influence of Hb on mitosis.

(4) metamyelocyte, (5) stab (band) cell, (6) polymorphonuclear – the mature granulocyte with lobed nucleus. Using electron microscopy Wetzel[29] classifies granulocytes into four stages. The first stage is the myeloblast (the precursor of all forms of granulocyte). The subsequent stages are early (pro-myelocyte), intermediate (myelocyte) and late granulocytes (metamyelocyte, stab cell and polymorph).

Granulocytes are classified by granule content into three groups (Plate 7.9), – neutrophils, eosinophils and basophils. (1) Neutrophils have three types of granules. Primary (azur) granules are first seen in pro-myelocytes. They are followed by secondary and tertiary granules as differentiation proceeds. (2) Eosinophils have two types of granule –

homogeneous and crystalline, (3) Basophils have only one form of granule.

Features of differentiation
The essence of differentiation in granulocytes is the development of the various types of granules and the organelles associated with their production. Also glycogen appears which is used in the anaerobic metabolism of neutrophils.

During development cell size and N:C ratio decrease. Nuclear chromatin becomes increasingly condensed and nucleoli, although initially prominent, eventually disappear. Free and attached ribosomes (granular E.R.) together with a Golgi zone are initially very prominent since they are associated with the production of the various types of granules. When a full complement of granules is reached the Golgi apparatus and ribosomes recede. Mitochondria decrease with differentiation. Histochemical evidence of differentiation includes the appearance of several enzymes (e.g. peroxidase and alkaline phosphatase) in the developing granules, an increasingly positive P.A.S. reaction as glycogen accumulates and increasing positivity to Sudan black B as lipo-proteins develop in the granules.

Morphology and histochemistry of granulocytes
Myeloblast. Light microscopy. This is a large cell (15–20 μm diameter), with a round leptochromatic nucleus. Although usually central, the nucleus is sometimes eccentric with flattening or slight indentation ('hof') at one edge. The cytoplasm here is characteristically pale (Plate 7.10) and marks the position of the developing Golgi-complex and impending granule production. The N:C ratio is quite high but cytoplasm is more abundant and less basophilic than in pro-erythroblasts. The myeloblast is easily confused with its close relative, the monoblast.

Ultrastructure. Nuclear chromatin is fine and dispersed. The large nucleoli present indicate RNA synthesis which is generating protein for general and specific use in enzymes and granules. Mitochondria are numerous, as are ribosomes and polyribosomes. The Golgi zone is not quite fully developed.

Histochemically myeloblasts are negative or react only minimally for enzymes or to P.A.S. and Sudan black B reactions.

Development of neutrophil granulocytes

(i) *Pro-myelocyte*. Light microscopy. The pro-myelocyte equals the myeloblast in size. The nucleus is round but often eccentric with a distinct indentation. The nuclear chromatin is still quite leptochromatic and the nucleoli are large (Plate 7.10). The cytoplasm is less basophilic, with a distinct pallor in the area close to the nuclear indentation. Here the Golgi apparatus lies together with the primary non-specific azurophilic granules which it forms. These granules are specific to neutrophils. Thus the pro-myelocyte is the earliest form of neutrophil. The N:C ratio is not quite as high as in the myeloblast.

Ultrastructure. Large (0·4 μm), homogeneous electron dense granules are present in the cytoplasm. They are the azurophilic granules visible using light microscopy. They originate from smaller very dense granules in vacuoles close to the Golgi sacs. These vacuoles fuse together to form the primary granules. The new granules then become enclosed by membranes.

Histochemically the granules are primary lysosomes containing hydrolytic enzymes, including acid phosphatase, aryl-sulphatase, α-galactosidase, β-glucosamidase, β-glucuronidase, α-mannosidase, esterase, 5-nucleotidase, and lysozyme. *Peroxidase* is also present and is *specific* for *primary granules*. Azurophilia of the granules is due to their sulphated mucopolysaccharide which is also P.A.S. positive. The granules are sudanophilic and this may be especially evident in the nuclear-hof where granules are concentrated.

(ii) *Myelocyte*. Light microscopy. Myelocytes are slightly smaller than pro-myelocytes. The N:C ratio is lower and nuclear chromatin is distinctly clumped. Nucleoli are indistinct. The nucleus is usually eccentric and flattened or indented on its deep side. Cytoplasmic basophilia has practically disappeared and *secondary granules* now appear. Although called neutrophil they in fact stain pink with Romanowsky dyes. Eosinophil and basophil granules are easily distinguished from them by their distinctive colour and large size.

Ultrastructure. Many of the features seen with the light microscope are confirmed, including condensation of nuclear chromatin and inconspicuous nucleoli. Ribosomes, polyribosomes and mitochondria are less evident. Glycogen granules now begin to appear and also the characteristic 'neutrophil' or *secondary granules*. These are smaller and less dense than the primary ones and are formed within Golgi saccules. When secondary granule formation begins primary granule production ceases. Primary granules will therefore decrease with any subsequent mitosis, but some do persist into the mature stage.

Secondary granules contain most of the lysozyme present in neutrophils, and *alkaline phosphatase* is exclusive to them. Peroxidase and most of the acid hydrolases are absent. The advent of secondary granules is reflected by increasing sudanophilia. Increasing amounts of glycogen are demonstrated by the P.A.S. reaction.

(iii) *Metamyelocyte*. Light microscopy. Metamyelocytes are slightly smaller (14 μm) than myelocytes. Their nuclear chromatin is more condensed and the N:C ratio diminished. Nucleoli are obscure. It is clearly distinguished from the myelocyte by its deeply indented, kidney-shaped nucleus. Secondary granules predominate over the primary by more than 10:1.

Ultrastructure. Advancing maturation is shown by further reduction in ribosomes and polyribosomes and a diminishing Golgi zone. Glycogen becomes even more conspicuous in the cytoplasm.

In addition to reacting histochemically to all the enzymes present in primary and secondary granules there is a marked reaction to Sudan black B. Increasing glycogen is shown by a strong P.A.S. reaction.

(iv) *Band (stab) cell*. Light microscopy. Since it is an almost mature neutrophil, its diameter approaches that of a polymorph (10–12 μm). The nucleus is 'U' or horse-shoe shaped but sometimes the limbs are folded so that the shape is distorted. The abundant cytoplasm contains numerous orange-pink granules (the secondary and tertiary granules visible with the electron microscope).

Ultrastructure. Besides a small number of primary granules and numerous secondary granules, a *third type* of granule appears which is unique to *mature* neutrophils (Plate 7.11). Its electron density is greater than that of the secondary granule. It may arise in the Golgi zone after secondary granules cease to be produced or may be formed from secondary granules themselves. Glycogen particles are abundant.

Histochemically the reactions are similar to those in metamyelocytes. In addition the reactions of the *tertiary* granules resemble those of the primary except for peroxidase and lysozyme. The large numbers of primary, secondary and tertiary granules are reflected by an intense reaction to Sudan black B. Also the P.A.S. reaction is very strong due to the very high glycogen content.

Polymorphonuclear cell. Light microscopy. In general detail polymorphs closely resemble stabcells except for segmentation of the nucleus which has 2–5 lobes connected by thin chromatin strands.

About 5 per cent of polymorphs in females have a short (1–2 μm) drum-stick appendage. This is the Barr body and represents chromatin of the female sex chromosome. Chromatin in the lobes is coarse and nucleoli are not visible.

Ultrastructure. All the features of maturation are seen. Nuclear chromatin is condensed and organelles associated with granule production have regressed. Thus, nucleoli are absent, ribosomes depleted and Golgi zone atrophic. An abundance of glycogen explains the paucity of mitochondria.

Biochemical maturation, namely a full content of granules and abundant glycogen, is reflected by intense sudanophilia, and strong reactions to P.A.S. and the various enzymes associated with the different granules.

Development of Eosinophils. Their development follows the general pattern already outlined. The *myelocyte* is the earliest recognizable stage when characteristic eosinophil granules first appear. They later become so numerous that they may partially obscure the nucleus. Eosinophils pass through the usual stages of maturation including metamyelocyte and stab cell but the polymorphonuclear has only 2–3 lobes.

(i) *Granules.* These contain enzymes elaborated in cisternae of the endoplasmic reticulum. They are 'packaged' in the very conspicuous Golgi zone and accumulate there in vacuoles distending the saccules. The vacuoles appear to condense into granules. Eosinophil granules are 0·6–1·0 μm in diameter and much larger than those of neutrophils (0·3 μm) or basophils (0·4 μm). There are two basic types of granule (Plate 7.11) and also intermediate forms. Type 1 are round, homogeneous and very electron dense. Type 2 are angular with a crystalloid internum consisting of thin lamellae. Type 1 granules predominate in immature eosinophils whereas type 2 form the bulk of granules in circulating eosinophils and are almost absent in the early forms. Although each type of granule might arise separately, intermediate forms suggest that type 2 granules may arise by maturation of type 1. The total number of granules remains fairly constant during development and granule production overlaps. (c.f. neutrophils in which production of primary granules ceases before secondary granule production commences.)

(ii) *Histochemistry.* The strong affinity of the granules for acid dyes results from their high content of Zn-containing protein, and P.A.S. positivity reflects their content of sulphated mucopolysaccharide.

They also show intense sudanophilia. Both types of granules contain enzymes. Peroxidase levels are even higher than those in neutrophils but acid phosphatase levels are lower. There are traces of ribonuclease, β-glucuronidase, arylsulphatase, deoxyribonuclease, nucleotidase and lipase. Alkaline phosphatase is absent (cf. neutrophil), also lysozyme, chloroacetate esterase and non-specific esterase.

Development of basophils. Like the eosinophil, this cell becomes recognizable at the *myelocyte* stage when its typical granules first appear. These stain intensely purple with metachromatic stains. Their size and number vary with species and functional state. Human basophil granules are relatively small but those in guinea-pigs may be so large and numerous that the nucleus is obscured. Sometimes in smears a purple 'halo' surrounds the cell. It is caused by heparin diffusing from the granules.

(i) *Granules.* There is only *one basic type of granule* (cf. eosinophil and neutrophil). Its diameter is 0·3 μm (range 0·15–0·5) and it is extremely electron dense (Plate 7.11). Each granule consists of many tightly packed particles (0·01–0·02 μm) within a double-layered limiting membrane. Morphological variations such as vacuolation and a thread-like appearance in granules may be caused by fixation. Precursors of granules appear in the Golgi zone in the form of small dense spheres. Mitochondria are somewhat larger and more numerous than in neutrophils.

(ii) *Histochemistry.* The purple staining of the granules by basic dyes is due to high levels of heparin, a sulphated acid mucopolysaccharide. The granules also contain glycogen (less than neutrophils) and the vaso-active compounds, histamine and serotonin. Their enzyme content is very much lower than that of other granulocytes.

Functional aspects of granulocyte differentiation
Functional and biophysical criteria[30] can now be added to morphological and histochemical evidence of maturation. They result from changes in the cell periphery including development of contractile actinomyosin-like proteins and decreases in density of negative surface charges on the cell membrane. The negative charge is associated with carboxyl groups of n-acetyl neurominic acid molecules. Sequelae of these physical changes include: (1) enhanced pseudopodia formation and easier deformation, thus facilitating penetration of the sinusoidal wall and emigration of granulocytes from the peripheral circulation into the tissues; (2) increased adhesiveness which assists margination

of blood neutrophils, an important preliminary to diapedesis into tissues; (3) increased motility; (4) increasing phagocytic ability. Phagocytosis is evident even in myelocytes as soon as secondary granules appear and it then increases progressively until the mature polymorph stage is reached.

Proliferation, differentiation and reserves in the developing leucon
The total turnover of granulocytes reflects granulocytopoietic capacity. In man it is 1600–3200 $\times$ 10^6 cells per kg body weight, and approximately 3000 $\times$ 10^6 in the dog. Essentially, the leucon has two compartments (Fig. 7.8): a pool of dividing and differentiating cells extends from the progenitor cell to the myelocyte stage. Recent evidence suggests that metamyelocytes ought also to be included in this pool since they actively synthesize DNA. At least five mitotic events occur during differentiation of granulocytes through this pool, one being in the myeloblast, one in the pro-myelocyte and three in the myelocyte. Thus, it is during the myelocyte stage that granulocyte production is principally expanded (Fig. 7.8). In dogs there is evidence for myelocyte overproduction by a factor of two, as estimated from the metamyelocyte population. The excess die and constitute 'ineffective granulopoiesis'. Small myelocytes are thought to be in the G_1 or early S phase in the mitotic cycle, and larger myelocytes in late S or G_2 phase. In dog myelocytes DNA synthesis lasts for 5 h, G_2 for 0·5 h and mitosis for 1–2 h. In rats, myelocyte production declines with advancing age being 12 times greater at 3 weeks of age as compared with 3 years. The second compartment of the leucon is the non-proliferating reserve of virtually mature cells, the stab and segmented forms. The reserves are very considerable as shown quantitatively in Fig. 7.8 and by the technique of leucopheresis which involves removal of granulocytes from the circulation, the reserves can be mobilized rapidly within 3–6 h. Neutrophil reserves alone are particularly large and at 6–12 $\times$ 10^9 cells per kg body weight are nearly 7 times greater than the total circulating granulocytes. In the guinea-pig, total discharge of marrow reserves could produce a granulocytosis of almost 100 000 mm³. In guinea-pigs about 70–140 $\times$ 10^6 eosinophils enter the blood daily and the marrow reserve probably exceeds 400 $\times$ 10^6. If suddenly discharged they could produce a blood oesinophilia of 13 000 mm³. The marrow reserve of basophils is 62 $\times$ 10^6 and their complete discharge would cause a basophilia of about 2500 mm³.

For many years it has been taught that a granulocyte's age is directly related to the number of lobes in its nucleus. Recent evidence suggests that the number of lobes is not a reliable index of the age of blood neutrophils. However, the non-lobed stab cell is certainly the youngest form of mature granulocyte.

Marrow transit time

During steady-state granulocytopoiesis the rates of inflow and outflow through the various compartments are balanced. In humans and dogs 4–6 days are required for full maturation of neutrophils, the last two days being spent in storage. In rats and mice transit from myelocyte to mature neutrophil takes only 16 h and 36 h respectively. There is less data for marrow transit times of eosinophils and basophils. In rats the generation time for eosinophil precursors is 40 h and emergence time is 36 h. Thus the total time for formation and release is 3 days. Irradiation studies in guinea-pigs suggest that 5–6 days elapse before eosinophils reach the circulation whilst basophils take 4 days. Quantitative data (Fig. 7.8) and labelling with radio-isotopes suggest that compartment transit times become longer as cells mature. In man, transit from myelocyte to metamyelocyte takes 3 h, from metamyelocyte to band cell takes 9 h, and from band to polymorph 12–24 h.

Blood phase and intra-vascular survival

Total blood granulocytes are 700×10^6 cells per kg body weight. They consist of an actively circulating pool (45%) and a marginated pool (55%) in contact with vessel walls or sequestrated in sites such as spleen and lungs. Distribution between the two pools is determined by physiological and pathological conditions, e.g. exercise and acute inflammation. Transfer from marginated into circulating pool is shown by a leucocytosis. The intra-vascular life-span of human neutrophils is very short ($5\frac{1}{2}$–$9\frac{1}{2}$ h). Most disappear randomly from the circulation (i.e. unrelated to age) but a few remain in the circulation for up to 2 days where they die of old-age. Having left the circulation most granulocytes remain extra-vascular but recent evidence in humans suggests that small numbers do return to the circulation.

Functional aspects of neutrophil development

Glycogen content. Mature neutrophils contain large amounts of glycogen since their metabolism has to be anaerobic for them to survive in damaged hypoxic tissues and phagocytose bacteria.

Congenital enzyme deficiency. Phagocytosis does not by itself destroy bacteria. After ingestion, bacteria become enclosed within the membrane bound granules where they are exposed to a wide spectrum of lethal enzymes. In human chronic granulomatous disease, recurrent bacterial infections are associated with quantitative deficiency of the enzyme NADPH oxidase. This is essential for H_2O_2 production by the hexose-monophosphate shunt. The H_2O_2 is utilized by peroxidase in primary granules to facilitate lethal iodination of bacteria.

Mechanisms for increasing blood granulocyte and RBC levels

Two mechanisms are alerted to increase blood granulocyte and RBC levels. Initially, *functionally mature cells* are *discharged from* the substantial *marrow reserves* of granulocytes and reticulocytes. Thus, almost 70% of marrow erythroid cells are reticulocytes and they number 250 000 mm³. Their total discharge would produce a reticulocytosis of about 1·5%. The granulocytic reserves have already been considered. After discharge of reserves there is a *'follow-up' response* initiated by humoral haemopoietic factors. The response consists of expansion of production (almost tenfold in the case of RBC) and earlier discharge from the marrow. The sequence is as follows: (1) More multipotent stem cells differentiate into progenitor cells of the required type. (2) More mitoses occur in progenitor cells, blasts, myelocytes and early erythroblasts. (3) To hasten maturation, mitoses may be omitted, especially in the erythroid series. Mitosis is switched-off prematurely due to rapid Hb accumulation, and macrocytes result. Macrocytosis tends to occur more in rodents than man, possibly because there is less reserve space for expansion of erythropoiesis in rodent marrow as compared with human. (4) Since cell release is hastened the more immature cells with proliferative ability have less contact inhibition, and this may allow their mitotic rates to increase. Anaemia causes release into the circulation of reticulocytes and even nucleated erythroblasts, whilst infections cause release of numerous stab cells and even metamyelocytes. Up to two days may be saved by earlier release of cells from the marrow, maturation being completed in the bloodstream.

Monocyte

Origin

Monocytes originate from multipotent stem cells which are acted upon by colony stimulating factor to differentiate into a common precursor of monocytes and granulocytes. A common ancestor seems logical since both monocytes and macrophages are phagocytic. Neutrophils become the *microphages*, and monocytes become the *macrophages* of inflammatory reactions. Both groups contain primary lysosomes (azurophil granules) and have similar enzyme spectra, their differences being quantitative rather than qualitative.

It has long been claimed that small lymphocytes transform into monocytes but direct proof is lacking. However, recent studies suggest that monocytes in exudates arise by local differentiation of small *lymphocyte-like* cells. These actively synthesize DNA and resemble small-sized multipotent stem cells. They are probably released from the marrow as part of a cell migration stream which is boosted by inflammatory reaction. They enter the tissues from the bloodstream and over a period of 48 h transform into monocytes. The transformation requires the presence of viable neutrophils. These probably provide CSF or other differentiating factors.

Stages of development and morphology
The multipotent stem cell transforms into a progenitor cell. Further differentiation produces a monoblast and this finally matures into a monocyte.

Monoblast (pro-monocyte). Light microscopy. This is the earliest recognizable form of monocyte. It is a large cell (diameter, 15–20 μm) with a large round nucleus containing several conspicuous nucleoli. The monoblast closely resembles the myeloblast but the nuclear contour tends to be more irregular and infolded. Cytoplasm is fairly abundant and moderately basophilic except for an area of pallor adjacent to the nucleus where a Golgi complex is located.

Ultrastructure. There are all the features of a relatively undifferentiated cell. Nuclear chromatin is homogeneous with minimal condensation on the nuclear membrane; nucleoli are prominent and mitochondria large; ribosomes and polyribosomes are numerous. There is a conspicuous Golgi zone with small electron dense granules in the cisternae. These are the precursors of primary lysosomes which eventually mature into the azurophil granules of monocytes.

Monocyte. Light microscopy. The mature monocyte is distinctive in its size, its cytoplasm and its nucleus. It is the *largest* of the leucocytes (diameter, 15–20 μm). The nucleus is kidney or horse-shoe shaped, and even semi-lobulated but it is never frankly lobed as in the neutrophil. Nucleoli are indistinct. The cytoplasm is extremely abundant. It is pale blue-grey in colour due to the presence of many primary lysosomes the majority of which are so small they can only be seen with the

251

electron microscope. A few are visible under the light microscope as small purple azur granules but they are not as large or as numerous as those in neutrophils.

Ultrastructure. The nucleus is horse-shoe shaped and one or two nucleoli are usually present. Roughness and irregularity of the cell membrane reflects potential motility of the cell. The cytoplasm contains moderate numbers of small mitochondria. There is abundant rough ER where lysosomal enzymes are segregated and a conspicuous Golgi zone where enzymes are packaged into granules. Two types of primary lysosomes are described in relation to cell development. Type I are formed during early development in the marrow and also whilst the monocyte circulates in the blood. Type II are produced only after the monocyte has entered the tissues. The largest of the type I lysosomes are visible under the light microscope as azur granules. They number between 50–100 and have a diameter 0.1–0.5 μm. They are uniformly electron dense. Type II lysosomes appear as small coated vesicles associated with the Golgi zone. Although phagocytic vacuoles are not normally seen in marrow or blood monocytes, those in the blood are capable of phagocytosis.

Enzyme content. Lysosomal enzymes, aryl sulphatase and acid phosphatase, are present in several species (rabbit, guinea-pig, human). Peroxidase is present in monocytes of mouse, guinea-pig and human but not the rabbit. Although peroxidase appears to be less than in neutrophils, the difference may merely be one of technique. Monocytes have a very high content of non-specific esterase and a very low content of chloroacetate esterase, but the converse is true of neutrophils.

Monocyte kinetics. When increased numbers of macrophages are required, not only do monoblasts divide but also mature macrophages themselves may divide. In the transit from primitive stem cell to monocyte, 3–4 mitotic cycles occur and differentiation takes 2–3 days. Some monocytes may enter the blood only 24 h after one division of a promonocyte. About 3.6×10^6 monocytes are produced daily and the marrow reserve is about twice the number circulating in the blood. Discharge from the marrow is random and maturation, including lysosomal development, still continues in the blood. Intra-vascular life-span is about $1\frac{1}{2}$ days and monocytes leave blood randomly (half-life 22 h). They may remain for up to two months in the tissues and having phagocytosed material some may return to the blood.

The megakaryocyte

The megakaryocyte produces blood platelets. It has two distinguishing features, (1) it is one of the largest cells in bone marrow (diameter, 80–100 μm), (2) it has a very large multi-lobed nucleus. In smears the overall shape is round but the edge of the cell membrane is poorly defined. The megakaryocyte is a free cell and it is also amoeboid. This is important in the mechanism of cell release and under certain circumstances megakaryocytes may themselves enter the blood stream.

Structure

The nucleus in the megakaryocyte may have 4, 8, 16 or even 32 lobes. It should not be confused with the *multinucleated* osteoclast which is also a very large cell. The megakaryocyte is unique in the *polyploidy* of its nuclei. Unlike normal diploid (2n) somatic nuclei, they are either 8, 16 or even 32 n. The number of lobes is not related to the stage of cytoplasmic differentiation or degree of polyploidy. The nuclear chromatin is pale and homogeneous, an appearance which suggests a more immature type of marrow cell. In smears stained by Romanowsky stains the nucleus appears to be dense.

The cytoplasm is abundant and its amount is related to DNA-concentration (degree of ploidy). The cytoplasmic volume is important since it determines the number of platelets which can be formed. In smears the cytoplasm is pink and its edge is indefinite. The cytoplasm may contain purple granules some of which appear to be detaching themselves from the surface. These are platelets. There are three zones in the cytoplasm and their detailed structure is revealed by the electron microscope[31]. The zones are inner (perinuclear), middle (intermediate), and outer (marginal). The three zones may vary in width even in the same megakaryocyte. In some places the outer zone appears to be entirely absent.

The inner zone contains all the organelles associated with intensive protein formation. These include numerous ribosomes, rough ER mitochondria and an extensive Golgi apparatus which produces granules and membranes. The granules are membrane bound and electron dense. They are also present in platelets and contain factors used in blood clotting, including platelet Factor 3 (phospholipid) which activates prothrombin into thrombin, serotonin (5-hydroxytryptamine) a powerful vaso-constrictor, and lysosomes containing hydrolytic enzymes. The membranes which first appear in the perinuclear zone become more obvious in the middle zone. They are called 'demarcation membranes' and play a vital role in platelet formation.

It is in the middle zone that there is the first evidence of the cytoplasm transforming into platelets. This zone contains all the organelles found in platelets including mitochondria, electron dense granules and microtubules. The outer zone contains very few organelles apart from fibrils composed of thrombosthenin. Thrombosthenin is a specific contractile protein, and is also present in platelets. Amoeboid movements occur in the outer zone of the cytoplasm and may play a part in platelet release.

Platelet formation and release
Wright[32] first described the origin of platelets from megakaryocytes nearly 70 years ago. More recently electron microscopy has provided detailed analysis of the mechanism of platelet formation[33, 31]. The sequence of events is shown in Plate 7.12. In mature megakaryocytes the cytoplasm disintegrates into small fragments, each fragment becoming a platelet. Thus the various organelles found in the cytoplasm of megakaryocytes will also be present in platelets. Platelet formation resembles the behaviour of mud in a river-bed which is drying up during a drought. Fissures appear in the mud and progressively coalesce until separate flakes are formed. The first sign of platelet formation is the appearance of double-layered 'demarcation membranes'. At first the membranes are poorly developed and take the form of vesicles which then line-up into a lattice formation. The lattice causes the cytoplasm to break-up into discrete areas which become the future platelets. At first the areas inter-communicate but as the membranes link together the cytoplasm becomes isolated into discrete areas which are then shed as platelets. The demarcation membranes which originate in the Golgi complex, develop in several stages[34]. Initially small vesicles appear which then coalesce into tubules (some of these persist as canaliculi). The tubules become flattened into double-layered smooth demarcation membranes, each layer of which forms the surface membrane of two adjacent platelets. Where demarcation membranes reach the surface they are continuous with the plasma membrane of the megakaryocyte, and the lumen of the tubule between the double membranes opens onto the cell exterior.

The megakaryocyte is actively amoeboid and this activity particularly involves the outer zone of the cytoplasm. Thus, although megakaryocytes are extra-sinusoidal, by their amoeboid activity they can extend their cytoplasmic processes up to the sinusoidal walls and project them through the small gaps in the wall. From these projections plate-lets can be easily shed into the sinusoids. In some megakaryocytes, platelets demarcated in the cytoplasm of the middle zone appear to flow through the cytoplasm to enter a pseudopodium projecting through a sinusoidal gap. They are then shed from the pesudopodium into the sinusoid. Although there is a basic method of detachment, platelets may be shed individually, or in short chains, or as several platelets adhering together in a small group. Under certain circumstances when there is an increased demand for platelet production (e.g. after a severe haemorr-hage) 'giant' platelets may be released. These consist of groups of in-completely separated platelets which complete their separation having entered the circulation.

Formation of the megakaryocyte
Stages in the formation of megakaryocytes are shown in Fig. 7.10 (see also The Lancet[35] and Ebbe[36]). Megakaryocytes are fully differentiated cells and therefore are unable to divide. There are four phases in their development and only in the last phase is their highly characteristic morphology apparent. *Phase 1.* This is the earliest stage when the multipotent stem cell is activated to start differentiating along the path-way of megakaryocyte production. In *phase 2* the fully committed progenitor cell is reached. This has a much higher rate of proliferation than the primitive stem cell and is irrevocably committed to form a megakaryocyte. *Phase 3* is the principal stage of *polyploidy development*. This involves DNA replication and levels of 4, 8, 16 or even 32 n are achieved. It should be noted that this DNA synthesis is not associated with actual cell division. Mitosis is confined to phases 1 and 2. *Phase 4* is the stage when full maturation occurs and the highly characteristic morphology of the megakaryocyte emerges. Three cell types can be defined, each linked by transitional forms.

Megakaryoblast. About 20% of identifiable 'megakaryocytes' are in this category. They rapidly incorporate ^{3}H-thymidine into their DNA and this represents the 'tail-end' of the ploidy development which is so prominent in phase 3. Ploidy can still increase from 4 to 32 n even at this stage. The megakaryoblast itself is not able to divide. It is a very large cell (diameter 25–30 μm). The rounded nucleus occupies most of the cell and contains several nucleoli. The cytoplasm is deeply baso-philic and contains many ribosomes, also numerous mitochondria. There is very little evidence at this stage of granules or demarcation membranes as seen in mature megakaryocytes.

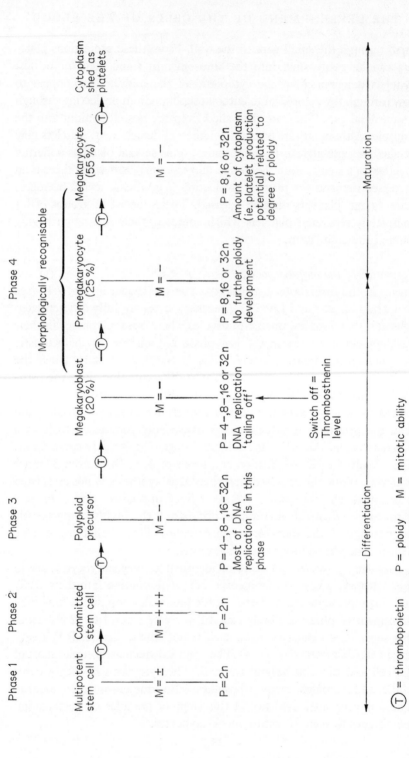

Fig. 7.10 Development of megakaryocytes

(T) = thrombopoietin   P = ploidy   M = mitotic ability

Phase 1   Phase 2   Phase 3   Phase 4

Multipotent stem cell
M = ±
P = 2n

Committed stem cell
M = +++
P = 2n

Polyploid precursor
M = –
P = 4-, 8-, 16-32n
Most of DNA replication is in this phase

Megakaryoblast (20%)
M = –
P = 4-, 8-, 16 or 32n
DNA replication 'tailing off'

Switch off = Thrombosthenin level

Promegakaryocyte (25%)
M = –
P = 8, 16 or 32n
No further ploidy development

Megakaryocyte (55%)
M = –
P = 8, 16 or 32n
Amount of cytoplasm (ie. platelet production potential) related to degree of ploidy

Cytoplasm shed as platelets

Morphologically recognisable

———— Differentiation ————   ———— Maturation ————

Pro-megakaryocyte. This is the stage intermediate between the megakaryoblast and the mature megakaryocyte. About 25% are in this form. Cytoplasm is more abundant and less basophilic, also a few azur granules (platelets) may appear. Ploidy does not increase any further at this stage.

Megakaryocyte. About 55% of megakaryocytes are in this final stage of maturation. It is a mature cell whose function is to discharge pieces of its cytoplasm into the circulation as platelets. There is no change in ploidy at this stage.

Significance of ploidy and its relation to platelet production.
A unique feature in the differentiation of the megakaryocyte is the development of polyploidy. This is a vital process since the DNA content of the nuclei determines the amount of cytoplasm and this in turn determines the number of platelets which can be formed. Platelet production only begins when DNA replication (i.e. ploidy development) ceases. The mechanism which shuts-off DNA synthesis is unknown but the specific protein *thrombosthenin*, present in the cytoplasm of megakaryocytes, may play a part. When it reaches a critical level in the cell, DNA replication ceases.

Maturation time of the megakaryocyte
In rodents it has been estimated that differentiation as far as phase 3 occurs in under 24 h, but maturation through phase 4 is appreciably slower, taking a further 2–3 days. In *man* there is evidence that it requires about 10 days to produce platelets.

Control of megakaryocyte formation and platelet production
The hormone *thrombopoietin* which is present in plasma and excreted in urine appears to mediate in a feedback mechanism which controls megakaryocyte production. There is a reciprocal relationship between blood thrombopoietin and blood platelet levels. The feedback may operate by platelets phagocytosing thrombopoietin.

Composition and origin of thrombopoietin. It is a glyco-protein present in the β-globulin fraction of plasma. Its origin is uncertain. Kidney and spleen have been investigated but platelets are still produced after their removal.

Mechanism of action. Thrombopoietin has two effects: (1) by increasing the number and size of megakaryocytes it increases the numbers of platelets produced, (2) it accelerates platelet release into the circulation.

It appears to act along several points in the pathway of megakaryocyte formation: (a) more multipotent stem cells are induced to differentiate into megakaryocyte precursors (b) mitotic activity is increased in these precursors (c) maturation of precursors is hastened by curtailing mitotic and inter-mitotic times and shortening the duration of polyploidy. This results in accelerated release of platelets. (d) By increasing polyploidy cytoplasmic volume is increased, thus making more available for platelet formation.

Assay of thrombopoietin. Thrombopoietin levels can be assayed either by counting blood platelets or by estimating blood isotope levels following injection of ^{35}S [37] or ^{75}Se- seleno- methionine[38]. These isotopes are incorporated into the cytoplasm of developing megakaryocytes and enter the blood when the cytoplasm is shed as platelets.

Anti-thrombopoietic factors. There is evidence that the spleen produces a factor which inhibits platelet production and that blood platelets themselves release it. The inhibiting factor may be a chalone.

The platelet

General features

Platelets are anucleate and the smallest cells in the blood (diameter 1–2 μm). Their average count is 250 000 mm^3 and they have a relatively short life-span of 4–5 days in the rat. Freshly liberated platelets are larger and more adhesive than older platelets. They contain large amounts of glycogen which is their principal energy source and also much ATP. They are rich in glycolytic and oxidative enzymes, and have moderate numbers of mitochondria.

Contents of platelets and their origin

Platelets contain various organelles and substances which were originally present in the cytoplasm of megakaryocytes and which are vital in the mechanism of clotting. They also contain other substances which they appear to take up themselves from the bloodstream. Whether they synthesize some of the compounds themselves is doubtful. They certainly do not possess protein synthesizing apparatus.

Ultrastructure

Detailed observations have been made by Zucker Franklin[31]. The plasma membrane is trilaminar. The two outer layers are protein and electron dense. The innermost layer is phospholipid and electron translucent. It may be a source of platelet Factor 3 used in clotting. There is also a *surface coating factor* which may be derived from platelets themselves or from plasma. The factor may contribute to the connecting bridges which form between platelets when they are aggregated by ATP. Thus it is probably concerned in platelet aggregation and adhesion to vascular endothelium.

Canaliculi and vacuoles are present in platelets and may be formed by infolding of the plasma membrane. This may explain how platelets 'ingest' viruses, chylomicrons, endotoxin and serotonin. Having been ingested they become membrane-bound. It is debatable whether the uptake of materials is by active phagocytosis (platelets do contain acid hydrolytic enzymes) or is merely passive, coinciding with canaliculi formation. Membrane-bound granules are present in platelets. They appear to be a heterogeneous group which vary in electron density. Each particular type of granule may contain a different factor used in blood clotting. These factors include (1) serotonin which is an important constituent of platelets and by its action on smooth muscle causes powerful vaso-constriction; (2) platelet Factor 3 which activates prothrombin, and (3) acid hydrolytic enzymes which will break-up fibrin and cell debris.

Platelets contain the protein *thrombosthenin* but they do not appear to produce it themselves. Its contractile properties resemble those of actomyosin in muscle. The microtubules and microfibrils visible in

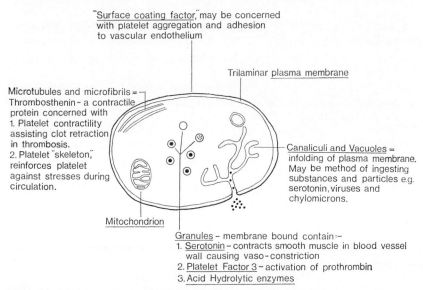

"Surface coating factor" may be concerned with platelet aggregation and adhesion to vascular endothelium

Trilaminar plasma membrane

Microtubules and microfibrils = Thrombosthenin – a contractile protein concerned with
1. Platelet contractility assisting clot retraction in thrombosis.
2. Platelet "skeleton", reinforces platelet against stresses during circulation.

Canaliculi and Vacuoles = infolding of plasma membrane. May be method of ingesting substances and particles e.g. serotonin, viruses and chylomicrons.

Mitochondrion

Granules – membrane bound contain:–
1. Serotonin – contracts smooth muscle in blood vessel wall causing vaso-constriction
2. Platelet Factor 3 – activation of prothrombin
3. Acid Hydrolytic enzymes

Fig. 7.11 Platelet ultrastructure in relation to function.

electron micrographs of platelets may be interpreted in relation to thrombosthenin in two ways. Either they may represent two different structural states (polymerization), the form presenting being related to platelet activity at the time of fixation, or thrombosthenin may have two basic components. Microtubules tend to be concentrated in layers just beneath the plasma membrane whereas microfibrils are scattered throughout the platelet. Fibrils appear to be thick in the microtubules but thin in the microfibrils (cf. skeletal muscle with its thick myosin and thin actin filaments). Microtubules lying beneath the plasma membrane may act as a cytoskeleton strengthening the platelet against stresses whilst it circulates in the bloodstream. The deeper microfibrils may comprise the contractile mechanism which contributes to clot-retraction during blood coagulation. The correlation between structure and function is shown in Fig. 7.11.

References and Bibliography

1. MOORE, M. A. S. and OWEN, J. J. T. (1967), 'Stem-cell migration in developing myeloid and lymphoid systems', *The Lancet, ii*, 658–659.
2. TILL, J. E. and McCULLOCH, E. A. (1961), 'A direct measurement of the radiation sensitivity of normal mouse bone marrow cells', *Radiation Research*, **14**, 213–222.
3. METCALF, D. and MOORE, M. A. S. (1971), 'Haemopoietic cells', Research Monograph, *Frontiers of Biology*, Vol. 24, North-Holland: Amsterdam, London.
4. ANDREW, W. (1971), 'The anatomy of aging in man and animals', 118–120, Heinemann: London.
5. BROOKES, M. (1971), 'The blood supply of bone. An approach to bone biology', Chap. 6, *Blood vessels in bone marrow*, Butterworth: London.
6. WEISS, L. (1965), 'Functional interrelationships of vascular and hematopoietic compartments in experimental haemolytic anaemia: an electron microscopic study', *Journal of Morphology*, **117**, 467–538.
7. HARRIS, P. F., HAIGH, G. and KUGLER, J. H. (1963), 'Observations on the accumulation of mononuclear cells and the activities of reticulum cells in bone marrow of guinea-pigs recovering from whole body gamma irradiation'. *Acta Haematologica* (Basel), **29**, 166–179.
8. BESSIS, M. and BRETON-GORIUS, J. (1962), 'Iron metabolism in the bone marrow as seen by electron-microscopy: a critical review', *Blood,* **19**, 635–663.
9. CUNNINGHAM, R. S., SABIN, F. R. and DOAN, C. A. (1925), 'The development of leucocytes, lymphocytes and monocytes from a specific stem cell in adult tissues', *Contributions to Embryology*, **16**, 227–276.
10. DOMINICI, H. (1902), 'Polynucleaires et macrophages', *Archives de médecine expérimentale et d'anatomie pathologique*, **14**, 1–72.

11. MAXIMOW, A. (1909), 'Der lymphozyt als gemeinsame Stammzelle der verschiedenen Blutelemente in der embryonalen Entwicklung und im post Fetalen heben der Saugetiere', *Folia haematologica* (Leipzig), **8**, 125–134.

12. BOND, V. P., FLIEDNER, T. M., CRONKITE, E. P., RUBINI, J. R., BRECHER, G. and SCHORK, P. K. (1959), 'Proliferative potentials of bone marrow and blood cells studied by in vitro uptake of H³-thymidine', *Acta haematologica* (Basel), **21**, 1–15.

13. HARRIS, P. F. and KUGLER, J. H. (1971), 'Unusual mononuclear cells in guinea-pig peripheral blood during anaemia', *Journal of Anatomy* (London), **108**, 1–12.

14. MALININ, T. I., PERRY, V. P., KERBY, C. C. and DOLAN, M. F. (1965), 'Peripheral leukocyte infusion into lethally irradiated guinea-pigs', *Blood*, **25**, 693–702.

15. CAFFREY-TYLER, R. W. and EVERETT, N. B. (1966), 'A radioautographic study of haemopoietic repopulation using irradiated parabiotic rats', *Blood*, **28**, 873–890.

16. MICKLEM, H. S. and LOUTIT, J. F. (1966), *Tissue grafting and radiation*, Academic Press: New York.

17. SELLER, M. J. (1968), 'Transplantation of anaemic mice of the W-series with haemopoietic tissue bearing marker chromosomes', *Nature* (London), **220**, 300–301.

18. YOFFEY, J. M. (1966), *Bone marrow reactions*, Arnold: London.

19. HARRIS, P. F. and KUGLER, J. H. (1964), 'The use of regenerating bone marrow to protect guinea-pigs against lethal irradiation', *Acta haematologica* (Basel), **32**, 146–167.

20. CUDKOWICZ, G., UPTON, A. C., SMITH, L. H., GOSSLEE, D. G. and HUGHES, W. C. (1964), 'An approach to the characterisation of stem cells in mouse bone marrow', *Annals of the New York Academy of Sciences*, **114**, 571–585.

21. MORRISON, J. H. (1967), 'Separation of lymphocytes of rat bone marrow by combined glass-wool filtration and dextran-gradient centrifugation', *British Journal of Haematology*, **13**, 229–235.

22. DICKE, K. A., VAN NOORD, M. J., MAAT, B., SCHAEFER, U. W. and VAN BEKKUM, D. W. (1973), 'Attempts at morphological identification of the haemopoietic stem cell in primates and rodents'. In *Haemopoietic stem cells*, Ciba Foundation Symposium 13 (new series), Associated Scientific Publishers: Amsterdam.

23. TILL, J. E. and McCULLOCH, E. A. (1961), 'A direct measurement of the radiation sensitivity of normal mouse bone marrow cells', *Radiation Research*, **14**, 213–222.

24. McCUSKEY, R. S., MEINEKE, H. W. and TOWNSEND, S. F. (1972), 'Studies of the hemopoietic microenvironment. I. Changes in the microvascular system and stroma during erythropoietic regeneration and suppression in the spleens of CF1 mice', *Blood*, **39**, 697–712.

25. BRADLEY, T. R. and METCALF, D. (1966), 'The growth of bone marrow cells in vitro', *Australian Journal of Experimental Biology and Medical Science*, **44**, 287–300.

26. BURWELL, R. G. (1963), 'The role of lymphoid tissue in morphostasis', *The Lancet, ii*, 69–74.

27. LAJTHA, L. G. and OLIVER, R. (1960), 'Studies on the kinetics of erythropoiesis: a model of the erythron'. In *Haemopoiesis–cell production and its regulation*, Ciba Foundation Symposium, Eds. Wolstenholme, G. E. W. and O'Connor, M., Churchill: London.

28. STOHLMAN, F. (1970), 'Regulation of red cell production'. In *Formation and destruction of blood cells*, Lippincott: Philadelphia.

29. WETZEL, B. K. (1970), 'The fine structure and cytochemistry of developing granulocytes, with special reference to the rabbit'. In *Regulation of hematopoiesis*, Vol. 2, Ed. Gordon, A. S., Chap. 33, pp. 769–817, Appleton-Century-Crofts: New York.

30. LICHTMAN, M. A. and WEED, R. I. (1972), 'Alteration of the cell periphery during granulocyte maturation: relationship to cell function', *Blood*, **39**, 301–316.

31. ZUCKER-FRANKLIN, D. (1970), 'The ultrastructure of megakaryocytes and platelets'. In *Regulation of hematopoiesis*, Vol. 2, Ed. Gordon, A. S., Chap. 55, pp. 1553–1586, Appleton-Century-Crofts: New York.

32. WRIGHT, J. H. (1906), 'The origin and nature of blood platelets', *Boston Medical and Surgical Journal*, **154**, 643–645.

33. HUHN, D. and STICH, W. (1969), *Fine structure of blood and bone marrow*, Lehmann's Verlag: Munchen.

34. YAMADA, E. (1957), 'The fine structure of the megakaryocyte in the mouse spleen', *Acta anatomica* (Basel), **29**, 267–290.

35. LEADING ARTICLE, 'Life history of platelets', *The Lancet, ii*, 393–394, 1968.

36. EBBE, S. (1970), 'Megakaryocytopoiesis'. In *Regulation of hematopoiesis*, Vol. 2, Ed. Gordon, A. S., Chap. 55, pp. 1553–1586, Appleton-Century-Crofts: New York.

37. HARKER, L. A. (1970), 'Regulation of thrombopoiesis', *American Journal of Physiology*, **218**, 1376–1380.

38. PENINGTON, D. G. (1969), 'Assessment of Platelet Production with [75]Se Selenomethionine', *British Medical Journal*, **4**, 782–784.

39. GOSS, C. M. (1928), 'Experimental removal of the blood island of Ambyostoma punctuatum embryos', *Journal of Experimental Zoology*, **52**, 45–63.

8 Functional Histogenesis of the Lymphoid Organs

MARGARET J. MANNING
and J. D. HORTON

Introduction

The lymphoid organs play a major rôle in the defence mechanisms of the body. Their lymphoid cells have a special function in immune responses, the small lymphocyte being a carrier of immunological information[1]. The production pathway of lymphocytes is from a large 'blast' cell, through successive mitotic divisions, to small lymphocytes which are mobile cells with little cytoplasm. At one time small lymphocytes were thought to be end cells but they are now known to be capable of further differentiation following contact with antigenic (immunogenic) material.

The lymphoid organs can be divided into two categories: primary lymphoid organs, such as the thymus and the bursa of Fabricius, which are the first to develop in ontogeny and which produce lymphocytes for functional expression elsewhere in the body, and secondary lymphoid organs (the spleen, lymph nodes and lymphoid follicles of Peyer's patches etc.) in which lymphocytes, on contact with an appropriate antigen, enter into a further phase of proliferation and differentiation. In the primary lymphoid organs, lymphopoiesis (production of lymphocytes) is independent of exogenous stimulation whereas, in the secondary lymphoid organs, it is predominantly antigen-induced. Secondary lymphoid organs have the executive rôle of producing antibody and initiating cellular immune responses.

The extreme sensitivity of lymphocytes to stimulation by minute amounts of antigen is of special interest to embryologists. Moreover, this lymphocyte reactivity forms the basis of the vertebrate immune system[2]. Functionally, this system enables the animal to respond to and destroy antigenic material through the production of humoral antibody (immunoglobulin molecules) and/or cell-mediated reactions, such as graft rejection, in which antibody does not appear to play an essential rôle. The hallmarks of the immune response are its specificity (any one lymphoid cell can only produce antibody to one antigen or a few closely related antigens) and its memory (a heightened response results if the same antigen is contacted a second time). Initial recognition of antigen by the lymphoid cell is currently believed to involve selection by the antigen of antibody-like receptors on the cell surface[3] rather than an inductive phenomenon in which antigen actually enters the cell. Contact of antigen with cell receptor leads to proliferation which yields a clone of cells[4] reactive with (and/or able to produce antibody to) that antigen. The seemingly endless diversity of the immune response can be traced to unique amino acid sequences in a variable portion of the immunoglobulin molecule (see p. 284), but its genetic basis remains an unsolved problem.

Although the specificity of the immune response is attributable to the lymphoid cells, other cell types are also implicated in immunological reactions. Macrophages may process antigen into highly immunogenic fragments, or retain antigen in immunogenic form and keep it in an appropriate location in the lymphoid system[5]. Although *in vitro* tests have indicated that macrophages may be essential for an immune reaction to take place, current opinion favours their rôle as non-specific helper cells, since antibody induction can take place in their absence[5]. Other cells, too, play a rôle in defence mechanisms; granulocytes, for example, may be attracted to sites of immune reactions. In addition, although the specificity of antigen-antibody combination lies at the level of the immunoglobulin molecule, this combination can trigger events involving other substances which are non-specific in their actions, such as those of the complement system which effect cytolysis and chemotaxis and facilitate phagocytosis[6]. It is also probable that some of the small cells termed 'lymphocytes' on morphological criteria are not part of the immune system but have some different rôle, for example, as stem cells in other lines of differentiation[7]. In the present chapter, however, we shall confine our discussion to the cells which are concerned with the specific immune responses of the vertebrate lymphoid tissues.

Development of the Lymphoid Organs

Sequence of development and growth of organs

The fact that primary lymphoid development is initiated in the thymus has been known for many years. As early as 1894, Beard[8], from his studies on the skate, suggested that the thymus is a source of cells which ultimately become distributed to the peripheral lymphoid organs. Many subsequent descriptions of the histogenesis of the lymphoid system confirm that lymphocytic differentiation in the thymus precedes that in other lymphoid organs[9-13]. Table 8.1 illustrates the way in which development of the thymus occurs before that of the spleen in the amphibian, *Xenopus laevis*[14]. A similar sequence occurs in the mammalian fetus[15]. Thus, in the fetal mouse, an epithelial thymus is present by 10-11 days and lymphocytic proliferation is active in the thymus by day 12-13. At this latter time, the spleen is present only as an anlage; splenic lymphopoiesis begins later, at 17 days. Similarly, in the lamb,

Table 8.1 *Chronology of histogenesis of the thymus and spleen in an amphibian,* Xenopus laevis[14].

	State of differentiation at larval stage[87] of development					
	42	45	47	48	49	50
Thymus						
anlage	+					
active lymphopoiesis			+			
many small lymphocytes					+	
Spleen						
anlage		+				
active lymphopoiesis				+		
many small lymphocytes						+

the thymus is seen in the 30 day old fetus and is producing lymphocytes by day 43, while the spleen first appears at about day 40, lymphopoiesis beginning at about day 65. In man, the thymus is present at day 40 and is a lymphopoietic organ by day 48, while the spleen appears at about day 70 and does not actively produce lymphocytes until much later, at about day 140. In mammalian lymph nodes, lymphocytic proliferation starts at times roughly comparable with those of the spleen, but in gut-associated lymphoid organs, such as Peyer's patches, it usually commences later and often is not apparent until after birth. However, an exception to the general pattern of lymphocytic differentiation has been reported in the fetal cat where small lymphocytes were observed in the lymph nodes before they were apparent in the thymus; presumably these came from extra-thymic sources[16]. Birds have a distinctive lymphoid structure in the cloacal region (the bursa of Fabricius) which, like the thymus, is a primary lymphoid organ. Lymphopoiesis occurs in the chicken thymus at about the ninth to tenth day of incubation and in the bursa of Fabricius around 15-16 days. Both organs are lymphocytic at hatching and are the only fully differentiated lymphoid organs at this time[17].

The primary lymphoid organs reach their maximum size early in life[18], their growth pattern differing from that of other organs of the body. In birds, both the thymus and the bursa of Fabricius are large until about the second or third month. Thereafter the bursa disappears while the thymus persists, although reduced in size[17]. In many mammals, the thymus reaches its maximum weight at about the time of puberty. It then undergoes 'age involution', a process involving a marked decrease in size and the progressive development of adipose tissue. Atrophic changes with advancing age also occur in the secondary lymphoid organs. This reduction in lymphoid tissue is associated with a decline in immunological function and is particularly significant in reactions involving cell-mediated immune responses[19]. However, the old animal still retains some functional lymphoid tissue; in aged humans, for example, areas of normal thymic parenchyma are still readily identifiable.

Structure and histogenesis
Thymus
The thymus first appears in the developing animal as a bud from the pharyngeal epithelium (see Plate 8.1). It is present in all jawed vertebrates but probably absent in the hagfish. In the lamprey, lymphoid accumulations in the larval pharynx have received various interpretations and are accepted as thymic homologues by some authors[20, 21]. In most mammals, the thymus migrates to the thoracic region where the adult organ is situated. In the guinea pig, however, it remains in the neck and both ectoderm and endoderm contribute to its formation[22]. Thus, during development, the ectoderm of the cervical sinus first forms a deep groove at the base of the neck then isolates and, together with the endoderm of the third pharyngeal pouch, forms the definitive thymus. In the bat, which has thymic lobes in both neck and thorax, the cervical thymus is formed from ectoderm of the cervical sinus, while the thoracic lobes are endodermal in origin[23]. In birds, Hammond[24] has demonstrated experimentally that the thymus is derived from ectoderm. Ruth, Allen and Wolfe[25] review the evidence on the relative importance of ectoderm and endoderm in the formation of the thymus and the bursa of Fabricius. They conclude that ectodermal/endodermal interactions probably play a significant part in the early induction of the primary lymphoid organs.

In the ancestral vertebrate each visceral cleft region may well have been capable of yielding a thymic bud and primitively this probably originated from the dorsal epithelium near the junction of the out-pushing endodermal pouch and the ingrowing ectodermal groove. This may perhaps explain both the variable origin of the thymic epithelial cells from ectoderm, endoderm or both, and the fact that the definitive thymus forms from different visceral regions in different vertebrates. Thus there are six pairs of thymic buds in selachian fishes while, in urodeles, five pairs of buds appear but only three of these pairs, those from visceral clefts III, IV and V contribute to the adult thymus. In anuran amphibians, the thymus develops in the region of the second visceral cleft, in lizards from the second and third, in snakes from IV and V, while in turtles, birds and mammals the thymic bud is associated with pouches III and IV, the major contribution coming from pouch III. In mammals, unlike other vertebrates, it is the ventral part of the pouch, not its dorsal region, which gives rise to the developing thymus (see Jolly[26], for a review).

Further differentiation of the thymic epithelium occurs as blood-borne cells invade the anlage (see p. 274 and Fig. 8.1). These force the epithelial cells apart but the latter remain held together by desmosomes and thus come to form a branching network[27]. The epithelial cells give rise to a number of derivatives such as thymic corpuscles and cystic structures; these are more conspicuous in the medulla than in the cortex. The thymic epithelium is responsible for directing the pathways of proliferation of the blood-borne cells which enter the organ and for influencing their differentiation into antigen-reactive small lymphocytes (T-cells – see p. 275). Secretion of humoral substances by the thymic epithelial cells is probably involved in this process[28, 29].

The small lymphocytes of the thymus are the progeny of a series of mitotic divisions which starts with the large lymphocyte and yields, first, medium-sized then, small lymphocytes. Mitotic activity is brisk, particularly in the thymic cortex which is densely packed with small lymphocytes (Plate 8.2). In mice it has been calculated that the small lymphocyte population of the cortex is replaced every 3–4 days. However, there is some doubt as to the percentage of lymphocytes that actually leave the thymus[30, 31]: Metcalf believes that is as low as 1% and that many cells die *in situ*[30]. The occurrence, but not the extent, of lymphocytic movement out of the thymus has been demonstrated by intra-thymic injection of tritium-labelled thymidine[32]. In these studies, labelled thymic lymphocytes could be traced to other tissues.

Most of the cells in the thymus are still incapable of reacting to antigens as fully competent T-cells. However, some lymphocytes of the thymic medulla give responses similar to those of the T-cells in the peripheral circuit and have, perhaps, undergone a further step in differentiation towards immunocompetence. These cells form some 5% of the total

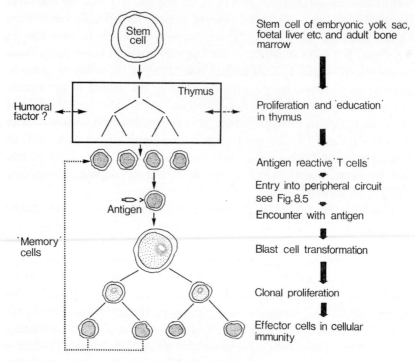

Fig. 8.1 Functional differentiation of the T-cell system[15].

population in the mature thymus. They show a resistance to the action of corticosteroids which is higher than that of the majority of thymus lymphocytes.

Gut-associated lymphoid organs

Bursa of Fabricius. The bursa of Fabricius forms as a dorsal diverticulum of the cloaca in birds. The surface epithelium of the bursal anlage is thrown into folds and then thickens in places to form buds which project into the surrounding connective tissue. Each epithelial bud forms the medullary portion of a bursal follicle and is surrounded by a basement membrane which is continuous with that of the epithelium. Outside this, there develops a mesenchymal condensation which gives rise to the cortex of the follicle[26]. Production of lymphocytes starts within the epithelial bud following the entry of blood-borne stem cells (Fig. 8.2); later small lymphocytes become numerous in the surrounding connective tissue. A distinct cortex and medulla appear after hatching. The bursa of Fabricius plays a special rôle in initiating the B-cell line of

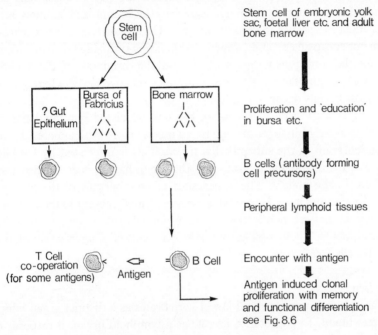

Fig. 8.2 Functional differentiation of the B-cell system[15].

lymphocytic differentiation which yields the precursors of antibody-forming cells (see p. 275). Since similar B-cell lymphocytes occur in mammals, efforts have been made to identify a mammalian equivalent of the bursa of Fabricius. However, it now seems likely that at least some of the mammalian gut-associated structures are secondary lymphoid organs and that the bone marrow plays the 'bursa-equivalent' rôle in the adult mammal (Fig. 8.2).

Other gut-associated organs. Lymphoid nodules occur in the gut wall and in the walls of the pulmonary, urinary and genital tracts. A simple gut-associated lymphoid organ from the intestinal wall of an amphibian is shown in Plate 8.3. An accumulation of lymphocytes can be seen in the sub-epithelial connective tissue and lymphocytes can also be observed lying within the gut epithelium. Lymphocytes similarly occur within the gut epithelium of other vertebrates[33]. In mammals, for example, they are numerous in the villus epithelium of the intestine. Other lymphoid tissues, such as the tonsils, Peyer's patches, appendix and the solitary lymphoid nodules, have a more complex structure which includes lymphoid follicles with germinal centres similar to those found in the spleen and lymph nodes. Germinal centres are rich in blast cells and show intense mitotic activity, especially in immunized animals after secondary exposure to certain antigens. Organs such as Peyer's patches are not encapsulated and antigenic stimulation probably normally occurs from the gut lumen. Some of the cellular products of the follicles possibly accumulate in the surrounding connective tissue where groups of plasma cells occur.

Embryological development of the gut-associated lymphoid tissues, such as the rabbit appendix and tonsil iliaca, begins with a mesenchymal condensation in the subepithelial tissue. This is infiltrated by lymphocytes which probably migrate into the organ having been formed elsewhere in the body[34]. The appearance of lymphocytes in the gut epithelium occurs later. Germinal centres are usually absent in the newborn mammal and are poorly developed in animals reared under germ-free conditions[35]. They apparently form as a result of antigenic stimulation of B-cells (see below).

Spleen

The spleen is present in all jawed vertebrates as a distinct organ which filters blood. In addition to its rôle as a lymphoid organ, it disposes of effete erythrocytes and it stores and produces erythrocytes, platelets and granulocytes. Its phagocytes take up foreign material from the blood. The structure of the mammalian spleen is shown in Fig. 8.3 (see also Diener[36]). Around the central arterioles of the white pulp are periarteriolar sheaths of T-lymphocytes. Other lymphocytic regions of the white pulp contain follicles with germinal centres. These regions are separated by their marginal sinuses from the marginal zones; the latter blend imperceptibly into the red pulp. The marginal zones receive most of the blood first entering the spleen: here, lymphocytes and macrophages are held in a specialized reticular network and blood-borne antigens are trapped. The red pulp contains nests of antibody-secreting plasma cells. These are conspicuous in the stimulated spleen.

The spleen starts its development as a condensation of mesenchymatous tissue in the dorsal mesogastrium (see Plate 8.4). This is vascularized and immature blood-forming cells appear in the vascular spaces and in the perivascular mesenchyme. Differentiation into red pulp and white pulp begins as the reticular sheaths around the splenic arterioles become more conspicuous and lymphocytes infiltrate into the area to form the periarteriolar lymphocytic sheaths. Blood cell formation in the spleen of the developing mammal at first involves the formation of erythrocytes and granulocytes, only later does lymphopoiesis become important. There is some evidence that the micro-environment of

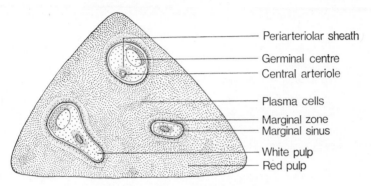

Periarteriolar sheath

Germinal centre

Central arteriole

Plasma cells

Marginal zone
Marginal sinus

White pulp
Red pulp

Fig. 8.3 Schematic diagram of the rat spleen (T.S.)[27, 36].

different regions of the spleen provides a stimulus which determines the future line of differentiation of primitive stem cells (towards erythropoiesis, granulopoiesis, lymphopoiesis, etc.) (see Moore and Metcalf[15]).

Lymph nodes
Lymph nodes are secondary lymphoid organs which filter lymph. They are situated along routes of lymphatic drainage and their organogenesis is related to that of the lymphatic system. This is clearly seen in the rudimentary lymph nodes of some birds which develop by simple invasion of the lumen of a lymphatic vessel by mesenchyme[26]. The mammalian lymph node is a more complex structure but its origin is essentially similar. It forms as a plexus of anastomosing lymphatic vessels interspersed by condensations of mesenchymatous cells.

Lymph nodes are encapsulated organs consisting of an outer cortical region, a mid-cortex (paracortical zone) and an inner medulla (Fig. 8.4). Foreign material reaches the node by way of the afferent lymphatic vessels and is phagocytosed by macrophages, particularly those of the medullary sinuses. Some antigen, however, is dealt with in a more specialized way, being trapped extracellularly on dendritic processes of reticular cells within the lymphoid follicles of the cortical region. This form of antigen retention also occurs in the lymphoid follicles of the spleen, Peyer's patches etc. It probably depends upon the presence of antibody. Functionally, it serves to bring antigen into intimate contact with the lymphoid cells of the follicles and this elicits active proliferation of B-lymphocytes with the formation of germinal centres. Some of the cells produced as a result of antigenic stimulation appear in the efferent lymph leaving the node but some apparently remain within the organ

Afferent lymphatic

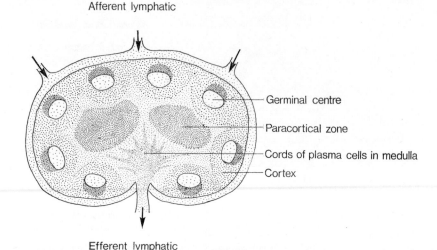

Germinal centre

Paracortical zone

Cords of plasma cells in medulla

Cortex

Efferent lymphatic

Fig. 8.4 Schematic diagram of a mammalian lymph node[27, 37].

and, in stimulated lymph nodes, the medulla becomes richly populated with antibody-secreting plasma cells.

The paracortical zones of the mammalian lymph node are areas which, like the periarteriolar lymphocytic sheaths of the spleen, contain T-cells and respond in cell-mediated immune responses. The cortex and the medullary cords, on the other hand, respond to antigens which elicit humoral antibody production. Many antigens are complex, however, and stimulate both regions of the node[37].

While the majority of mammalian lymph nodes filter lymph, in some species, particularly in ruminants, there are a number of so-called haemolymph nodes whose sinuses are filled with blood. The rudimentary lymph nodes of amphibians are also of this nature (see Plate 8.5 and Baculi and Cooper[38]).

Within the secondary lymphoid tissues (lymph nodes, spleen, etc.), in both the B-cell and the T-cell lineages, encounter of a lymphocyte with an antigen to which it can respond leads first to enlargement and then to a series of mitotic divisions which produce (i) effector cells of the immune system (see p. 275) and (ii) sensitized lymphoctyes which form memory cells.

Other haemopoietic tissues

Some of the organs which we have so far considered have clearly defined functions within the immune system. Thus the thymus is a primary lymphoid organ, responsible for initiating the line of lymphocytes concerned in cell-mediated responses (T-cells). The bursa of Fabricius is a primary lymphoid organ essential to the development of B-lymphocytes which are the precursors of antibody-forming cells. The lymph nodes, spleen and the gut-associated organs such as Peyer's patches are secondary lymphoid organs. These have a complex structure, with arrangements for trapping and processing antigen and ensuring that the latter is presented in suitable form to cells capable of reacting to it. They include areas for circulation of cells of the immune system within an anatomical framework which increases the opportunities for the chance meeting of antigen and antigen-reactive cell. They accommodate the proliferating lymphoid populations which result from antigenic stimulation, house the developing effector cells and provide for the release of their humoral and cellular products.

There are other vertebrate lymphoid tissues whose function within the immune system is less clear, partly because in the case of poikilotherms, we lack detailed knowledge and partly because these tissues also house precursor cells of other blood cell lines, as well as primitive stem cells which are probably still uncommitted as to their future line of development. The incomplete separation of myeloid and lymphoid components is particularly evident in poikilotherms and in the immature stages of all vertebrates.

In the ancestral vertebrate, these blood-forming elements were probably housed in convenient 'unoccupied' spaces. Thus the present-day lamprey has a typhlosole-like structure which serves to increase the absorptive surface of the gut; this also accommodates blood-forming tissue. The lamprey also has haemopoietic cells in the soft tissues of the protovertebral arch, an area which lies above and cushions the otherwise poorly-protected spinal cord. Terrestrial vertebrates have hollow bones and, at this evolutionary stage, haemopoietic tissue is regularly found occupying parts of the bone marrow cavity. This is an available location which is absent from the thinner bones of fish, although in some fishes blood-forming tissue is occasionally associated with the cranial skeleton (for example, in some holosteans).

During ontogeny, the blood-forming tissue arises from mesenchymal cells associated with the extra-embryonic splanchnic mesoderm of the yolk sac. Later, in mammals, the liver, spleen and ultimately the bone marrow assume haemopoietic functions. In adult mammals the bone marrow is the definitive blood-forming tissue. It, alone of the organs, contains all the necessary progenitor cells to restore the full blood-forming capacity of an irradiated animal. In addition to other lines of blood cells, the bone marrow contains many lymphocytes. These bone marrow lymphocytes are a somewhat problematic population. They are replaced about every three days, being produced partly from progenitor cells from within the marrow itself and also possibly by entry of cells from the blood. Small lymphocytes probably leave the marrow for the blood circulation and may perhaps be progenitor cells for other cell lines[7]. In anamniotes, the kidney also is a blood-forming organ. It contains many lymphocytes and, in some amphibians and fishes, has been shown to play an important part in immune responses[39].

Migrations of Lymphoid Cells

Establishment of the system

The establishment of the immune system in the embryo begins with the development of the primary lymphoid organs and the entry into these organs of blood-borne stem cells derived from blood-forming tissue which, at the first stages or organogenesis, is that of the yolk sac. This was demonstrated by Moore and Owen[15] who traced the chromosome markers of male and female chick embryos joined in parabiosis. This work, together with experiments using tritium labelling, showed that the lymphocytes of the thymus are derived from stem cells which colonize the epithelial thymic bud. Similarly, in the bursa of Fabricius, invading blood-borne cells are the ancestors of the bursal lymphocytes. This contrasts with the views of some earlier workers who believed that the lymphoid cells of the primary lymphoid organs were direct descendants of their epithelial components. This view was based on histological studies (for a review see Sainte-Marie and Leblond[40]) and supported, more recently, by experiments *in vitro*[41]. It now seems probable that the thymic epithelial tissue cultured in these latter experiments already contained immigrant stem cells. It appears that not only do the epithelial cells of a primary lymphoid organ have an essential inductive influence on exogenous stem cells but also that the type of epithelium profoundly influences later functional differentiation. Thus stem cells which lodge in the thymus proceed along a future line of development which is quite distinct from that of cells which develop within the bursa of Fabricius.

This rôle of the primary lymphoid organs in influencing the future differentiation of cells of the immune system is important not only in early ontogeny but also in maintaining a working system. In the adult mammal, the stem cells which populate the thymus are derived from the bone marrow[42]. This was demonstrated in mice using a chromosome marker to trace cellular migrations. The bone marrow is the only definitive blood-forming tissue with a self-sustained population of stem cells. All other lymphoid tissues require an input of immigrant cells to maintain their cellular composition.

Thymus-dependent and thymus-independent populations

In current terminology, cells processed by the thymus are called T-cells; these form a population of lymphocytes which are responsible for cell-mediated immunity (see p. 282). Those which develop independently of the thymus are the B-cells; these are progenitors of antibody-forming cells. Although in the mammal, lymphocytes of the T-cell and B-cell populations cannot be separated by histological criteria, it is possible to distinguish between them on other grounds[43]. In B-cells, surface immunoglobulins can easily be demonstrated using a heterologous anti-serum to the immunoglobulin (see p. 282), while a useful marker of T-cells has been described in some mice. This is the antigen theta, an antigenic marker on the surface of T-cells which can be identified by means of an anti-theta antiserum raised in mice of a different strain. Most of the long-lived lymphocytes which circulate in the blood and lymph bear the theta antigen. Theta-bearing cells are also found in the lymph nodes and spleen where they occupy, respectively, the para-cortical zones and the splenic periarteriolar sheaths. These areas are depleted of lymphocytes by neonatal thymectomy. T-cells are less 'sticky' than B-lymphocytes (they adhere less readily to surfaces), and they are more negatively charged. B-cells, but not T-cells, can bind the immunoglobulin of immune complexes. *In vitro* certain mitogens, such as phytohaemagglutinin, stimulate only T-cells while other antigens, such as *Escherichia coli* endotoxin, stimulate only B-cells.

T-cells can produce pharmacologically-active substances (the so-called lymphokine factors) which are non-specific mediators of cellular immune responses[44]. T-cells also play an important rôle in co-operating with B-cells in antibody production to many antigens (see p. 283). They are not, however, progenitors of antibody-forming cells. This is the responsibility of B-cells.

Recirculation pathways

The cellular movements described on pp. 274–275 are those which first establish and then maintain the working population of antigen-reactive lymphocytes. Further migrations occur as these cells circulate around the body, through blood, lymph and lymphoid tissues. The use of radioactive tracers has enabled these circulation pathways to be studied in some detail[45,46]. Early experiments of Gowans and his co-workers showed that small lymphocytes injected intravenously could be traced to the lymph, whence they were returned to the blood. They also appeared in lymph nodes, Peyer's patches and spleen but did not enter the thymus. A route for the passage from blood to lymph was identified in the paracortical areas of the lymph nodes where specialized blood vessels, the post-capillary venules, occur. These vessels have a high-walled endothelium. Electron microscope studies have shown that small lymphocytes actively penetrate the cytoplasm of the cuboidal endothelial cells and move out of the blood vessel by passing through, rather than between, these cells. Most of these circulating lymphocytes are long-lived cells. In rodents, tritiated thymidine incorporated into the nuclei of blood lymphocytes persists for up to one year. In the rat, it has been calculated that about 90% of circulating small lymphocytes have a life span of several months, the remaining 10% being short-lived cells of only a few days. Large lymphocytes are also present in the circulation in relatively small numbers. It is believed that they are produced by the lymph nodes and migrate to gut-associated lymphoid tissues.

The depletion of lymphocytes from the peripheral circuit of neonatally thymectomized animals suggests that many of the long-lived recirculating small lymphocytes are thymus-dependent cells (see Fig. 8.5) and evidence from experiments involving injection of radioactively-labelled thymus cells[47, 48] and from local intra-thymic labelling[32] indicates that thymus-dependent lymphocytes localize exclusively in areas which are involved in recirculation. Thus thymus cells entering the lymph nodes from the blood move into the paracortical zone. In the spleen, thymus cells first reach the marginal zone, which receives most of the afferent blood, but later 'home' into the periarteriolar sheaths. On the other hand, bone marrow cells injected intravenously 'home' into thymus-independent areas of the secondary lymphoid organs, such as the lymph node medulla and the splenic red pulp. This ability of lymphocytes to move into appropriate areas apparently does not rely on external factors but is an attribute of the lymphocytes, possibly depending on their surface properties[48].

Lymphocytes differ from other cells such as polymorphs and macrophages which play a less specific rôle in defence mechanisms. The transcellular movements of lymphocytes across the endothelium of the postcapillary venules is a peculiar process which may well involve selective

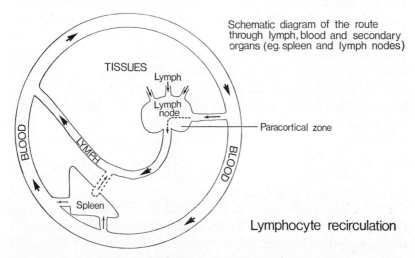

Fig. 8.5 The pathway of recirculation of small lymphocytes[7, 46].

passage of cells. Polymorphs and blood macrophages, on the other hand, move out of vessels in a less specialized way by squeezing between the endothelial cells and are influenced in their movements by a number of pharmacological substances which are non-selective in their action, in the sense that they can affect the movements of all cells of the type concerned. Lymphocytes, on the contrary, have a complex migratory pattern which is not obviously dependent on external factors (chemotactic agents, etc.) and whose 'raison d'être' probably relates to the highly specific nature of their immune responses. Thus continual recirculation of lymphocytes increases the possibility that an antigen localized in a particular part of the body will encounter the infrequent cell of the immune system which can respond specifically to it. In this way the lymphocytes can perform their task of immunological surveillance.

Functional Maturation of the Lymphoid System

Appearance of immunological competence

Immunological immaturity and tolerance

The amniote embryo obtains protection from a potentially hostile environment as it lies within the egg or within its mother's womb. In this respect, anamniotes offer particular advantages for analysing the ontogeny of immunocompetence. The embryo is free-living and so delicate manipulations (such as transplantation of tissues and cells and introduction of antigens) are readily feasible. Moreover, many mammals receive passively-transmitted immunity from the mother before and/or after birth. Likewise, birds and reptiles obtain maternal antibodies via the egg yolk before hatching[49]. Such transmission from mother to young in amniotes plays a vital rôle in the defence of the developing animal by providing protection of the embryo before its own system of active immunity has matured. However, the presence of maternal antibody suppresses or even abolishes the induction of active immunity during early ontogeny and may well modify the basic processes involved in the development of the vertebrate immune system.

Concomitant with development of the animal's lymphoid tissues is the differentiation of antigenic expression by maturing cells throughout the body. These cells must not be rejected but rather tolerated by the developing immunocompetent cells. If an animal meets foreign material during early development this may similarly be tolerated: thus, after vascular sharing *in utero*, cattle twins accept erythrocytes from each other, even though these may be of a different blood group[50]. The concept of a developmental period when all antigens are treated as 'self' and a later stage of differentiation when 'non-self' antigens are rejected, thus emerged[51]. Experimental documentation of this concept soon followed when Billingham, Brent and Medawar[52, 53] showed that mice or chickens inoculated with foreign cells early in development tolerated a skin graft applied from the same donor in adult life. Grafts from unrelated donors were rejected in normal fashion thus demonstrating the specificity of acquired immunological tolerance.

Immunological tolerance was, then, first demonstrated in relation to foreign cells. However, we now know that tolerance following embryonic administration of foreign proteins can also be induced[54]. Moreover, the concept that immunological tolerance relates only to the immature, immunologically incompetent animal must also be dismissed in the light of research which indicates that tolerance can be achieved in the adult[55].

In addition, rather than looking upon immunological tolerance as an unreactive state, it is now suggested that this phenomenon is a specific immunological reaction that parallels the immune response in sensitivity and degree of specificity[56, 57]. Although data on ontogeny of tolerance induction are not readily available such information would seem to be of great importance in determining the onset of differentiation of immunocompetent cells. Thus the ontogeny of immunity is not only dependent on maturation of antigen-reactive cells but, unlike tolerance induction, is also dependent to a certain extent on the appearance of a mechanism that can first process or present the antigen in an immunogenic form. This might involve development of enzyme systems that can suitably degrade antigen for recognition by host lymphoid cells. Tolerance may ensue in the neonate because its reticulo-endothelial system is not well developed which results in poor trapping and retention of antigen, i.e. the deficiency here may not be in the immunological reactivity of the lymphoid cells themselves, but in the immature antigen-trapping mechanisms.

Ontogeny of active immunity
As mentioned in previous sections, the primary lymphoid organs are essential to the maturation of immune capacity of the lymphoid cells. As demonstrated by Miller[58] in his work on mice, the thymus clearly governs the maturation of cell-mediated responses. Thus neonatal thymectomy in the mouse resulted in severe impairment of the ability to reject foreign grafts, whereas thymectomy later in life had relatively little effect on this response. These findings have now been substantiated in many mammalian, avian and amphibian species[59]. One of the interesting points to emerge from such research is the variable effect that neonatal thymectomy has on subsequent immune reactivity in different mammalian species. Why is this so? The answer possibly lies in the varying extent of lymphoid tissue maturation displayed in these animals at the time of the thymectomy operation – animals possessing more mature lymphoid tissues being relatively little affected by thymectomy whereas those with immature lymphoid organs at the time of this operation display severe immunological defects[59]. Congenital absence of the thymus similarly leads to severe crippling of the cellular immune system[60, 61]. In contrast to the development of cellular immunity, the maturation of antibody production is dependent upon other primary lymphoid organs – in birds, the bursa of Fabricius. This is readily displayed by bursectomy which severely abrogates antibody production

but has relatively little affect on graft rejection[62]. Experiments of this kind on the early removal of the thymus or the bursa of Fabricius first demonstrated the dichotomy of function within the immune system between, on the one hand, the thymus-dependent (T-cell) system and, on the other, the thymus-independent (bursa-dependent) system of B-cells.

The maturation of immune responses may appear at the same physiological age in all vertebrates[63]. The critical relationship here is not such processes as birth, hatching or metamorphosis, but rather morphological events such as first appearance of small lymphocytes. Let us first consider the maturation of the immune system in the lamb, an animal with a long gestation period (150 days), and which apparently receives no passive immunity before birth[49]. Silverstein and his colleagues[64, 65] have worked extensively on this species and have revealed a stepwise development of the immune response. Thus antibody to a bacteriophage occurred as early in gestation as was technically feasible to immunize and bleed the fetus (about 40 days – the thymus is just becoming lymphoid at this time), whereas antibody to ovalbumin was only produced at 120 days and to *Salmonella typhosa* only after birth. The ability to respond to foreign skin grafts develops at about 85 days gestation, by which age lymphocytes are present in the thymus, lymph nodes and spleen. The sequential ontogeny of immune responses may depend on the acquisition of clones of antigen-reactive cells responsive to the different antigens and/or it may reflect the immunogenicity of the antigen itself. Differences in inducibility of antibody production to different antigens may of course partly reflect the differences in extent of natural immunity that already exist prior to immunization. Also, in transplantation experiments, the antigenicity of the transplanted tissue may also vary as this tissue itself develops.

The opossum is another good experimental animal from the viewpoint of the developmental immunobiologist: it is born only 12 days after copulation and then crawls into the maternal pouch. In the opossum immunological capabilities develop gradually. This is in contrast to the lamb which can express adult levels of reactivity as soon as it becomes immunologically competent. Thus in the opossum chronic rejection (12–13 weeks) of foreign grafts occurs if these transplants are applied to 17 day old pouch young, whereas grafts placed on animals older than 3 months are rejected (as in the adult) in acute fashion (1–3 weeks)[66]. Moreover, Rowlands *et al.*[67] have shown that older opossums respond more vigorously to viral antigens than younger pouch animals and have demonstrated that 5 day old pouch young can produce antibody. In the opossum, lymphocytes are first seen in the thymus at 2 days, in the lymph nodes from 3–6 days and in the spleen from 17–20 days of pouch life[67].

The relationship between lymphoid organ development and the ability to respond to foreign grafts is readily apparent in the Amphibia. Hildemann and Haas[68] showed that, in the bullfrog, transition from the period when larvae accept foreign tissue to the time when they display immune reactivity to transplants is directly correlated with the appearance of small lymphocytes in the circulation. Horton[69, 70] has related the maturation of transplantation immunity to specific stages of lymphoid organ histogenesis in two anuran species. The first sign of immune responsiveness is seen when only the thymus contains many lymphocytes: subsequent increase in complexity of the response to grafts is concomitant with a rapid phase of lymphoid organ maturation. The work of Volpe and his colleagues[71] suggests that the immunological outcome of embryonically applied foreign grafts depends on the size of the transplant. Thus immunity follows if the embryonic grafts are small, whereas tolerance is induced if these grafts are large. In birds and mammals the relative ease with which premature exposure to foreign antigens induces tolerance may be related to the special embryonic conditions of amniotes. In free-living developmental stages (such as found in amphibians) early exposure to a foreign antigen may not necessarily preclude the possibility of an immune response to the same antigen in a more mature animal. This is perhaps not surprising since a free-living larva will naturally encounter many pathogenic organisms in its environment before its lymphoid system is mature enough to deal with them.

The ontogeny of the *in vitro* rather than *in vivo* capacity of lymphoid cells to display immunological reactivity is currently under investigation in many vertebrate species[72]. Thus the capacity of lymphoid cells at different stages of development, to respond to phytohaemagglutinin and other mitogens should yield invaluable information on the ontogeny of cellular reactivity. Using *in vitro* tests which detect antibody formation against sheep erythrocytes, Du Pasquier[73] has shown that antigen-reactive cells are present in the larva of the midwife toad when it possesses remarkably few lymphocytes (less than one million).

Rôle of antigen in lymphoid cell differentiation

Antigen recognition

The first step in the stimulation of the lymphoid cell by antigen is the encounter of the latter with an antigen-specific receptor site on the surface of the immunologically competent cell. It has been postulated that this site is immunoglobulin in nature. This is well substantiated for B-cells but, in the case of T-cells, the issue remains controversial. If a fluorescein-labelled antiserum to the immunoglobulin is applied to B-lymphocytes, fluorescence can be detected at the cell surface; at 37 °C this appears as a cap at one pole of the lymphocyte. This cap formation is attributed to a reaction between the receptor immunoglobulin and the specific antibody. Engagement of the receptor site alters the surface properties of the cell in such a way as to immobilize this part of the membrane. It has been suggested that a similar combination of receptor immunoglobulin with antigen may form part of the mechanism by which antigen triggers the next stage of differentiation, i.e. clone formation leading to antibody production (see Taylor et al.[74]).

Induction of tolerance

There is still much debate concerning the cellular mechanisms involved in tolerance induction. Initially it was suggested that tolerance resulted from a deletion of a clone of cells predestined to react with a certain antigen[4, 75]. Current opinion[5], however, favours the idea that induction of tolerance may simply depend upon the concentration of antigen that meets the antigen-reactive lymphoid cell: only the correct amount of antigen that reaches the cell results in an immune (rather than a tolerant) response. Although deficiency of macrophage activity may well be an important non-specific factor involved in tolerance induction, the basic cellular defect has been shown to reside in the lymphocyte population[2].

Induction of immunity

In cell-mediated immune reactions involving thymus-dependent lymphocytes (T-cells), the initial contact of antigen with the antigen-reactive cell may occur either outside the lymphoid tissue or within a secondary lymphoid organ. The stimulated T-cell enlarges to form a cell with pyronin-staining cytoplasm (the large pyroninophilic cell) which then undergoes a series of mitotic divisions (see Fig. 8.1). Plasma cells are not produced in this proliferative response, rather lymphocytes of progressively decreasing size result. These cells ('killer cells') may, by direct action, destroy the antigen. They may, for example, invade a skin graft[76] (see Plates 8.6 and 8.7). The reaction of sensitized lymphocytes with antigen may result in the production of non-immunoglobulin mediators of immunity (lymphokines). These are secreted specifically but expressed non-specifically, and have the ability to inhibit the migration of macrophages, attract granulocytes, etc.[44, 77].

The induction of humoral antibody synthesis to many antigens involves the co-operation of T-cells and B-cells[78, 79]. The mechanism of co-operation is not completely understood but may involve initial recognition (after processing of antigen into suitable immunogenic form by macrophages) by the T-cell of the carrier determinant of the antigen and the presentation of the inducing (haptenic) determinant to the immunological receptor of the B-cell[80] (see Fig. 8.2). Mitchison[80] suggests that the initial reaction of antigen with the T-cell may be thought of as a concentrating mechanism required for stimulation of the B-cell receptor. Thymus-independent antigens (i.e. those which can stimulate B-cells without requiring T-cell co-operation) seem to be polymeric forms with repeating identical determinants.

The cellular events occurring in humoral antibody production (see Fig. 8.6 and Nossal[81]) take place in the secondary lymphoid organs. Following antigen recognition, the resting nucleus of the lymphoid cell of the B-series becomes activated and the cytoplasmic organelles display enhanced activity. RNA synthesis and protein synthesis increase, as indicated cytologically by the intense pyronin-staining properties of the resulting blast cell (large pyroninophilic cell). Acceleration of RNA synthesis begins before the onset of DNA synthesis. Thus the time of maximal incorporation of tritiated thymidine into DNA occurs later than the time when the greatest number of large pyroninophilic cells are present[82]. DNA synthesis is followed by cell division and rapid mitotic division cycles of daughter cells then take place, producing cells intermediate in morphology between lymphocytes and the cells that are eventually produced in this proliferative event – the antibody-forming plasma cells. Although immunoglobulin production can occur to a small extent in lymphocytes, this process chiefly occurs in the plasma cell which has the organelles for the production and secretion of large amounts of protein (such as Golgi apparatus, polyribosomes and rough endoplasmic reticulum). Immunoglobulin is localized in the endoplasmic reticulum and, in the mature plasma cell, becomes concentrated in extended cisternae. Eventually, it is extruded from the cytoplasm and enters the circulation.

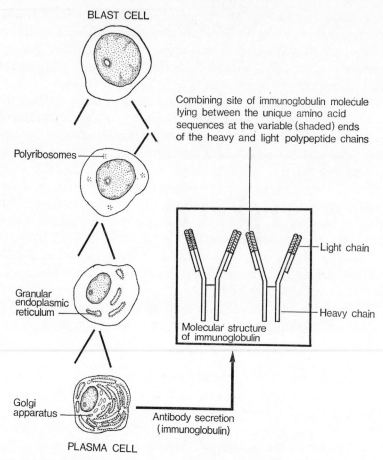

Fig. 8.6 Differentiation of antibody-secreting cells[27, 81]

Immunoglobulin structure

The immunoglobulin molecule is based on subunits comprising four polypeptide chains, two identical heavy chains and two identical light chains. The antigen-binding sites lie between the N-terminal sections of the light and heavy chains (see Fig. 8.6). Diversity among immunoglobulin molecules exists in the N-terminal portion of the heavy and light chains. Thus in each immunoglobulin so far studied, the N-terminal section has a unique sequence of amino acids. This is in contrast to the constant nature of the carboxy-terminal sections[83, 84]. The known variability is sufficient to account for the large number of antibodies which an animal can make. There is evidence that the constant and variable portions are coded by separate genes but considerable uncertainty

284

remains about the number of v (variable)-genes and about the possible integration process of v-genes with the c (constant)-gene. An even more important unsolved problem is how diversity is generated in the variable segments of the immunoglobulin chains. Does every immunologically competent lymphocyte have a large number of genes which are randomly expressed in different cells or are somatic mutations or recombinations involved? Various theories are reviewed by Talmage[85] and Jerne[86].

The immunoglobulins (Ig) fall into different classes (IgM, IgG, IgA etc.) according to pharmacological and biological properties which are governed by the structure of the heavy chains. The whole range of antibody specificities appears to be available in each class. Several workers have shown that IgM is the first type of immunoglobulin which can be detected in ontogeny and there is evidence that even cells which will eventually produce IgG, initially express IgM on their surface. The ability to synthesize IgM occurs very early in development: thus sensitive *in vitro* tests have shown its production in the chicken bursa of Fabricius only 6 days after histogenesis of the lymphoid system has commenced. *In vitro* studies also indicate that lymphocytes tend to be IgM secretors whereas plasma cells usually secrete IgG[27]. T-cells appear to be required for the full expression of IgG responses but it is not yet clear whether they act by inducing a cellular switch from IgM to IgG production or whether they selectively activate those B-cells which already bear appropriate IgG receptors.

Immunological memory

An initial (primary) response to an antigen usually increases the animal's capacity to respond again to that same antigen, although under some circumstances, as we have seen, negative memory (i.e. tolerance) may result. Experiments have shown that memory cells exist in the long-lived population of circulating small lymphocytes and also may occur in cells that are not depleted by drainage of lymph from the thoracic duct[46]. A heightened reaction to introduction of antigen a second time is probably due to the clonal expansion which occurred in the antigen-induced phase of the first response (see Fig. 8.1). Whether the memory cells acquire any special properties during this proliferation and thereby become specially sensitized lymphocytes which differ from virgin antigen-reactive cells is an open question. Memory could be due simply to their numerical increase but it is more likely that it involves conversion to a further level of differentiation. It is also possible that a decreasing requirement for co-operation with T-cells may play a part in the memory phenomenon with respect to humoral immunity.

The acquisition of memory cells during the life of the animal results in an increase in receptor sites for antigens previously encountered, which are predominantly those common in the environment. Thus the animal becomes adapted and under these circumstances, and in the limited space of the immune system, some restriction of the input of randomly-generated new cells may be desirable. This might account for the typical 'age involution' of the primary lymphoid organs. Some neo-lymphogenesis is however advantageous both to replace clones whose life span is finite[3] and to maintain sufficient representation throughout life of cells which can deal with new antigens.

Adaptive significance

In a system where, throughout life, the final stages of lymphocyte differentiation are governed by exogenous antigen, the insertion of internally regulated primary organs provides a necessary safeguard against exhaustive overstimulation by external factors. The peculiar variability of the immunoglobulin molecule permits the highly discriminatory recognition of large numbers of potential antigens. The number of different recognition sites involved (estimated as being of the order 10^5) means that each individual site can be only poorly represented in the lymphocytic population – especially if there is only one type per individual lymphocyte. Thus, although in ontogeny individual lymphocytes become committed to responding to one, or a relatively small number of antigens, the fact that antigen-triggering leads first to mitotic activity and only later to effector cell differentiation, allows clones of cells with the required specificities to be built up as necessary.

The elaborate circulatory pathways of the long-lived lymphocytes and the complex arrangements for trapping antigen from circulating body fluids in the secondary lymphoid organs ensure that fixed antigens are reached in most situations and that mobile antigens are held in tissues where antigen-reactive lymphocytes occur. The existence of mechanisms which enable cells with little or no specificity (e.g. phagocytes) to amplify the effector activities of lymphocytes is clearly advantageous and factors such as the lymphokines which are released from T-cells help to bring this about.

Summary

Primary lymphoid organs such as the thymus and the bursa of Fabricius are the first to develop ontogenetically. They originate as epithelial anlagen and have a framework of cells which, although branched and highly modified, are clearly epithelial in nature. They are responsible for 'educating' stem cells that have immigrated from embryonic yolk sac, fetal liver etc. by a process which possibly involves humoral substances.

At some stage during the development of immunological competence, individual lymphocytes acquire antigen-specific receptors on their surface and so become committed to responding to these antigens. They then enter the peripheral circuit as antigen-reactive lymphocytes capable of recognizing antigen and being triggered by it to a further phase of differentiation.

This encounter with antigen takes place in the tissues or in secondary lymphoid organs such as the spleen, lymph nodes and gut follicles, the secondary lymphoid organs being structurally adapted for antigen trapping.

The 'recognition site' of an antigen-reactive cell is believed to be a surface-associated immunoglobulin molecule which, in the precursors of antibody-forming cells, is similar to the secreted antibody (immunoglobulin) of their progeny. Engagement of this site by appropriate antigen probably leads to membrane changes in the antigen-reactive cell and results in transformation of the latter into a blast cell (large pyroninophilic cell) and initiation of a series of mitotic divisions. This produces a clone of cells of identical specificity.

Encounter with antigen usually leads to the production of effector cells of the immune system (either antibody-forming cells or lymphocytes involved in cell-mediated immunity) and to an increase in the number of cells able to respond to subsequent encounter with the same antigen (memory cells). On the other hand an encounter of lymphocyte with antigen may lead to specific tolerance (negative memory), this latter condition being more readily induced in the immature animal.

Morphologically similar cells of the lymphocyte population can be divided not only into the virgin antigen-reactive cells and cells which have already undergone antigen-induced differentiation but also into the thymus-dependent T-cells (i.e. cells educated by the thymus) and the B-cells which develop independently of the thymus, being educated by other primary lymphoid tissue.

B-cells are the progenitors of antibody-forming cells. T-cells effect

287

cell-mediated immune responses: they can produce pharmacologically active substances (lymphokines) but do not secrete immunoglobulins. However, co-operation of T-cells with B-cells occurs in humoral antibody synthesis to a wide array of antigens. T-cells are a mobile population, circulating between lymph, blood and secondary lymphoid organs in their function of immune surveillance.

The secondary lymphoid organs, although possessing a structural framework of reticular fibres etc., are made up of a changing and mobile population of cells (macrophages, granulocytes, lymphocytes) whose capacity to settle or to migrate depends partly on the intrinsic properties of the cells concerned (for example, the surface properties of T-cells and B-cells) and partly on local environmental factors such as the presence of antigen.

Some lymphocytes are long-lived, others have a life of only a few days, but ultimately the secondary lymphoid organs rely on the primary system to maintain their lymphocytic population. The primary lymphoid organs, in turn, require an input of blood-borne stem cells which in adult mammals comes from the bone marrow.

The maturation of immunological reactivity by the developing vertebrate is closely paralleled by and is dependent upon the lymphoid histogenesis of its lymphoid organs.

References

1. BURNET, F. M. (1969), *Self and Not-self*, Melbourne University Press: Melbourne.
2. MILLER, J. F. A. P. (1971), 'The thymus and the immune system', *Vox Sanguins*, **20**, 481–491.
3. WILLIAMSON, A. R. (1972), 'Clones of antibody-forming cells: natural and experimental selection', *Endeavour*, **31**, 118–122.
4. BURNET, F. M. (1959), *The Clonal Selection Theory of Acquired Immunity*, Cambridge and Vanderbilt University Press.
5. NOSSAL, G. J. V. and ADA, G. L. (1971), *Antigens, Lymphoid Cells, and the Immune Response*, Academic Press: London.
6. GEWURZ, H. (1971), 'The immunologic role of complement'. In *Immunobiology*. Eds. Good, R. A. and Fisher, D. W., pp. 95–103, Sinauer Associates, Inc.: Stamford, Connecticut.
7. YOFFEY, J. M. (1967), 'The fourth circulation'. In *The Lymphocyte in Immunology and Haemopoiesis*. Ed. Yoffey, J. M., pp. 1–10, Edward Arnold: London.
8. BEARD, J. (1894), 'The development and probable function of the thymus', *Anatomischer Anzeiger*, **9**, 476–486.

9. HAMMAR, J. A. (1905), 'Zur Histogenese und Involution der Thymusdrüse', *Anatomischer Anzeiger*, **27**, 23–30, 41–89.

10. GOOD, R. A. and PAPERMASTER, B. W. (1964), 'Ontogeny and phylogeny of adaptive immunity', *Advances in Immunology*, **4**, 1–115.

11. ABRAMOFF, P. and LA VIA, M. F. (1970), *Biology of the Immune Response*, McGraw-Hill: New York.

12. PAPERMASTER, B. W. and GOOD, R. A. (1962), 'Relatvie contributions of the thymus and bursa of Fabricius to the maturation of the lymphoreticular system and immunological potential in the chicken', *Nature, London*, **196**, 838–840.

13. HORTON, J. D. (1971), 'Histogenesis of the lymphomyeloid complex in the larval leopard frog, *Rana pipiens*', *Journal of Morphology*, **134**, 1–20.

14. MANNING, M. J. and HORTON, J. D. (1969), 'Histogenesis of lymphoid organs in larvae of the South African clawed toad, *Xenopus laevis* (Daudin)', *Journal of Embryology and Experimental Morphology*, **22**, 265–277.

15. METCALF, D. and MOORE, M. A. S. (1971), *Haemopoietic Cells*, North Holland Publishing Company: Amsterdam.

16. ACKERMAN, G. A. (1967), 'Developmental relationship between the appearance of lymphocytes and lymphopoietic activity in the fetal cat', *Anatomical Record*, **158**, 387–399.

17. WARNER, N. L. (1967), 'The immunological role of the avian thymus and bursa of Fabricius', *Folia Biologica*, **13**, 1–17.

18. WALFORD, R. L. (1969), *The Immunologic Theory of Aging*, Munksgaard: Copenhagen.

19. MACKAY (1972), 'Ageing and immunological function in man', *Gerontologia*, **18**, 285–304.

20. SALKIND, J. (1915), 'Contributions histologiques à la biologie comparée du thymus', *Archives de Zoologie Expérimentale et Générale*, **55**, 81–322.

21. GOOD, R. A., FINSTAD, J., POLLARA, B. and GABRIELSEN, A. E. (1966), 'Morphological studies on the evolution of lymphoid tissues among the lower vertebrates'. In *Phylogeny of Immunity*. Eds. Smith, R. T., Miescher, P. A. and Good, R. A., pp. 149–168, University of Florida Press: Gainesville.

22. KLAPPER, C. E. (1946), 'The development of the pharynx of the guinea pig with special emphasis on the morphogenesis of the thymus', *American Journal of Anatomy*, **78**, 139–179.

23. MASSART, C. (1940), 'Morfologia e sviluppo del timo con richerche originali nei Chirotteri (*Vesperugo pipistrellus*)', *Archivio Italiano di Anatomia e di Embriologia*, **44**, 489–550.

24. HAMMOND, W. S. (1954), 'Origin of thymus in the chick embryo', *Journal of Morphology*, **95**, 501–522.

25. RUTH, R. F., ALLEN, C. P. and WOLFE, H. R. (1964), 'The effect of thymus on lymphoid tissue'. In *The Thymus in Immunobiology*. Eds. Good, R. A. and Gabrielsen, A. E., pp. 183–206, Hoeber-Harper: New York.

26. JOLLY, J. (1923), *Traité technique d'hématologie*. A. Maloine, Paris.

27. WEISS, L. (1972), *The Cells and Tissues of the Immune System*. Prentice-Hall Inc.: Englewood Cliffs (N.J.).

28. METCALF, D. (1967), 'Relation of the thymus to the formation of immunologically reactive cells', *Cold Spring Harbour Symposia in Quantitative Biology*, **32**, 583–590.

29. MILLER, J. F. A. P. and OSOBA, D. (1967), 'Current concepts of the immunological function of the thymus', *Physiological Reviews*, **47**, 437–520.

30. METCALF, D. (1967), Lymphocyte kinetics in the thymus. In *The Lymphocyte in Immunology and Haemopoiesis*. Ed. Yoffey, J. M., pp. 333–341, Edward Arnold: London.

31. SAINTE-MARIE, G. and PENG, F. S. (1971), 'Emigrations of thymocytes from the thymus: a review and study of the problem', *Revue Canadienne de Biologie*, **30**, 51–78.

32. WEISSMAN, I. L. (1967), 'Thymus cell migration', *Journal of Experimental Medicine*, **126**, 291–304.

33. FICHTELIUS, K. E., FINSTAD, J. and GOOD, R. A. (1968), 'Bursa equivalents of bursaless vertebrates', *Laboratory Investigation*, **19**, 339–351.

34. ACKERMAN, G. A. (1966), 'The origin of lymphocytes in the appendix and tonsil iliaca of the embryonic and neonatal rabbit', *Anatomical Record*, **154**, 21–40.

35. THORBECKE, G. J. (1959), 'Some histological and functional aspects of lymphoid tissue in germfree animals. I. Morphological studies', *Annals of the New York Academy of Science*, **78**, 237–246.

36. DIENER, E. (1970), 'The primary immune response and immunological tolerance', *Handbuch der Allgemeinen Pathologie*, *VII*, part 3, 250–325. Springer-Verlag: Berlin.

37. TURK, J. L. (1967), *Delayed Hypersensitivity*, North Holland Publishing Company: Amsterdam.

38. BACULI, B. S. and COOPER, E. L. (1967), 'Lymphomyeloid organs of Amphibia. II. Vasculature in larval and adult *Rana catesbeiana*', *Journal of Morphology*, **123**, 463–480.

39. COWDEN, R. R., GEBHARDT, B. M. and VOLPE, E. P. (1968), 'The histophysiology of antibody-forming sites in the marine toad', *Zeitschrift für Zellforschung und Mikroskopische Anatomie*, **85**, 196–205.

40. SAINTE-MARIE, G. and LEBLOND, C. P. (1964), 'Thymus-cell population dynamics'. In *The Thymus in Immunobiology*. Eds. Good, R. A. and Gabrielsen, A. E., pp. 207–235, Hoeber-Harper: New York.

41. AUERBACH, R. (1961), 'Experimental analysis of the origin of cell types in the development of the mouse thymus', *Developmental Biology*, **3**, 336–354.

42. MICKLEM, H. S., FORD, C. E., EVANS, E. P. and GRAY, J. (1966), 'Interrelationships of myeloid and lymphoid cells: studies with chromosome-marked cells transfused into lethally irradiated mice', *Proceedings of the Royal Society of London*, *B*, **165**, 78–102.

43. RAFF, M. C. (1973), 'T and B lymphocytes and immune responses', *Nature, London*, **242**, 19–23.

44. DUMONDE, D. C., WOLSTENCROFT, R. A., PANAYI, G. S., MATHEW, M., MARLEY, J. and HOWSON, W. T. (1969), 'Lymphokines: non-antibody mediators of cellular immunity generated by lymphocyte activation', *Nature, London*, **224**, 38–42.

45. FORD, W. L. and GOWANS, J. L. (1969), 'The traffic of lymphocytes', *Seminars in Hematology*, **6**, 67–83.
46. GOWANS, J. L. (1971), 'Immunobiology of the small lymphocyte'. In *Immunobiology*. Eds. Good, R. A. and Fisher, D. W., pp. 18–27, Sinauer Associates, Inc.: Stamford, Connecticut
47. PARROTT, D. M. V. and DE SOUSA, M. (1971), 'Thymus-dependent and thymus-independent populations: origin, migratory patterns and lifespan', *Clinical and Experimental Immunology*, **8**, 663–684.
48. DE SOUSA, M. (1973), 'Ecology of thymus dependency'. In *Contemporary Topics in Immunobiology* (Vol. 2): *Thymus Dependency*. Eds. Davies, A. J. S. and Carter, R. L., pp. 119–136, Plenum Press: New York and London.
49. BRAMBELL, F. W. R. (1970), *The Transmission of Passive Immunity from Mother to Young*. North Holland Publishing Company: Amsterdam.
50. OWEN, R. D. (1945), 'Immunogenetic consequences of vascular anastomosis between bovine twins', *Science, N.Y.*, **102**, 400–401.
51. BURNET, F. M. and FENNER, F. (1949), *The Production of Antibodies*. McMillan: Melbourne.
52. BILLINGHAM, R. E., BRENT, L. and MEDAWAR, P. B. (1953). '"Actively acquired tolerance" of foreign cells', *Nature, London*, **172**, 603–606.
53. BILLINGHAM, R. E., BRENT, L. and MEDAWAR, P. B. (1956), 'Quantitative studies on tissue transplantation immunity. III. Actively acquired tolerance', *Philosophical Transactions of the Royal Society, Series B*, **239**, 357–414.
54. SMITH, R. T. and BRIDGES, R. A. (1958), 'Immunological unresponsiveness in rabbits produced by neonatal injection of defined antigens', *Journal of Experimental Medicine*, **108**, 227–250.
55. MITCHISON, N. A. (1963), 'Immunological paralysis in the adult', *Proceedings of the Royal Society of Medicine*, **56**, 937–940.
56. MITCHISON, N. A. (1967), 'Immunological paralysis as a problem of cellular differentiation'. In *Ontogeny of Immunity*. Eds. Smith, R. T., Good, R. A. and Miescher, P. A., pp. 135–138, University of Florida Press: Gainesville.
57. VOISIN, G. A. (1971), 'Immunity and tolerance: a unified concept', *Cellular Immunology*, **2**, 670–689.
58. MILLER, J. F. A. P. (1962), 'Effect of neonatal thymectomy on the immunological responsiveness of the mouse', *Proceedings of the Royal Society of London, B*, **156**, 415–428.
59. HESS, M. W. (1968), *Experimental Thymectomy: Possibilities and Limitations*, Springer-Verlag: Berlin.
60. LISCHNER, H. W., PUNNETT, H. H. and DIGEORGE, A. M. (1967), 'Lymphocytes in congenital absence of the thymus', *Nature, London*, **214**, 580–582.
61. WORTIS, H. H. (1971), 'Immunological responses of "nude" mice', *Clinical and Experimental Immunology*, **8**, 305–317.
62. GLICK, B. (1970), 'The bursa of Fabricius: a central issue', *Bioscience*, **20**, 602–604.

Plate 8.1 Early development of the thymus. The section passes through the pharyngeal region of an amphibian (*Xenopus laevis*) larva at a very early stage of development (stage 42[87] – approx. 3 days old). At this early maturative stage an epithelial bud has appeared from, although is still attached to, the pharyngeal epithelium (ph.e). This bud is the thymus anlage (th). (br.p) branchial pouch. *Scale:* 50 μm.

Plate 8.2 Differentiation of the thymus. At this much later stage of development the thymus has become differentiated into an outer cortex (densely populated with small lymphocytes) and a more centrally-located (paler-staining) medulla. The latter zone contains fewer lymphocytes but large numbers of epithelial cells and their derivatives, such as thymic corpuscles and cystic structures, are found. This section was taken from a young adult *Xenopus* toad, 16-weeks of age. The thymus is no longer attached to the pharyngeal epithelium: such detachment occurred early in larval life. *Scale:* 200 μm.

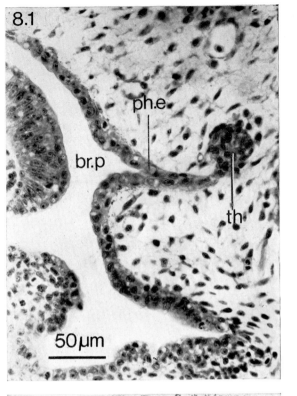

8.1

ph.e

br.p

th

50 μm

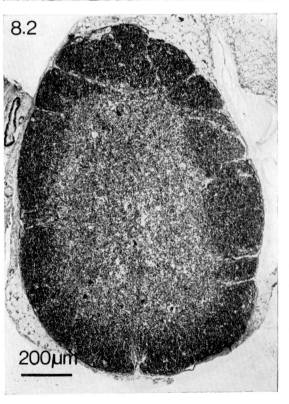

8.2

200 μm

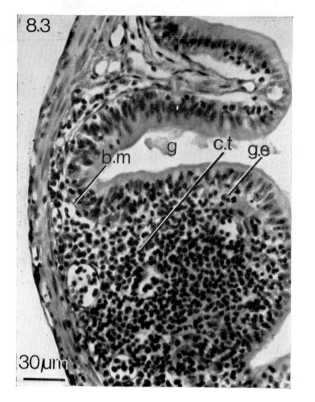

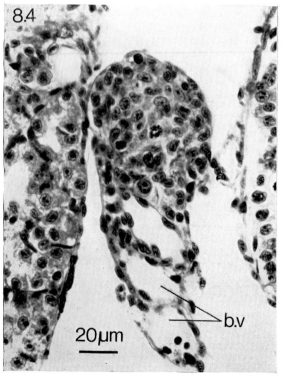

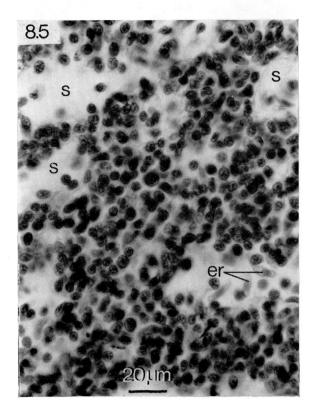

Plate 8.3 The development of a simple kind of gut-associated lymphoid tissue is demonstrated in this section which passes through the small-intestinal region of a 16 week old adult toad. Lymphocytes are seen in the subepithelial connective tissue (c.t) and in the overlying gut epithelium (g.e). The basement membrane (b.m) becomes difficult to detect in areas where infiltration of epithelium with lymphocytes is high. A more normal villus is seen at the top of the plate. (g) gut lumen. *Scale:* 30 μm.

Plate 8.4 Early development of the spleen. The section shows the development of the spleen anlage in an 8 day old *Xenopus* larva (early stage 48[87]). The spleen is apparent as a thickening of the mesenchymal tissue of the dorsal mesogastrium that is associated with the splenic blood vessels (b.v). Active lymphopoiesis will soon begin in the spleen. Gut tissue is seen to the right and to the left of the plate. *Scale:* 20 μm.

Plate 8.5 The sinuses of the rudimentary lymph nodes of amphibians are filled with blood. This is apparent in the section of a small region of the lymph gland (a lymphoid organ which develops in the lateral branchial region) of a leopard frog larva. Erythrocytes (er) and lymphocytes can be seen in the pale-staining sinusoidal regions (s). The densely-staining parenchyma of the lymph gland contains lymphocytes, reticular cells and macrophages. *Scale:* 20 μm.

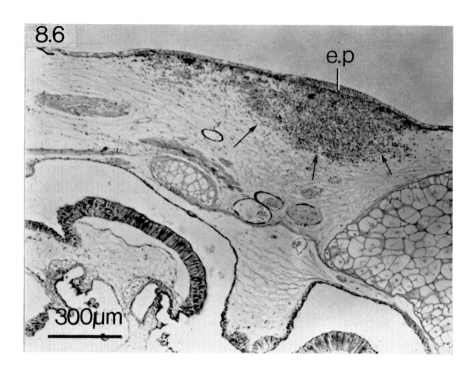

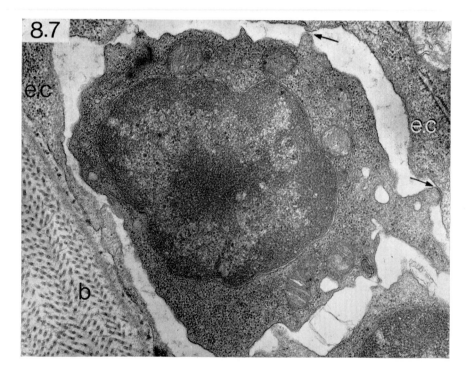

Plates 8.6 and 8.7. In graft rejection reactions, large numbers of host lympho-cytes attack and may, through cell-to-cell contact, destroy the foreign tissue. This is clearly demonstrated in transplantation experiments with amphibian larvae. In Plate 8.6 a large mass of cells (mostly lymphocytes – arrowed) have invaded the region where a larval foreign skin graft has been applied. This extensive immune reaction is seen in the second post-operative week in *Xenopus* larvae (with differentiated lymphoid organs) kept at 23 °C. Lymphocytes are found invading the sub-epidermal connective tissue and within the graft epidermis (e.p) itself. Abrogation of lymphocytic invasion of foreign skin grafts in *Xenopus* larvae can be obtained by very early removal of the thymus [76]. The branchial region of the host is seen to the bottom of plate 8.6. Plate 8.7 is an electron micrograph of the basal epidermal layer of another foreign skin graft (applied 8 days previously to a *Xenopus* larva). A lymphocyte is seen lying in the intercellular space between graft epidermal cells (e.c). The lymphocyte has a high nucleo-cytoplasmic ratio, margination of dense nuclear chromatin, and many mitochondria, free ribosomes and vacuoles in its cytoplasm. Note the close proximity (arrows) of the plasma membranes of the host lymphocyte and the epidermal cells of the graft in certain places. Area 6 is the basement lamella (cutis) of larval skin graft. Outer epidermal cells of the skin lie to the right (out of field). Plate 8.6 – Scale: 300 μm   Plate 8.7 – Magnification × 24 000

63. SOLOMON, J. B. (1971), *Foetal and Neonatal Immunology*. North Holland Publishing Company: Amsterdam.
64. SILVERSTEIN, A. M., PARSHALL, C. J. and PRENDERGAST, R. A. (1967), 'Studies on the nature of immunological immaturity'. In *Ontogeny of Immunity*. Eds. Smith, R. T., Good, R. A. and Miescher, P. A., pp. 143–147, University of Florida Press: Gainesville.
65. SILVERSTEIN, A. M. and PRENDERGAST, R. A. (1971), 'The maturation of lymphoid tissue structure and function in ontogeny'. In *Morphological and Functional Aspects of Immunity*. Eds. Lindahl-Kiessling, K., Alm, G. and Hanna, M. G., pp. 37–42.
66. LAPLANTE, E. S., BURRELL, R., WATNE, A. L., TAYLOR, D. L. and ZIMMERMANN, B. (1969), 'Skin allograft studies in the pouch young of the opossum', *Transplantation*, **7**, 67–72.
67. ROWLANDS, D. T. and DUDLEY, M. A. (1969), 'The development of serum proteins and humoral immunity in opossum "embryos",' *Immunology*, **17**, 969–975.
68. HILDEMANN, W. H. and HAAS, R. (1962), 'Developmental changes in leucocytes in relation to immunological maturity'. In *Mechanisms of Immunological Tolerance*. Ed. Hašek, M., Langerová, A. and Vojtišková, M., pp. 35–49, Czechoslovakia Academy of Sciences: Prague.
69. HORTON, J. D. (1969), 'Ontogeny of the immune response to skin allografts in relation to lymphoid organ development in the amphibian *Xenopus laevis* Daudin', *Journal of Experimental Zoology*, **170**, 449–466.
70. HORTON, J. D. (1971), 'Ontogeny of the immune system in amphibians', *American Zoologist*, **11**, 219–228.
71. VOLPE, E. P. (1972), 'Embryonic tissue transplantation incompatibility in an amphibian', *American Scientist*, **60**, 220–228.
72. GOLDSHEIN, S. J. and COHEN, N. (1972), 'Phylogeny of immunocompetent cells. I. *In vitro* blastogenesis and mitosis of toad (*Bufo marinus*) splenic lymphocytes in response to phytohemagglutinin and in mixed lymphocyte cultures', *Journal of Immunology*, **108**, 1025–1033.
73. DU PASQUIER, L. (1970), 'Ontogeny of the immune response in animals having less than one million lymphocytes: the larvae of the toad *Alytes obstetricans*', *Immunology*, **19**, 353–362.
74. TAYLOR, R. B., DUFFUS, W. P. H., RAFF, M. C. and DE PETRIS, S. (1971), 'Redistribution and pinocytosis of lymphocyte surface immunoglobulin molecules induced by anti-immunoglobulin antibody', *Nature, New Biology, London*, **233**, 225–229.
75. MEDAWAR, P. B. (1960), 'Theories of immunological tolerance'. In *Cellular Aspects of Immunity*, pp. 134–149. J. & A. Churchill, London.
76. HORTON, J. D. and MANNING, M. J. (1972), 'Response to skin allografts in *Xenopus laevis* following thymectomy at early stages of lymphoid organ maturation', *Transplantation*, **14**, 141–154.
77. PICK, E. and TURK, J. L. (1972), 'The biological activities of soluble lymphocyte products', *Clinical and Experimental Immunology*, **10**, 1–23.
78. DAVIES, A. J. S. (1969), 'The thymus and the cellular basis of immunity', *Transplantation Reviews*, **1**, 43–91.

79. PLAYFAIR, J. H. L. (1971), 'Cell cooperation in the immune response', *Clinical and Experimental Immunology*, **8**, 839–856.

80. MITCHISON, N. A. (1969), In *Mediators of Cellular Immunity*, ed. Lawrence, H. S. and Landy, M., pp. 73–80, Academic Press; New York.

81. NOSSAL, G. J. V. (1964), 'How cells make antibodies', *Scientific American*, **211**, 106–115.

82. LING, N. R. (1968), *Lymphocyte Stimulation*. North Holland Publishing Company: Amsterdam.

83. PORTER, R. R. (1967), 'The structure of antibodies', *Scientific American*, **217**, 81–90.

84. KABAT, E. A. (1968), *Structural Concepts in Immunology and Immunochemistry*. Holt, Rinehart and Winston Inc.: New York.

85. TALMAGE, D. W. (1969), 'The nature of the immunological response'. In *Immunology and Development*, ed. Adinolfi, M., pp. 1–26, Spastics International Medical Publications: London.

86. JERNE, N. K. (1973), 'The immune system', *Scientific American*, **229**, 52–63.

87. NIEUWKOOP, P. D. and FABER, J. (1967), *Normal Table of Xenopus laevis* (Daudin). North Holland Publishing Company: Amsterdam.

9 Cellular Ageing

D. BELLAMY

Introduction

In many organisms, death may be regarded as a programmed event, in that it is an essential part of the life cycle. For example, annual plants develop in such a way that food is transferred from parent to seeds during the terminal phase of the reproductive cycle, resulting in the rapid death of the parent plant. In many animals, both invertebrate and vertebrate, similar developmental processes have evolved. In the salmon, for example, during the up-river spawning migration, the irreversible utilization of post-mitotic tissues, mainly skeletal muscle, to provide amino-acids for the formation of eggs and sperm, culminates in rapid death after the gametes are shed and it is difficult to escape the conclusion that the reproductive process, with its associated endocrine changes is the major cause of death. In these organisms, death is an obvious outcome of the natural selection of an efficient reproductive process.

On the other hand, for many organisms, ageing does not appear to be directly connected with reproduction in this way. There is clearly a steady decrease in the ability of these organisms to resist the various traumas of life, which is expressed as a statistical concept of an increasing force of mortality. The underlying loss of vigour or resilience in the face of environmental factors, is manifest particularly at the more complex and integrated levels of biological organization, but it is questionable as to what extent failures at this level are a cause of ageing.

It may be that ageing and death result from the extension of a sequence of chemical events that was selected to establish the organism in a habitat where it was highly probable that death would occur at, or soon after, peak reproductive efficiency had been attained. That is to say, the organism 'drifts' away from the optimal state when allowed to live far beyond its evolutionary life-span. Alternatives to this random 'differentiation theory' of ageing are that errors occur progressively at various key points in biochemical reactions that cannot be corrected, or that accumulated damage may arise resulting from random changes in

297

nuclear biochemistry, such as mutation, or the chemical deterioration and accumulation of non-renewable macromolecules. These latter theories encompass events at the cellular level and raise the question as to the importance of cellular ageing in relation to the ageing of the whole organism.

There can be little doubt that a common feature of ageing at all levels of animal organization is the loss of cells. This phenomenon, known as 'age involution', is correlated with an increase in the proportion of extracellular inert material of the body, such as the connective tissue protein collagen. It appears to be responsible for the decline in the mass of effectors, such as muscle, and the loss of integrative units, such as endocrine cells and motor end plates. Malfunctions leading to cellular death cannot be corrected through evolution if they are manifest beyond the time of peak reproductive efficiency of the population. This is because events characteristic of the post-reproductive period, which are advantageous in prolonging life, cannot be perpetuated in the gametes through the forces of natural selection.

Cell loss begins in some organs before the animal is adult. Indeed, it is an essential feature of differentiation that entire organ systems be dismantled to make way for more advantageous structures. During postnatal development of vertebrates, post-mitotic cells begin to disappear from key organs well before the onset of sexual maturity, and the rate of loss continues steadily throughout life. The existence of this phenomenon may be explained, in that organisms can tolerate a certain degree of imperfections in function which is in keeping with the maintenance of the correct degree of reproductive efficiency in relation to selection pressure. The random appearance of imperfections seriously interferes with the life process only when the organism is allowed to live beyond the natural life-span selected through evolution. This applies to animals in captivity and modern man.

One view on the biological significance of ageing holds that the individual must age because there is a need for the reproductive replacement of a certain number of individuals, but it is difficult to see how death rate alone serves a positive selection purpose. On an evolutionary plane, reproduction tends to create novelty in the form of new combinations of genes as well as tending to preserve novelty in the form of mutant genes. Thus, a fixed life-span may be advantageous in conferring genetic flexibility, although the mechanism by which it evolved is obscure.

Cellular Ageing in General

Working on the assumption that the ultimate control of the direction and force of development resides in the cells, it is logical to investigate cell function for the underlying causes of ageing. It is also sensible to set up models of the ageing process at the cellular level, although it is no exaggeration to say that most models at the present time pose questions of fact for which we must find the answers before the model may be tested.

The various theories have been classified according to the nature of the postulated primary defect (Table 9.1). The three broad categories are

Table 9.1 *Classification of theories of ageing at the cellular level*

A. Accumulation of inert material (Collagen theory; age-pigment theory)
B. Chemical ageing of macromolecules

    Intracellular reactions involving:
      1. Small reactive molecules (Free-radical theory)
      2. Other polymers (Cross-link theory; nucleo-histone theory)

C. Random metabolic changes:
      1. Gene deletion (Somatic mutation theory)
      2. Errors in DNA duplication ⎫
      3. Errors in DNA transcription ⎬ Various error theories
      4. Errors in RNA translation ⎭
      5. Nuclear differentiation (Random drift theory; clonal selection theory).

concerned with: the accumulation of inert material, either within the cells or between cells; chemical interactions which reduce the efficiency of irreplaceable polymers; and the change in cellular phenotype occurring through random metabolic alterations. All of these processes may eventually lead to cellular death.

These theories are generally discussed in relation to the existence of two major types of cell, defined according to the possibilities for further division and replacement.

Cells that are capable of multiplication and replacement may be classed as 'mitotic'. Other highly specialized tissues, such as the central nervous system, together with cardiac and striated muscle, are composed of cells which have lost the power to divide. These 'post-mitotic' cells make up the bulk of living tissue.

Cellular Ageing in Mitotic Cells

For many years, it was held that dividing cells were potentially immortal. Recently, however, experimental work has clearly shown that populations of vertebrate somatic cells cultured *in vitro* cannot be maintained indefinitely. Cultures of human diploid embryonic fibroblasts undergo approximately fifty doublings then die out, often showing well-defined terminal histological changes with marked chromosome abnormalities. Similar conclusions on the finite life of cells come from experiments in which normal cell populations of skin, bone marrow and mammary tissue have been serially transplanted as their hosts age. There is as yet no clear indication of the mechanism involved in the eventual dying out of these cell populations. Diploid fibroblasts from adults grown *in vitro* show less than half the doubling potential of embryonic cells, suggesting that the cells change progressively from birth. This limitation of the life-span of cultured cells may not be directly related to the doubling potential, but may be a function of total metabolic time of the cell strain *in vitro*. There is also evidence that the post-mitotic life-span of human amnion cells *in vitro* may be extended by modifying the environment, particularly by increasing the concentration of steroid hormones. However, these findings do not alter the fact that cell strains in tissue culture are not immortal. There are several possible explanations for the gradual deterioration of cell cultures. There may be a progressive loss of some vital self-duplicating cellular entity – death occurring when the concentration of this entity falls below a critical level. The loss of such an entity could occur either through the cumulative effects of random losses or because the entity was itself dividing at a rate somewhat slower than that of the cells. On the other hand, the cells may be progressively changing by the operation of processes akin to differentiation. Workers in this field have suggested that the evidence for a finite lifetime of diploid cells *in vitro* may be the cellular manifestation of ageing *in vivo*.

Normal cell populations which duplicate *in vitro* have been described as cell 'strains' to distinguish them from cell 'lines' which may arise during the culture of normal cells. Cell lines, such as the HeLa cultures, are capable of indefinite proliferation *in vitro*, but have many abnormal properties. In particular, they may behave like cancer cells on inoculation into living organisms. The existence of cell lines suggests that mechanisms exist for perfect control of cell metabolism. Indeed, one would expect this from the fact that a continuous series of perfect cell division lineages has been maintained through evolutionary time to the germ cells of all present-day organisms. It is also a strong possibility that cell proliferation in other tissues, such as the intestinal epithelium, is a perfect process.

As an alternative to perfect duplication and self-maintenance, it is possible that mechanisms exist to counteract imperfections which inevitably occur. Elimination of imperfections may take place through the presence of error-correcting codes. Cell lines may appear in tissue culture because of the unmasking of a self-correcting code which is normally only expressed in germ cells. With regard to errors occurring in mitotic cells, it is possible that these could be brought about by any of the well-defined mechanisms such as mutation, non-disjunction, translocation, inversion, deletion, and repetition. In general, these processes are likely to result in malfunction of the nucleus, either through loss of chromosome material or by changes in relative gene-activity; it may be predicted that malfunctions would be greatest in the most rapidly dividing cells.

The rate at which mutations arise has been calculated as follows. If there are about 10^{14} somatic cells in the human body, about 10^{10} are known to be expended every day. Assuming a mutation rate of 10^{-6} per cell generation per locus, with 10^4 loci in each cell, then the replacement cells arising daily would carry 10^8 new mutations. The question that cannot be answered at the moment is how many of these accumulate and cause increasing damage?

Although much work has been carried out on mitotic cells of higher organisms in the context of ageing, in no case has it been shown that any of the cells become old. Ageing can only apply to the stem cell line, but even here there is a possibility of 'regeneration' occurring at each division. Amoeba are only 'immortal' when allowed to proliferate. Also, with regard to age-involution in mitotic tissues, it is likely that any random deleterious change in the cellular constitution of a daughter cell, which leads to a greater probability of this cell dying, would be masked by continuous proliferation of normal cells. In these terms, it is unlikely that age involution of mitotic tissues takes place by the accumulation of random errors. Mitotic tissues are particularly useful for studying age-changes that favour the proliferation of a new line, or that progressively alter the relationship between division and cellular death. However, in both situations, it is difficult to determine to what extent cellular age-changes are due to ageing processes at a more complicated level. From this point of view, studies directed towards primary cellular changes should be studied in cells that are either stem cell lines or are not proliferating.

The clonal selection theory

The fact that there are cells in many organisms that are capable of division raises the possibility that a process of selection could occur within the body to favour the multiplication of cell-variants which, during the lifetime of the individual, might prove harmful to the organism. This idea forms the basis for the 'clonal selection' theory of ageing. A particular development of this theory states that, through random mutation or other kinds of nuclear malfunction, cells which generate antibody-forming lymphocytes could change their properties and give rise to a lymphocyte population that would not be able to distinguish its own cells as being 'self'. This could result in the progressive loss of cells from a wide range of organs by a phenomenon termed 'auto-immunity'. It is believed by some workers that auto-immunity is the chief process which leads to a gradual deterioration of the vertebrate organism.

Evidence has been obtained for a decline in iso-antibodies and for auto-immune reactions developing in old mammals, and the view has also been expressed that adsorption of certain antigen-antibody or hapten-antibody complexes could result in ageing-phenomena. Where an increased titre of auto-antibodies has been determined in humans, the incidence in males has always been lower than that in females. If auto-immunity was a major factor in determining longevity, the incidence in shorter-lived males should be higher in the female. Also, the situation in invertebrates, where the classical aspect of immunity has no application, raises questions as to the general validity of the autoimmune theory. Immuno-suppressive drugs exert only a marginal beneficial effect on mammalian longevity, but the consequences of such treatment with regard to the destruction of normal immunological reactions make it difficult to make a valid test of the autoimmunity theory by this means.

Ageing in Protozoa

Although cultures of micro-organisms cannot be regarded as analogous to cell cultures, they have been examined from time to time in an attempt to throw light on cellular ageing. Early work predicted that because Protozoans were pure germ plasma, inevitably they must be immortal. However, it was soon found that individual ciliates underwent a series of temporal changes akin to sexual maturation, followed by senescence; they aged and died. Also, in the laboratory, a particular cell clone will eventually die out after passing through this life-cycle a certain number of times. New clones may arise by the processes of conjugation and autogamy which entail, respectively, the introduction of individuals of a different clone and a kind of internal fertilization. The nuclear apparatus is replaced during both the processes. In conjugation, male nuclei are reciprocally exchanged between organisms and a fertilized nucleus is formed in each male; in autogamy, haploid male and female micronuclei from the same individual unite to form a fertilization nucleus. Both kinds of reorganized cells become the starting points for new clones. Progeny of autogamy produced early or late in the senescent period of an individual develop in several possible ways: to give fully viable clones, which begin new life cycles; to give clones which are either non-viable or which divide at markedly reduced rates and soon die out; to give clones with intermediate characteristics. Despite this variability of response, it is clear that the probability of obtaining a new viable clone from a senescent clone is inversely proportional to the age of the parental cells. In extreme old age, new clones either cannot be formed at all, or if formed, always die off after a few generations have passed.

The earliest evidence for ageing in *Paramecium aurelia* is the appearance of abnormal numbers of micronuclei. In young cells, the norm is 2; in old cells, the range is from 1 to 4. Old cells also have abnormalities in the micronuclear spindles and chromosomes. Other cell components, such as cilia and the gullet, also experience characteristic age-changes. These abnormalities increase in number as the clone ages. It has been claimed that old clones accumulate many detrimental and even lethal mutations which are expressed as both dominant and recessive characters. These genetic changes in the macronucleus appear rapidly when conjugation is not allowed to occur; lethal changes are noted after only 80 fissions and by about 200 fissions, all of the progeny are non-viable. Under circumstances where a culture has the opportunity of periodic nuclear reorganization, it may be maintained for over 1000 fissions and 'clonal' old age is avoided.

There is much evidence pointing to the macronucleus as the central organelle which controls the activities of the micronuclei – via the intervening cytoplasm. Late in the life-span, various structural abnormalities appear in the macronucleus which are not inconsistent with a central hypothesis that ageing involves progressive changes in the function of macronuclear genes. There is also ample data indicating that ageing in ciliates is the result of differential gene expression. The two questions which are central to the phenomenon of ageing are: What is the mechanism by which cells with a common mitotic lineage, cultivated in a constant environment, develop an orderly sequence of diverse phenotypes? Also, how does nuclear reorganization halt the progress of ageing, transforming a degenerating cell into a totipotent individual? The answers to these questions are also pertinent to ageing in higher organisms which resemble protozoans in the progressive manifestation of two inevitable signs of ageing – phenotypic instability and loss of reproductive function.

Despite the obvious age deterioration in the nucleus, which is a common feature in a wide range of organisms, it is by no means certain that the key to ageing phenomena lies primarily in this organelle. There is no reason why we should not turn our attention to the cytoplasm or even the cellular environment for causes of ageing, because extra-nuclear processes influence nuclear function. It may be also wise not to overlook the fact that highly differentiated cells can be derepressed and revert to an unspecialized form, so that the ageing process may begin all over again. Neglected material in this context may be found in the invertebrates (de-differentiation of planarians), vertebrates (conversion of iris epithelial cells to lens cells) and plants (conversion of leaf epithelial cells into meristematic tissue). 'Rejuvenation' in this context appears to be initiated by virtue of the cells finding themselves in a new environment.

Cellular Ageing in Post-mitotic Cells

Because post-mitotic cells, such as muscle fibres and neurons, cannot be replaced if there is a failure in function, changes in non-dividing cells with the certainty of accumulated faults are likely to be an important cause of ageing. This aspect was highlighted by early work on flatworms, where the separation of anterior tissues from the posterior portion of the animals showed, through the subsequent process of regeneration, that senescence was a feature only of the anterior tissues which had a preponderance of post-mitotic cells. There are three views as to the way in which post-mitotic cells could deteriorate with time.

Accumulation of inert material

One of the most commonly observed age-deteriorations accepted widely by histologists is the progressive accumulation of lipofuscin granules in certain cells of the body. This 'age pigment', as it has been termed, appears in the light microscope as brown granules in nerve cells, seminal vesicles, adrenal cortex, interstitial cells of the gonads and is particularly prominent in cardiac muscle. The accumulation of lipofuscin in the myocardial cells results in a large displacement of muscle volume, and one would anticipate that this would result in a loss of efficiency of the contractile elements. However, there is a clear absence of correlation with cardiac disease or heart failure.

Age pigment apparently occurs in increasing amounts in a variety of non-dividing cells in long-lived animals, but in general is not so marked in the cells of those with a short life-span. Nevertheless, the view has been expressed that the presence of lipofuscin is the only constant cellular alteration that can be correlated with age in human subjects and laboratory mammals.

Age pigment has been isolated and attempts made to characterize it in chemical terms. It has a blue-green fluorescence and is partially soluble in aqueous and lipid solvents. Chemical analysis indicates that it is predominantly composed of lipid and protein. The lipid factor is similar to the constituent lipid of the tissue of origin – some of the components appear to be oxidized products of unsaturated fatty acid residues which are probably responsible for the colour, the fluorescence and the general resistance to enzymic degradation. The protein component is relatively rich in glycine and valine.

Another accumulation, termed 'an amyloid deposit', has been observed in old tissues from a range of animal types. There is no firm evidence as to its rôle in the ageing process, although one group of workers has suggested its appearance indicates the development of auto-immune reactions.

Chemical ageing

Recent refinements of the 'accumulated waste' theory hold that there are interactions between macromolecules and metabolites or highly reactive intermediates of metabolism, such as transient free-radicals, which result in the cross-linking together of proteins into large inert aggregations. It is postulated that many of these links are formed by non-enzymic mechanisms, particularly through oxidative reactions. These theories await the firm chemical identification of cross-linked polymers, the establishment of a rise in concentration with age, and the identification of causal relationships. Taking the viewpoint that ageing may be due in part to deleterious side effects of free-radicals normally produced in metabolism, mice have been treated with compounds, such as anti-oxidants, which are capable of reacting preferentially with free-radicals, so reducing their effective concentration. Results show that treatment with anti-oxidants from weaning does increase mean life-span in some, but not all, strains (maximum 26% increase). A difficulty in this type of study is that control diets have to be formulated and administered with great care in order to connect changes in life-span with a particular additive. This problem of experimental design arises because life-span is known to be connected with nutritional status.

Somatic mutations

One of the most widely held theories on ageing holds that spontaneous chemical changes occur in the DNA of somatic cells which are analogous to mutations in germ cells, in that they give rise to proteins with abnormal amino-acid sequences. Somatic mutations would be expected to alter the function of those organs composed largely of post-mitotic cells as the tissues gradually become occupied by malfunctioning cells, and of course, mutations could be held responsible for any kind of deterioration in cells, giving this theory a dominant position in gerontology.

Alternatively, mutations could result in rapid cell death with organ function deteriorating through the loss of its parts. Major problems surround the identification of mutations and, at the moment, there is no way of characterizing them in post-mitotic cells. Chromosome abnormalities definitely appear in histological sections of cells in regenerating liver. The frequency of the aberrations increases with age; and these aberrations have been, perhaps unjustifiably, equated with mutations.

There are serious difficulties in accepting the theory of somatic mutations. The frequencies of aberrations found in regenerating liver from different strains do not always fit observed differences in life-span. In order to demonstrate them, the mitotic rate of hepatic cells has to be greatly stimulated by drastic surgical measures. Also, the life-shortening action of radiation, which has been taken to involve somatic mutations, may be explained in other ways. For example, gamma irradiation of *Drosophila* imagos at specific times between 1–20 days of age, results in death at a constant time post-exposure. Age at irradiation influences survival time only when the flies are irradiated 30–90 days after eclosion. These results are taken to mean that the life-shortening effects of irradiation result from a radiation-induced sickness which, at least for 1–20 day old flies, is unrelated to ageing. The same conclusions apply to irradiated mice where, with some strains, the shortened life-span is associated with an abnormal endocrine syndrome. This highlights the shortcomings in using life-span as a measure of the rate of ageing.

It is established that ionizing radiations may cause specific mutations in the skin pigment cells of mice at the same rate as in spermatogonia. On the other hand, irradiation of experimental animals at doses that would be expected to result in somatic mutations, does not affect the rate of ageing. This applies particularly to reproductive capacity, ageing of collagen and the timing of the incidence of cancer in mice.

For human subjects, evidence against the mutation theory is somewhat indirect and comes mainly from the reports of the Atomic Bomb Casualty Commission of the U.S.A. These reports cover more than two decades of longitudinal studies on the survivors of the atomic bomb explosions at Hiroshima and Nagasaki. The relevant evidence arises from a comparison of a group who were within two kilometres of the explosion of one of the bombs and had survived acute radiation symptoms, indicating that they had been subjected to about one-third of the acute lethal radiation dose, with a matched non-exposed group. There is no evidence in the reports of any inter-group differences in cardiovascular function, sensory acuity and the cosmetic aspects of ageing. Immunological tests gave no indication that the irradiated group behaved in any respect as if they were older than their chronological age.

In summary, whilst it is highly probable that somatic mutations occur with age, the available evidence is inadequate to establish that this process is responsible for a general deterioration in function. Even in Protozoa, where both dominant and lethal mutations appear to be produced on a fairly large scale in very old individuals of 'old clones', there is some considerable doubt as to whether this is a cause or an effect of ageing.

Gene-suppression

Another theory which has been proposed also uses the central theme of inactivation of genes. The theory rests on the assumption that there are age-related losses in the ability of cells to synthesize RNA and protein, and further postulates that these losses in synthetic ability are due to the irreversible binding of repressors, such as histone-proteins, to corresponding structural or operator genes. This theory falls into the 'chemical ageing' category.

Incorrect translation

Failing a primary change in the chemical state of DNA through mutation or the irreversible binding of repressors to genes, abnormal protein patterns could arise due to a loss of competence of the cytoplasmic apparatus to correctly translate the genetic code into the polypeptide specified by the nucleus. Such an error would concern the specificity of information handling enzymes and lead to the synthesis of malfunctioning enzymes by virtue of the substitution of abnormal amino acids. Following treatment of micro-organisms and invertebrates with amino acid analogues which give rise to abnormal proteins through amino acid displacement, the organisms show premature senescence and a reduced life-span. These results suggest the probable outcome of disturbing the process of translation, but say nothing about effects on normal ageing processes. Experiments with cell-cultures of fibroblasts have shown that with increasing age of the culture, changes occur in the physical properties of some enzymes indicative of a change in primary structure, but the crucial test of the incorrect translation theory awaits the isolation of proteins from old organisms with abnormal amino acid sequences. These changes are inferred from differences in rates of denaturation.

If either mutation or faulty translation were the main cause of ageing, there would most likely be a steady decline in the biological activity of a large number of tissue proteins and a change in their physical properties. As will be made clear in the next section, this is not a general feature of ageing at the molecular level. Although there is evidence indicative of a fall in activity of some enzymes in old animals, this may be explained by a partial denaturation occurring because enzymes synthesized by old animals are released into a sub-optimum cellular environment, rather than by errors in synthesis of the protein.

Programmed death

Once a mitotic cell has differentiated into a post-mitotic cell, it has entered upon an irreversible developmental pathway, and there is a certain degree of probability that this pathway will result in cell death. An extreme form of this theory states that there is an ageing pathway in all post-mitotic cells which is controlled by a special group of 'ageing genes' and that in this sense, all specialized cells are dying cells. The origins of this idea stem from a consideration of the proliferation of epidermal cells which, in the process of dying, are converted into the important protective covering of the body. There are also numerous other examples from mitotic tissues with a high rate of cell turnover, where the steady state organ mass is maintained as a balance between cell division and cell death.

It is one thing to talk about programmed cellular death in mitotic tissues, often measured in periods of a few hours or days, but quite

308

another to discuss this phenomenon in relation to age-involution. It is an even bigger and less certain step to extrapolate these ideas, derived from mitotic tissues, to post-mitotic tissues in general. At the moment, the biochemical evidence does not support the idea that there is a progressive deterioration of post-mitotic cells according to a set pattern. Cellular death appears to be a random process, and this is difficult to reconcile with a programme theory.

Unselected differentiation

A 'random drift' theory of ageing is a reasonable alternative to the mutation theory as an explanation for a shift in the direction of metabolism. This theory holds that changes in the relative activities of various genes would produce progressive alterations in the concentrations of intra-cellular enzymes by the operation of normal biochemical reactions. At no period in life is there a perfect steady-state expressed as a stationary body weight and body composition. The corresponding shift in enzyme pattern need not involve special ageing genes and would then be no different in principle from those events responsible for differentiation. That is to say, a pattern of genetic information could develop in post-mitotic cells with time, which was influenced by events in the extra-nuclear phase. New protein patterns would differ from those of development, in that the changes would be quantitative rather than qualitative. Ageing, by this mechanism, differs from development because the process leads to a loss of efficiency and there can be no selection against random disproportionate alterations in enzyme activity which, on the time scale of ageing, may cause inefficiency at the organ level and possibly random death of cells.

Up to a certain point in time, it may be postulated that these changes are potentially reversible. In other words, there has been no change in the chemistry of DNA, only a progressive opening and closing of information channels leading to other parts of the cell. From this viewpoint, the regeneration of protozoa after autogamy may well be the key to an understanding of the nature of ageing, in that 'regeneration' could occur by a process which was, in effect, the uncoupling of DNA from cyto-plasmic events and its re-establishment in a 'youthful' environment, so allowing the specific developmental programme to unfold once more.

There is no shortage of evidence that changes in enzyme pattern occur in post-mitotic tissues, but only a few of these changes are obviously connected with a failure in tissue function, and many of them do not take a downward trend. On the whole, the age-dependent variations are not very marked, although few experiments have been designed to follow age-changes in enzyme activity throughout the life-span. Data from laboratory rodents suggests that the period of most rapid change occurs during the first few months of life and by the end of the first year of life, the activities have reached a stable level.

From the few comprehensive comparative studies that have been accomplished, it would appear that a high proportion of post-mitotic cells in the body maintain enzyme concentrations in old age which are close to those found in the mature adult. As is the case for physiological and morphological processes, it appears that biochemical processes involving enzyme synthesis exhibit ageing mostly in terms of a less effective response to a given stimulus.

Conclusions as to the stability of enzyme concentration patterns rest, for the most part, in the lack of statistical evidence for change. Many times, a large difference in mean values has been found in two-point experiments, but the deviation from the mean is often a large fraction of this difference. Some of the differences in means may become significant if more individual measurements were made (n is usually not more than 10 and often less than 10). More useful data could also be obtained by sampling at more than 2 points in time, e.g. consistent trends might be established. Thirdly, by analysing for many enzymes simultaneously instead of the usual one or two, patterns might be established.

As an indication of the biochemical constancy at the cellular level, the basal oxygen uptake of humans declines with age, but when expressed per unit of intracellular water, does not show any regression. From this it is concluded that there is no general impairment in the respiratory metabolism of old cells. The whole body respiratory response to the ingestion of a standard meal is essentially the same in old and young men, although older subjects show a slower response. In rats, the basal whole-body oxygen consumption actually increases with age between 10% and 13%. This discrepancy between rats and human subjects has not been resolved, but it is likely that the decrease in muscle mass of rats is counteracted by a higher rate of respiration of cells in other tissues. On the other hand, no age-changes have been detected in aerobic metabolism of tissue preparations from rat liver. Similar experiments with kidney indicate only a small age-decrement in aerobic metabolism, possibly due to a fall in the number of mitochondria per cell. The maintenance of aerobic metabolism at the cellular level is borne out by results obtained for guinea-pig tissues.

Little is known about ageing of anaerobic metabolism. Indirect evidence is suggestive of a decrease in the demand of the ageing brain for anaerobically produced energy. Also, some glycolytic enzymes in rat kidney have been observed to decrease with age by between 15% and 20%, but these changes are not associated with an impairment of glycolysis measured *in vitro*.

Ageing rat liver and brain are remarkable for their metabolic stability. Liver does not undergo age-involution and there is no alteration in protein turnover. It also appears that the enzyme pattern of liver and brain is largely independent of age. Despite this, there is a decrease in nitrogen content of liver, attributable mainly to losses of mitochondria and other membrane structures, and a decrease in the RNA content per cell. In humans, there is a rise in the volume-ratio of nucleus to cytoplasm; the actual quantity of RNA per nucleus is increased and there is an alteration in the base composition. These latter changes may be connected with the marked increase in gross RNA turnover, which is observed in nuclei of a number of tissues in old rats. Despite an alteration in RNA metabolism, cytophotometric measurements of Feulgen-stained nuclei in neurones of the cerebellum and liver cells indicate that there is no loss of DNA from individual cells during ageing. However, despite this uniformity, there are marked differences in the detailed ageing pattern of different tissues.

Most work on the biochemistry of ageing has been carried out using the laboratory rat, and it could be argued that the biochemical features of ageing in the laboratory rat are a consequence of inbreeding. This objection has been overcome with regard to some of the phenomena described above, in that they are also found in the ageing wild rat, although there are quantitative differences that have been interpreted as a slower rate of ageing in the wild strain, when both strains are housed in the laboratory.

Taken together, the biochemical work suggests that each tissue is characterized by a particular pattern of ageing, but to support this idea it is desirable that comprehensive studies be made on each tissue. So far, only one such analysis has been carried out on human blood vessels which indicates that there is a considerable variation in both the extent and pattern of change in enzyme activity. Some enzymes increase in activity (maximum, 3·5-fold rise), others hardly change and some at first increase and then decrease in activity. They may be divided into four groups according to the time when the activity reaches a maximum, but there are no obvious similarities in function of enzymes within any group. Enzymes in group 1 show a steady decline in activity from the first decade, amounting to about -5% per decade. Those in group 2 have a peak activity in the second decade (mean 50% increase), then show a decline which, between the third and eighth decade, proceeds on average at a rate similar to that for enzymes of group 1. Enzymes in group 3 reach their maximum activity in the third decade (mean, two-fold increase), then decline at a rate similar to those in groups 1 and 2. Group 4 contains enzymes that reach their peak between the third and ninth decades. The mean rate of increase to the peak is about the same for all groups.

Taking all enzymes together, there is no evidence to support the idea that there is a general deterioration in metabolism with age. By the ninth decade, about 50% of the enzymes retain more than 90% of the activity characteristics of the first decade; only one enzyme retains less than 50% of its activity in the first decade.

Another study, although not so many enzymes were measured, has been carried out on mouse prostate. In this organ, there is no detectable change in total DNA content as the animals age, indicating that cell loss did not occur on a large scale. Enzyme activities, expressed per unit of DNA, fell into two basic patterns: starting at age 9 months, some enzymes showed a steady decline; others showed a rise, followed by a fall. At the age of 30 months, only 5 enzymes out of 11 showed an activity per unit DNA which was less than that at 9 months.

Available evidence, therefore, although stressing the biochemical constancy of cells, suggests that there is some kind of drift in enzyme activity which is not always towards a drop in activity. It is within the bounds of probability that genotypic or phenotypic influences would, given time, eventually lead to ratios of key enzymes in some cells which were incompatible with life. The expression of this phenomenon at the population level might be expected to take the form of, first an increase in variability, followed by a decline in variability in the relevant enzymes at the time of peak mortality. The shift in mean plus the lowered variability would be the result of selection against these cellular phenotypes with greatly disturbed enzyme patterns. Linking these changes with increased chances of mortality, a small fraction of the organisms in the population would survive into extreme old age with an enzyme pattern closer to that of youth.

It is not an essential feature of the random drift theory that all cells in a given tissue change in the same way at the same time. There are variations in the environment of cells dependent upon local differences in the rate of blood flow and the presence of adjacent extracellular structures and other cell types. Thus, basic positional differences probably exist for cells to take up an individual ageing pattern. Indeed, there is clear histological evidence for age-dependent random variations in cells. For example, hepatic parenchymal cells, scattered either individually or in small groups, show variations in the morphology of cell organelles and functional differences are apparent through histochemical tests. Changes in enzyme pattern of this kind, once set in motion, may lead to an early death of the cells and may prove the explanation for the apparent random nature of cell death in post-mitotic tissues from an early age.

Conclusions

The readily observable deterioration in physiological adaptability after maturation, with the concomitant increasing incidence of mortality that comes with age, allows us to define ageing but it does not tell whether the phenomenon results from accumulated disease, repeated trauma, progressive primary cellular damage, or is inherent in the common chemical plan upon which living organisms are constructed. Only in the latter situation is there something inherent in the total organism itself which would strictly justify the use of the term 'process' to describe a well-defined intrinsic sequence of events controlled largely by genotype and environment

Gerontologists, in trying to distinguish between these various alternatives, are dealing with a particularly intractable area of biology which resists the efforts of the research worker for several unique reasons. Many experiments in the area of cellular ageing are, of necessity, whole animal experiments, with the attendant limitation that it is not possible to determine effects of relationships between cells from those due to primary cellular events. On the other hand, this difficulty cannot be overcome by using *in vitro* systems. It is not possible to obtain cells in simple culture that are identical with those in the multicellular organisms, and this is particularly true for the highly differentiated post-mitotic cells, where cellular changes would be expected to have important repercussions. Looked at in this way, we are far from establishing primary causal agents in ageing, and most of the current work is purely descriptive of the mean metabolic state of complex systems.

Another difficulty concerns the limitation of experimental techniques for studying the variations in the metabolic states of cells at a given point in time. Cellular death in the context of total life-span is a rare event and can only be studied using methods which allow the characterization of individual cells. This means that in the future, more emphasis has to be placed upon, on the one hand, developing high resolution microscopical techniques that allow cellular morphology and biochemistry to be inter-related and, on the other, to the evaluation of methods for cell separation in order to measure biochemical heterogeneity.

Cellular death is the worst thing that could happen to a cell that changed its behaviour with age, and all comparative evidence points to cellular death being a very important general characteristic of ageing. However, we know very little about cellular death and the factors that control it, even in mitotic tissues, where death is an essential component of normal functional organization. Nevertheless, cellular death in post-mitotic tissues is a measurable end point, with obvious effects on the organism, whereas if ageing results from a quantitative diminution of cellular efficiency, it will be correspondingly more difficult to assess in relation to the total organism.

With so many possibilities for failure in functional organization, it would be foolish for future workers to exclude any of the aspects of ageing presented in this review. It may well be that all play a part in the increasing incidence of mortality. We are still at a very elementary stage of investigation and although there can be little doubt as to the existence of cellular ageing, we must still consider the possibility that a large pro-portion of phenomena responsible for age changes occur at the super-cellular level.

Bibliography

1. EXTON SMITH, A. N. (1955), *Problems of Old Age*, Williams and Wilkins: Baltimore.
2. ALLISON, R. S. (1962), *The Senile Brain*, Arnold: London.
3. BRUES, A. M. and SACHER, G. A. (1965), *Ageing and levels of biological organization*, University of Chicago: Chicago.
4. KROHN, P. L. (1966), *Topics in the Biology of Ageing*, Interscience: New York.
5. ISAACS, B. (1965), *An introduction to geriatrics*, Baillière, Tindall and Cassell: London.
6. BIRREN, J. E. (1964), *The psychology of ageing*, Prentice Hall: New Jersey.
7. WOLSTENHOLME, G. E. W. and O'CONNOR, C. M. (1957), *Methodology of the study of ageing*, Churchill: London.
8. COMFORT, A. (1965), *The Process of Ageing*, Weidenfeld and Nicolson: London.

Index